P9-CMK-597

The HUMAN BODY EXPLAINED

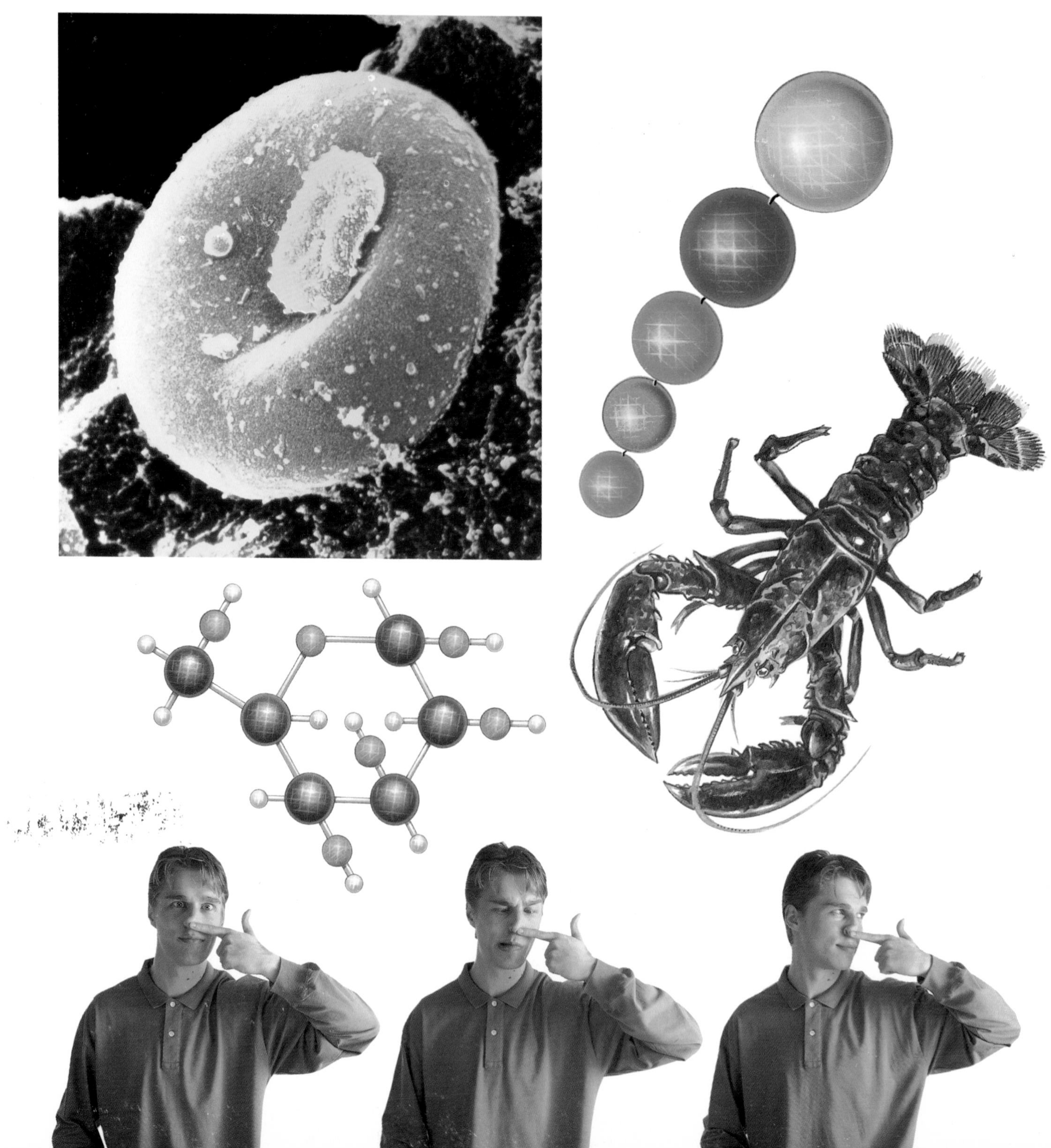

The HUMAN BODY EXPLAINED

A GUIDE TO UNDERSTANDING
THE INCREDIBLE LIVING MACHINE

Philip Whitfield, General Editor

A HENRY HOLT REFERENCE BOOK
HENRY HOLT AND COMPANY

NEW YORK

Contents

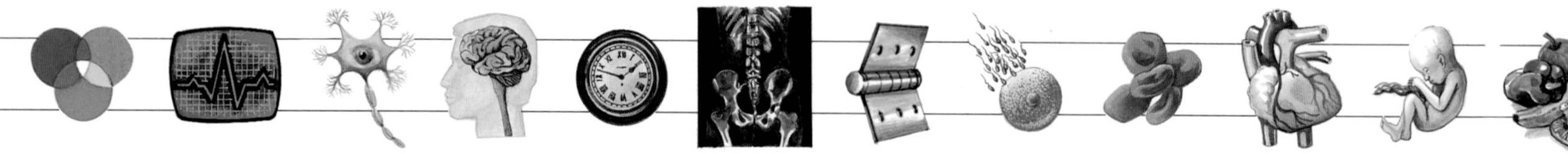

A Henry Holt Reference Book
Henry Holt and Company, Inc.
Publishers since 1866
115 West 18th Street
New York, New York 10011

Henry Holt ® is a registered trademark of
Henry Holt and Company, Inc.

Library of Congress
Cataloging-in-Publication Data

The human body explained: a guide to understanding the incredible living machine / Philip Whitfield, general editor. — 1st ed.
p. cm. — (Henry Holt reference book)
1. Body, Human—Popular works.
2. Human physiology—Popular works.
3. Human anatomy—Popular works.
I. Whitfield, Philip. II. Series.
QP38.H83 1995 95-86
612—dc20 CIP
ISBN 0-8050-3752-7

Henry Holt books are available for special promotions and premiums. For details contact: Director, Special Markets.

First Edition—1995

Conceived, edited, and designed by
Marshall Editions, London

Printed and bound in Italy by New Interlitho
Originated by HBM Print, Singapore

All first editions are printed on acid-free paper.∞

10 9 8 7 6 5 4 3 2 1

Project editor Jon Kirkwood
Art editor Simon Adamczewski
Assistant editor Jon Richards
Picture editor Zilda Tandy

Contributors Steve Parker
Dr. Philip Whitfield

Previous page (clockwise from top): *a red cell, part of blood supplying the body; linked amino acids make proteins for building bodies; a crustacean has its own way of staying in shape; soft cartilage, the nose's living framework; glucose – part of what you are and what you eat.*

Overleaf (clockwise from top): *all pulling together like the heart's special muscle; even when asleep the active brain keeps on working; the joints swing into action when you throw a ball; arteries and veins keep blood in circulation.*

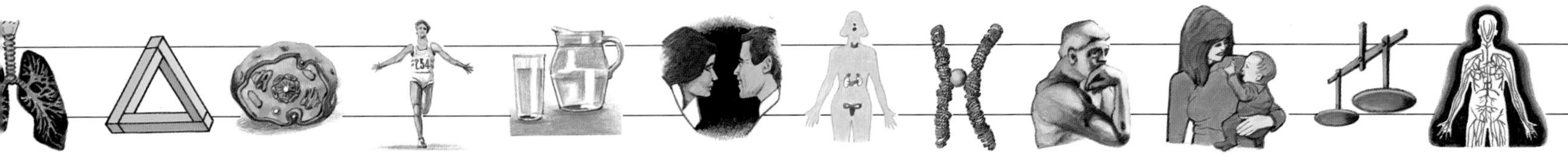

Energy

Circulation, Maintenance, and Defense

Reproduction and Growth

Foreword

Humans are inquisitive creatures. We investigate and explore. We are continually searching for new experiences, objects, and surroundings. When confronted with some new phenomenon or with something inexplicable, we want to understand how it works, what makes it tick. All the knowledge and understanding achieved in the many different branches of science and technology are the result of people down the ages asking the question "how does it work?"

Over many thousands of years, people have asked that question about the most complex living machine of all – the human body. Using the most up-to-date sources available, this book explains the ways in which the human machine works. The information in it has been gained by different types of specialists – anatomists, physiologists, neurologists, psychologists, pharmacologists, molecular biologists, and biochemists. Each area of specialism has provided knowledge of the way in which parts of the body function, from the way in which the heart beats and the digestive tract processes food to the way the brain thinks and babies grow.

In these and in all other facets of its working, the body can be fully understood only by knowing what happens at the three levels of organs, cells, and molecules. For instance, the movements of your eyes as they scan this line of text can be understood as the coordinated contractions of the set of six muscles (organs) that move each eye, as the actions of the specialized muscle cells from which those muscles are built, or as the interactions of the proteins actin and myosin (molecules) in the muscle cells which enable them to contract. This book is an illustrated manual of the workings of all parts of your body described at these three levels.

Phil Whitfield

Dr. Philip Whitfield

Introduction

This book's five interlinked sections, each looking at different aspects of the body's functions, reveal the body's workings and structure, how we interact with the world around us, and how we survive both individually and as a species. In this way, the knowledge of how the body machine is put together and how it performs its everyday tasks is made accessible and understandable.

SUPPORT AND MOVEMENT

A bony framework, along with other less rigid tissues, holds the body in shape, while muscles make it move. **Support and Movement** delves into muscle action – from the mechanics of getting around to the beating of the heart.

CONTROL AND SENSATION

Presiding over the body is the extraordinary human brain. **Control and Sensation** untangles the complex interactions between brain and body, looks at the systems that regulate the body, and offers explanations for how we think.

ENERGY

Among the countless billions of chemical reactions taking place inside us every second are those that provide power for the body. **Energy** keeps track of how we acquire, produce, and use energy both in cells and on a larger scale.

CIRCULATION, MAINTENANCE, AND DEFENSE

Blood – a complex fluid driven by a tireless pump – is in constant motion. **Circulation, Maintenance, and Defense** examines all the body's moving fluids and the phenomenal systems it uses to defend and repair itself.

REPRODUCTION AND GROWTH

The development of a person from the fertilization of an egg by a sperm to full adult maturity is a part of the amazing story of how humans populate the Earth. **Reproduction and Growth** explores human reproduction in the light of new facts about our genetic makeup.

It is relatively simple merely to describe the components and systems of the human body, but much more rewarding to understand how and why they work as they do. Using analogies with familiar items and events, and with the help of straightforward language, this book provides explanations of a kind not otherwise easily attainable.

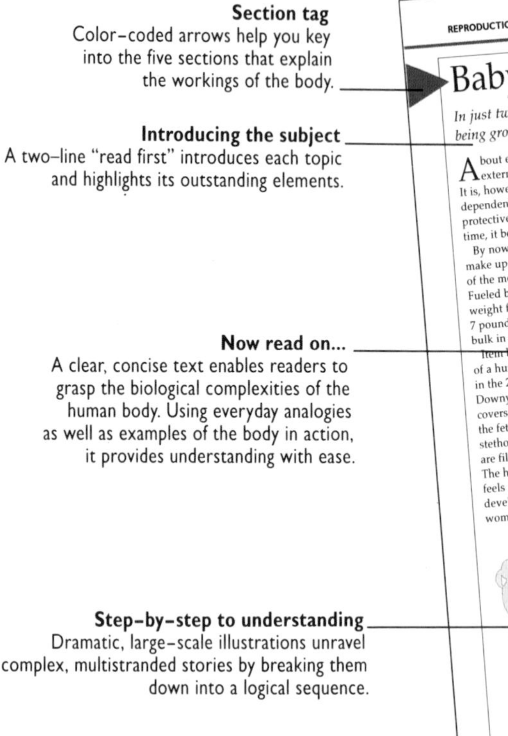

Section tag
Color-coded arrows help you key into the five sections that explain the workings of the body.
Introducing the subject
A two-line "read first" introduces each topic and highlights its outstanding elements.
Now read on...
A clear, concise text enables readers to grasp the biological complexities of the human body. Using everyday analogies as well as examples of the body in action, it provides understanding with ease.
Step-by-step to understanding
Dramatic, large-scale illustrations unravel complex, multistranded stories by breaking them down into a logical sequence.

HOW TO USE THIS BOOK

Photographs and illustrations are the starting points that lead to knowledge and an understanding of the nature and workings of the body. Words link closely to pictures to explain in depth the science of the human body. Each of the book's double-page spreads is a self-contained story. But since topics do not always fit neatly and completely under the headings superimposed on them, connections lead from the edge of each right-hand page to other topics, both within the same section and in different sections.

In order to promote flexibility of thought and depth of interest, the connections made are often deliberately wide-ranging. The connections can be used to make the book interactive and help forge links of understanding between different, yet complementary, aspects of human body functions.

Four sections of the book – **Support and Movement**; **Control and Sensation**; **Circulation, Maintenance, and Defense**; and **Reproduction and Growth** – are devoted to specific areas of the human body's function and performance. The other – **Energy** – deals with the overall process whereby the human machine powers its many activities.

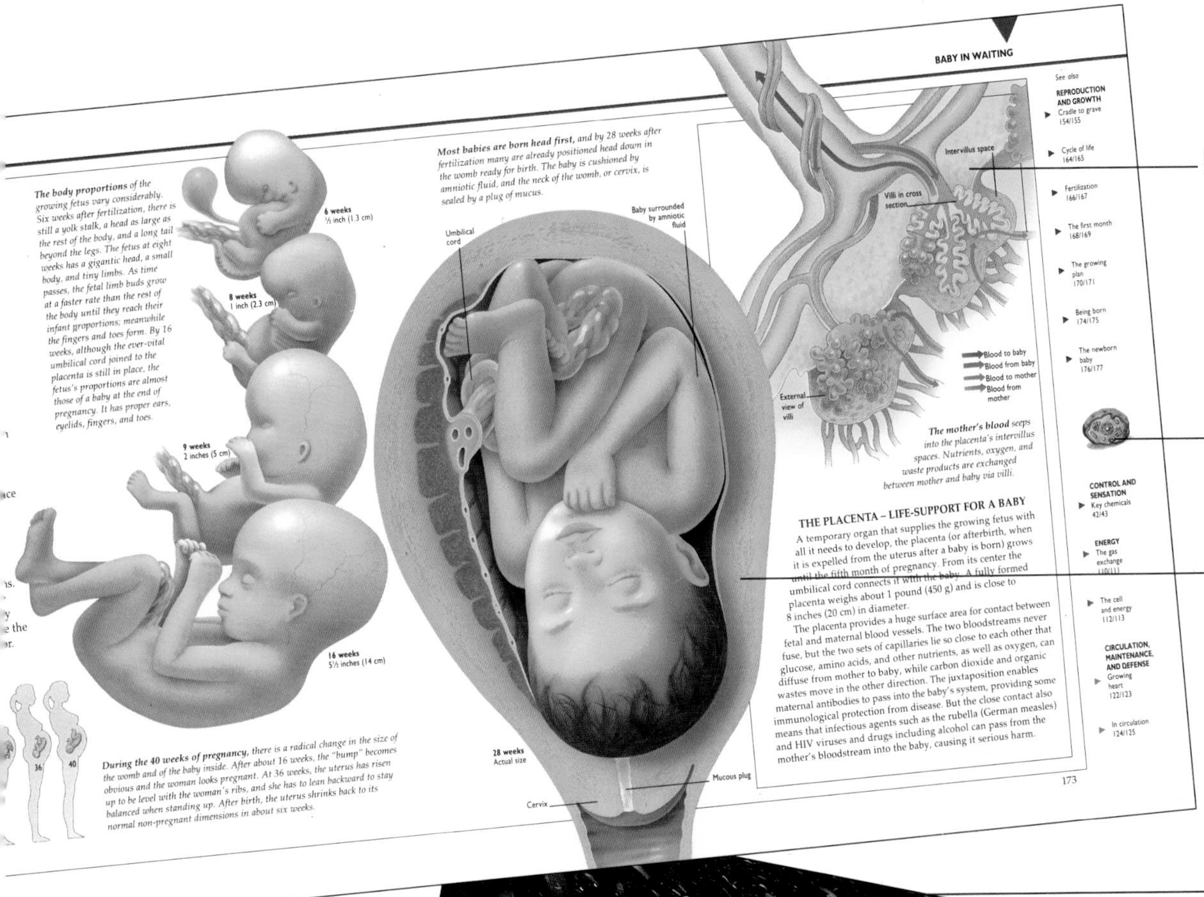

Feature box
To expand understanding, box features focus on the details of body systems and functions, giving contrasting or complementary examples. Here, the amazing nutritive interchange zone between mother and baby – the placenta – is explored and explained in detail.
Sometimes box features look at how knowledge from the field of medicine gives insights into the workings of the body. Boxes of this type are indicated by the medical symbol of a staff twined with snakes.

Connection icon
Graphic icons help you to make the link between a specific topic and connecting topics in the same or different sections.

Closeup on the body
Meticulous artworks explain by projecting the topic to a fuller, more complete level.
A step-by-step record of the all-important details of development shows a new human getting ready for birth.

Explaining with the familiar
Everyday objects, activities, and events provide vital clues to solving the mysteries of how the body works. For instance, a nighttime scene demonstrates the elements of visual perception in action. Details are dealt with by explanatory illustrations.

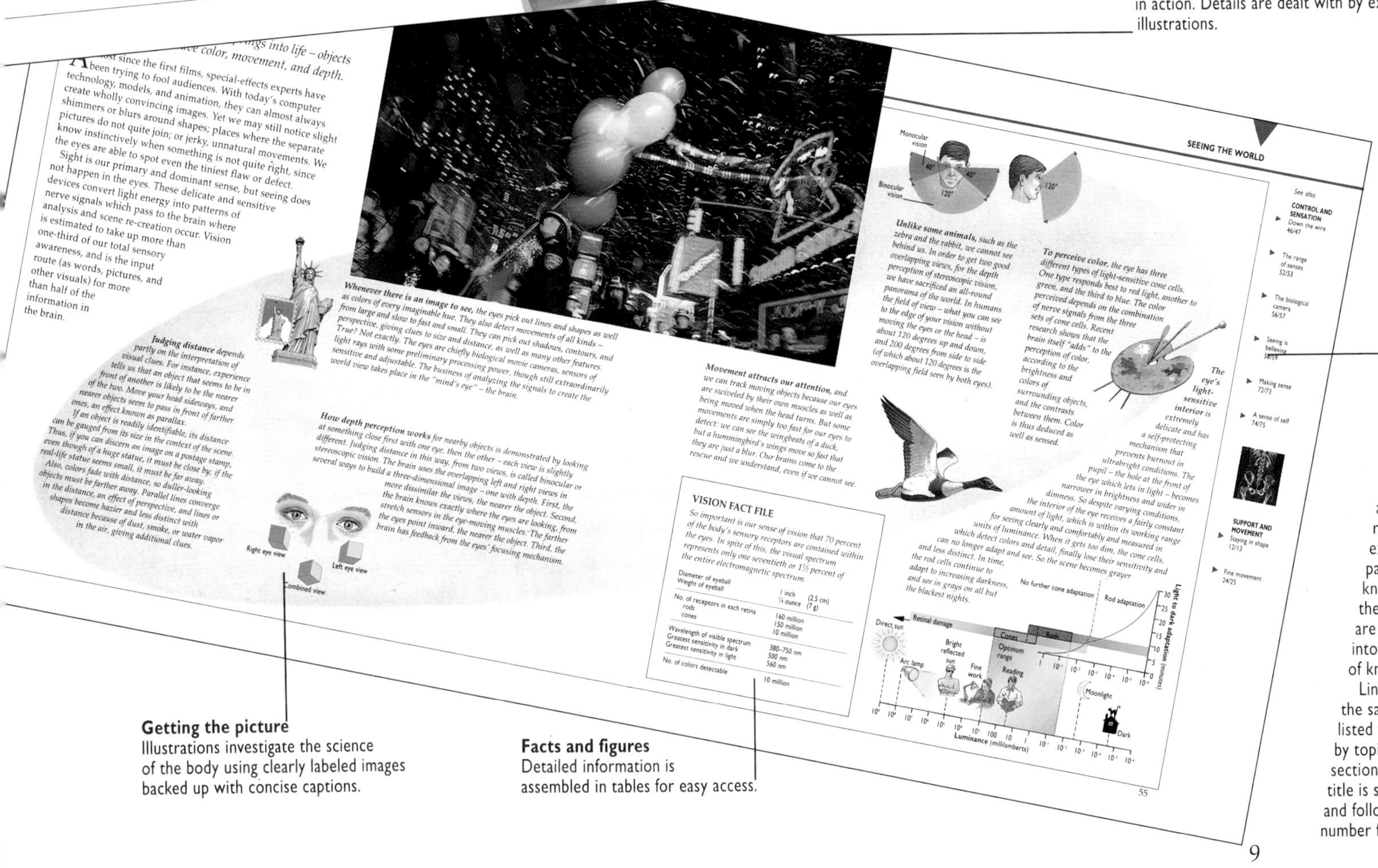

Connections
Follow the routes to suggested topics that contain backup facts to boost your grasp of each subject.
The connections also track down related ideas and explore parallel pathways of knowledge, so that the diverse themes are bound together into a coherent body of knowledge.
Linked topics in the same section are listed first, followed by topics in other sections. Each topic title is section-tagged and followed by its page number for easy access.

Getting the picture
Illustrations investigate the science of the body using clearly labeled images backed up with concise captions.

Facts and figures
Detailed information is assembled in tables for easy access.

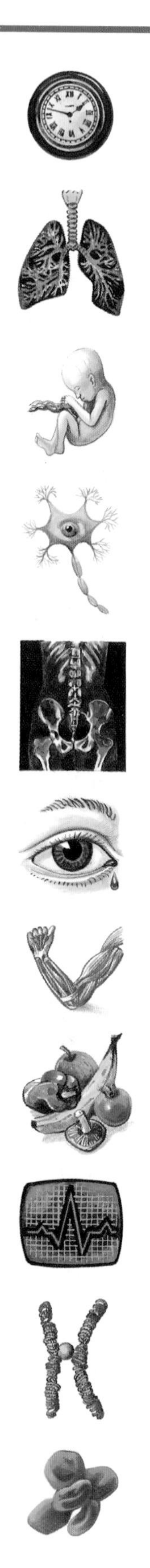

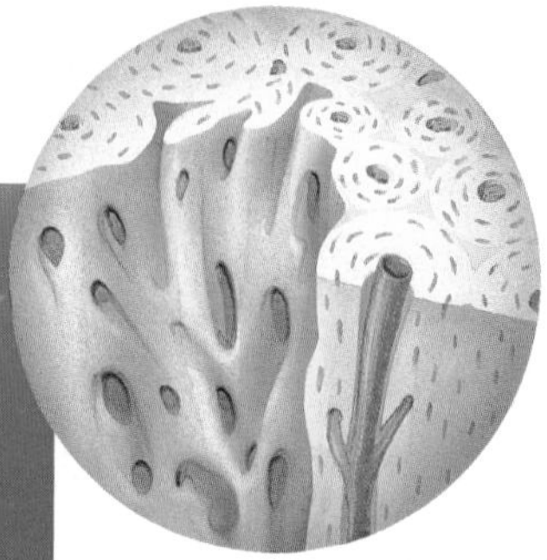

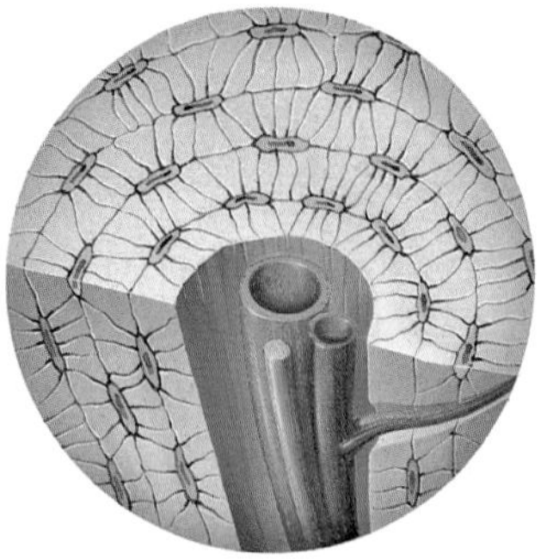

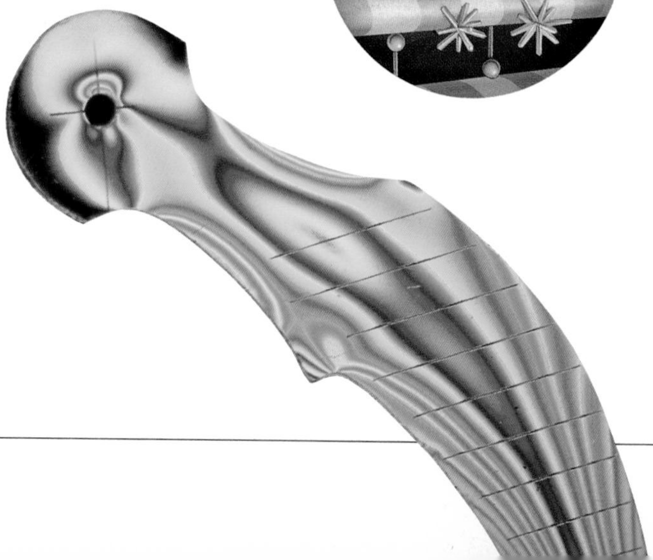

Support and Movement

A self-repairing internal framework stronger pound for pound than steel and a tireless pump controlled by electric signals from a powerful computer sound like some of the specifications for a superman. But they are just attributes of each and every human body – the skeleton and the bones that form it, and a muscle pump, the heart, whose rate of beating is timed by the brain.

Muscles and bones, together with the connective tissues, share the job of supporting the body and making it move. But muscles also play vital roles in other types of movements. Blood vessels with muscular walls act as circulatory control valves, while some muscles in the eyes focus the lens and others constrict the iris. Then there are the muscles in the digestive tract that force food through the intestine and others that keep body wastes contained.

Left (clockwise from top): *homing in on the living skeleton; a tough nut to crack, like the skull; a joint's stressful life; the watery way to stay in shape.* **This page (top):** *moving is easy with good connections;* **(left)** *resilient fibers.*

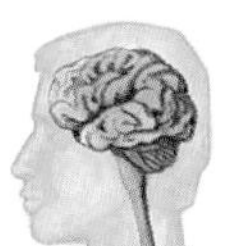
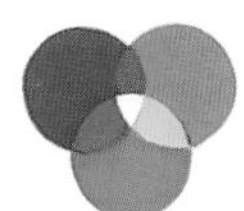
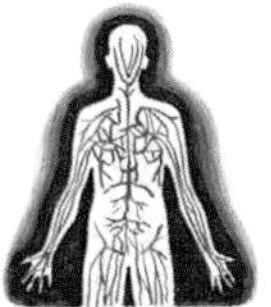

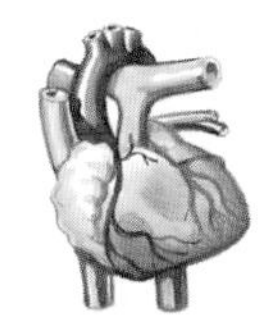

Staying in shape

Our shape is determined by evolution, but how is the familiar outline of the body supported?

All vertebrates, or backboned creatures (humans included), have endoskeletons – strong internal frames made of bone or cartilage. In fact, the shape of human beings reflects millions of years of evolutionary changes acting on the basic body plan of a land-living, four-limbed vertebrate. Fossil evidence shows how, some 350 million years ago, the fleshy-based fins of ancient fish developed into four limbs, each with five digits. The result was the earliest amphibians, which lumbered onto the land. Through the stages of early reptiles, mammal-like reptiles, and then the first shrewlike mammals over 200 million years ago, the basic skull-and-backbone-and-four-limbs design persisted.

When the dinosaurs disappeared some 65 million years ago, mammal evolution took off. We can trace our origins within the mammal group, through the first primates, to the early apes, and then via "ape-humans" *Australopithecus* to our own group, *Homo*, and the human species today, *Homo sapiens*.

The general plan of the human body is typically mammalian. The breathing and eating inlets, as well as the main senses of sight, hearing, smell, and taste, are sited at the head end. The respiratory and pumping sections of the circulatory system are packed into the chest. Most of the digestive, excretory (waste disposal), and reproductive parts are in the abdomen. The four limbs have five digits each.

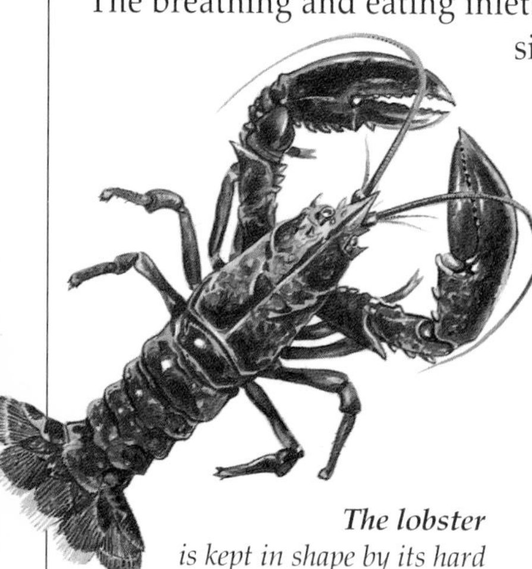

The lobster *is kept in shape by its hard shell. Other crustaceans, as well as insects and mollusks, also have rigid outer frameworks, known as exoskeletons, which give their soft internal tissues shape and support and provide solid anchorage points for their muscles.*

More mammalian characteristics feature both outside and within. Externally, the body has elastic, wear-resistant skin that protects it from knocks, germs, and rays from the Sun and prevents leakage of body fluids. The skin's strong fibers (similar to any mammal leather) give shape and support to the contours of muscles and softer organs below. Our strong endoskeleton, "the skeleton," supports and protects the softer fleshy parts and provides points for muscle anchorage. The only major anatomical features that make us unique among

The jellyfish *has neither solid endoskeleton nor exoskeleton, since it is a creature whose muscle groups do not need a rigid frame inside or out. Its form is maintained by the tension in its outer layers which take their shape when stretched by the soft internal tissues with their high water content. A shape maintained in this way is said to be supported hydrostatically.*

mammals are an unusually large and complex brain and our status as the only species that habitually stands and moves in an upright posture.

By contrast, most of the invertebrates – creatures without backbones, which make up the vast majority of animals and include crustaceans, mollusks, and insects – have rigid external shells, or exoskeletons, which support and protect the internal organs. Exoskeletons become impossibly heavy with increasing size, and the largest exoskeleton-clad animals live in the sea, their weight partly supported by water. They are virtually helpless on land.

Some invertebrates, such as jellyfish and worms, seem to have no body support at all and plainly do not have the sophisticated endoskeletons of the vertebrates. These creatures do, however, have a support in the form of a hydrostatic "skeleton," which relies on fluid under pressure. Their essentially fluidlike internal organs press against a flexible "skin," or outer layer, to keep them in shape, rather like a plastic bag full of water. This system works well for simple creatures, especially those that live in water where the surrounding fluid helps them to keep a shape.

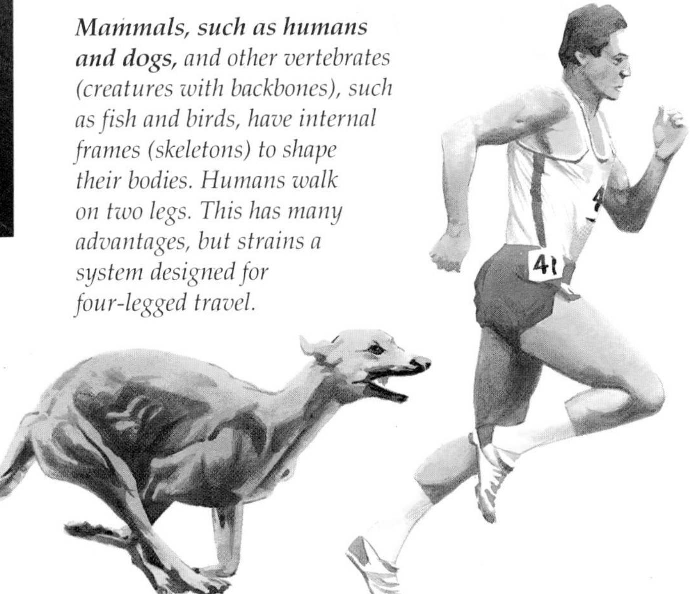

Mammals, such as humans and dogs, *and other vertebrates (creatures with backbones), such as fish and birds, have internal frames (skeletons) to shape their bodies. Humans walk on two legs. This has many advantages, but strains a system designed for four-legged travel.*

THE SHAPE YOU ARE IN

Most of us have similar body shapes, and variation between individuals is not great. Most obvious differences are in height and bulk. Laurel, on the left, is slim while Hardy is taller and stocky.

The body's size and shape depends on three main factors: skeletal size and proportions; muscular development; and stored fat. The relation of skeletal height and the lengths of limbs to the widths of skull and hipbone remains remarkably constant from the tall to the short. The skeleton determines our height and is the factor we can least influence.

But it is possible to add or lose muscle and fat. Almost anyone who exercises can build up muscle mass, thereby bulking up the shoulders, arms, torso, and legs. People who do not exercise, and who eat too much, round out in the face, neck, and abdomen, where the excess food is converted to stores of body fat.

See also

SUPPORT AND MOVEMENT

▶ Bearing the load 14/15

▶ The living framework 16/17

▶ Holding it together 20/21

▶ Making a move 22/23

CIRCULATION, MAINTENANCE, AND DEFENSE

▶ Transportation systems 118/119

▶ The water balance 132/133

▶ Outer defenses 142/143

REPRODUCTION AND GROWTH

▶ Building bodies 160/161

Bearing the load

The body's internal framework, the skeleton, is an extraordinary study in engineering design.

Without a skeleton, the body would collapse in a floppy heap, like a jellyfish on dry land. But more than 200 bones, linked at over 100 movable joints, provide rigid strength and support coupled with flexibility and agility.

Evolution has produced bones whose size and shape are perfectly matched to their roles. The primary roles are support for the softer parts, anchor points for muscles, and protection for delicate parts within. The long bones in the arms and legs function both as load-bearing beams and columns and as mechanical levers; the broad, flat bones in the shoulders and hips provide large areas for the attachment of many powerful muscles; and the rounded bones of the skull, ribcage, and pelvis shield the more delicate parts within, such as the brain, heart, lungs, and intestines, from knocks and damage. Certain bones are formed from several bones which have fused together during development. For example, each pelvic, or hip, bone has three fused elements; and the sacrum at the base of the spine has five elements.

__Like a walnut shell,__ the skull protects its contents within a hard covering. Both shell and skull are rigid containers that enclose and protect a rounded, wrinkled-looking object. In the skull's case, the object is the brain. The curved upper brain-box part is the cranium and consists of eight rounded bones which are fused strongly at wavy-line joints called sutures. Another 14 fused bones form the facial skeleton. Some of these create two bowl-shaped depressions – the orbits, or sockets, which house and shield the eyes.

__The body's backbone,__ or spinal column, is like a spiral staircase. All the treads and central rings of the staircase are anchored firmly together so the whole structure does not sway about like beads on a string. The weight of the staircase is carried by the central supporting column.

The spine has a similar segmented load-bearing structure – taking the entire weight of the upper body – but in addition, it is flexible. Each bone, known as a vertebra, is separated from its fellows by a tough "washer," the disk, which allows a limited amount of movement. The spinal column is supported and held in place by muscles and ligaments which allow movement but prevent motions that might damage the vertebrae or, crucially, the spinal cord that runs down its center. The cord contains nerves that control muscles and carry sensory messages.

__Bones have anatomical names,__ and most have everyday names as well (shown in brackets). Bones also lend their names to other body parts near them, such as blood vessels and nerves. Thus, the subclavian artery runs under the clavicle (collarbone).

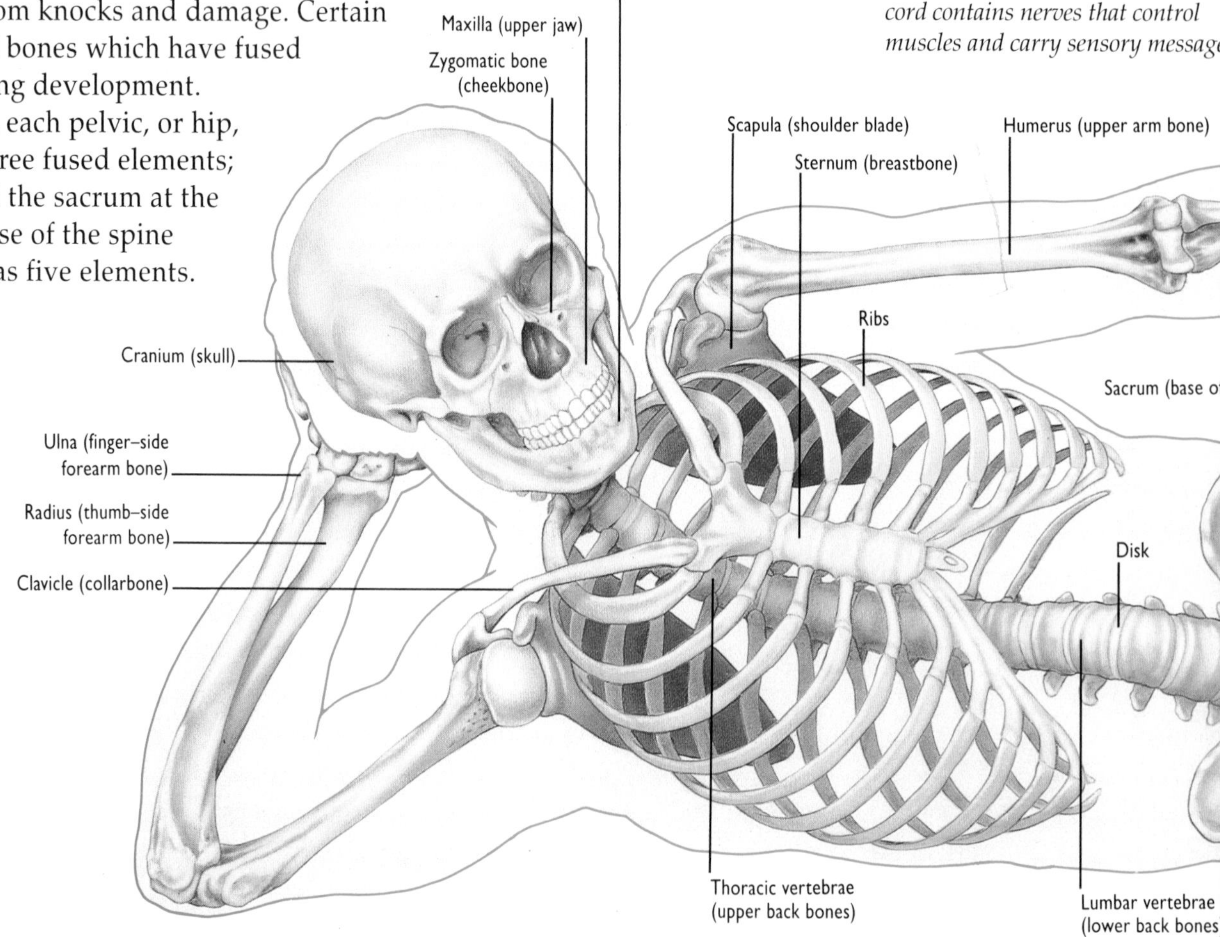

SKELETON FACT FILE

Some individuals have extra or fewer bones: one person in 20 has an extra pair of ribs, making 13 pairs instead of 12.

Total no. of bones	adult child	206 300*
No. of bones in	skull and neck each ear spinal column chest arm and hand hip, leg, and foot	23 3 26 25 32 31
Weight of skeleton in	110-pound adult (50-kg)	15½ pounds (7 kg)
Longest bone Widest bone Smallest bone	femur pelvis stapes	27% of height of an adult body width at hip ⅕ inch (0.5 cm) long

*Some bones fuse together as a child grows.

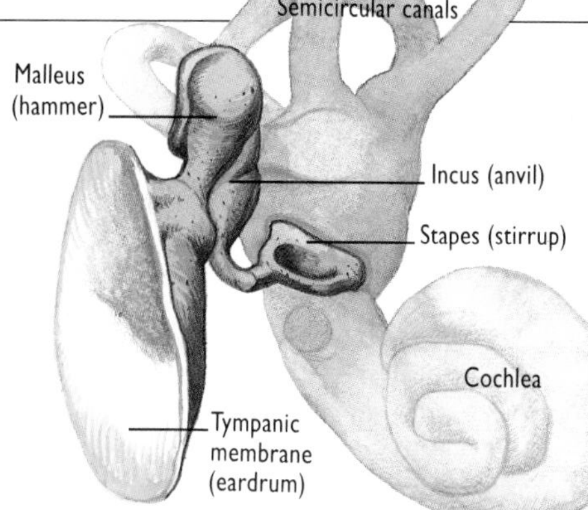

Three tiny bones, *known as the auditory ossicles (**left**), deep in the inner ear, are little larger than rice grains. They have no load-bearing or supporting function at all.*

Despite their minute size, they have all the usual bone features: a nerve and blood supply, lubricated joints with their neighbors, and muscle attachments.

The word "ossicle" comes from the Latin for "small bone." The bones' individual Latin names, and the translations, reflect their resemblance in shape to a blacksmith's hammer (mallet) and anvil, and to a stirrup used in horse riding. The function of the bones in hearing is to transmit and amplify the vibrations that represent sound waves from the tympanic membrane (eardrum) to the cochlea.

The leg and hip bones *work like the tower and jib of a tall crane. The tower is built to withstand both downward compression forces and sideways forces as the crane lifts a weight.*

In a similar manner, the thigh and shin bones not only support the body's downward weight but also resist sideways forces, since they are cantilevered out from the midline of the body by the width of the hips. Generally, though, the body's long bones are designed to bear compression forces along their long axes.

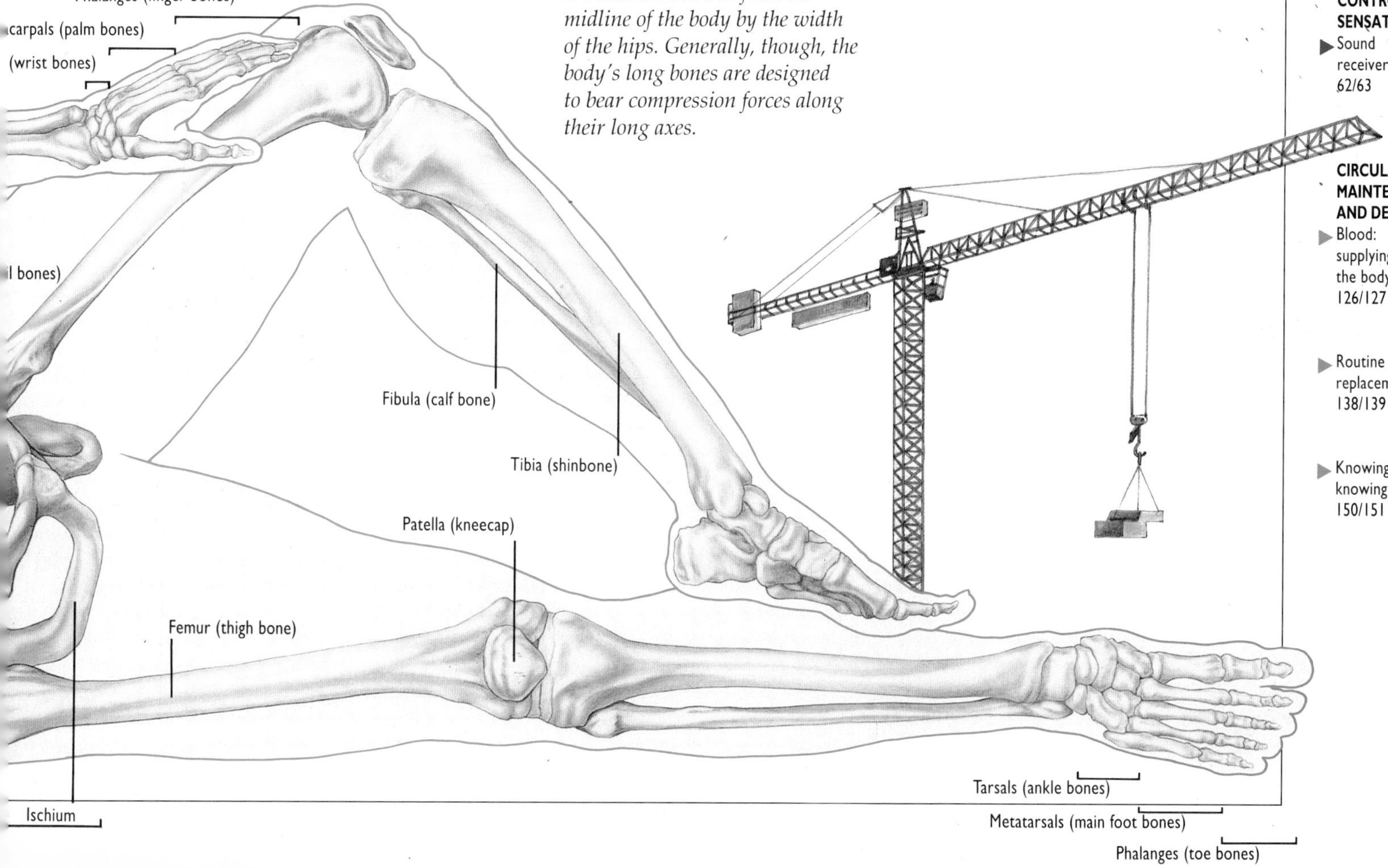

Phalanges (finger bones)
carpals (palm bones)
(wrist bones)
l bones)
Fibula (calf bone)
Tibia (shinbone)
Patella (kneecap)
Femur (thigh bone)
Ischium
Tarsals (ankle bones)
Metatarsals (main foot bones)
Phalanges (toe bones)

The living framework

Skeletons are made from incredibly strong material that is constantly changing, adapting, and rebuilding.

Bone is tough. It can withstand compression forces almost twice as well as granite, and stretching forces four times better than concrete. Yet despite this strength, bone is astonishingly light. If the skeleton were built of steel to an equivalent strength, it would weigh about five times as much. In the correct environment – the human body – bone can last for many years, even repairing itself when broken. For living bone is far from dry, brittle, and static. It is physiologically busy and is constantly exchanging minerals and other substances with other body parts via the bloodstream.

Like many other tissues, bone consists of cells surrounded by a substance called a matrix. Some 30 to 40 percent of this matrix is protein, chiefly fibers of the protein known as collagen. The rest is made up of minerals, mostly calcium and phosphate. These two components – collagen and minerals – are complementary, each performing a vital role. Without collagen, bone would be as brittle as glass. Without minerals, it would be as pliable as a rubber toy.

Embedded in the matrix are three main types of bone cells. In mature bone they are mostly spidery osteocytes. These maintain the matrix, removing or adding proteins and minerals as required. A typical osteocyte lives for decades deep in a prison of its own making. It was once an osteoblast, or bone-making cell, near the bone's surface. As the bone grew during body development, more bone tissue was added around it, and the osteoblast matured into an osteocyte. Osteoclasts, the third type of cell, can erode the matrix if the body requires minerals and other ingredients elsewhere. The architecture of bone is far from fixed. While the osteoclasts dismantle bone, osteoblasts rebuild it elsewhere, for example at regions where strain increases with age or injury.

__Cartilage__ is softer and more bendable than bone, but still strong and resilient. Eight curved cartilage plates form the squashable framework of the nose, with a ninth dividing the two nostrils.

Bone has an intricate microstructure of rods, each of which has a channel, or Haversian canal, at its center. When packed densely together, these rods form compact bone, which provides a strong outer layer or "shell" for the whole bone where stresses are greatest. Other rods are arranged in the open honeycomb structure of spongy bone, which can be trellislike or made up of plates and sheets. This type of bone forms where stress is lower. In the spaces inside spongy bone are various types of marrow, which produce blood cells or store fatty nutrient reserves.

Lacking the rigid mineral crystals found in bone, cartilage – the other main supportive tissue in the body – is a more flexible alternative. It is found in many organs, giving shape to the ears, nose, and trachea, helping you to sneeze, breathe, speak, and hear effectively.

AS GOOD AS NEW

Most parts of the body can mend themselves, like skin, which heals over with scar tissue. A fractured bone can also self-repair, provided its stresses are relieved by other bones and muscles, and the broken ends are still close to each other. This is why a broken limb needs a splint or a plaster cast in order to heal. First blood from the vessels inside the bone clots to stop its continued leakage (**1**). Then a callus of fibrous tissue forms around the fracture (**2**). The callus is manufactured by cells known as fibroblasts. These cells also make cartilage,

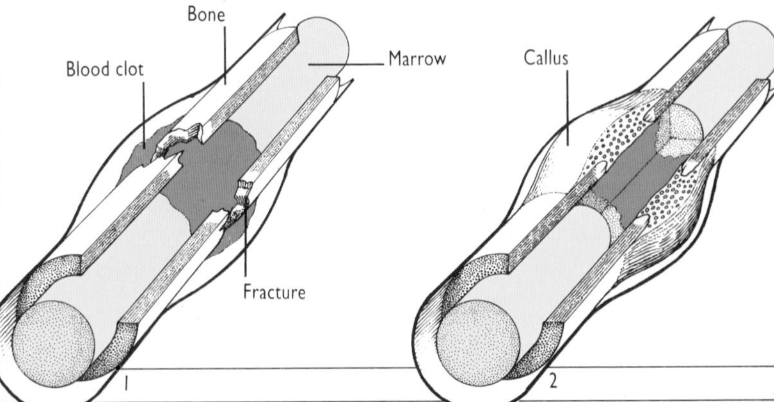

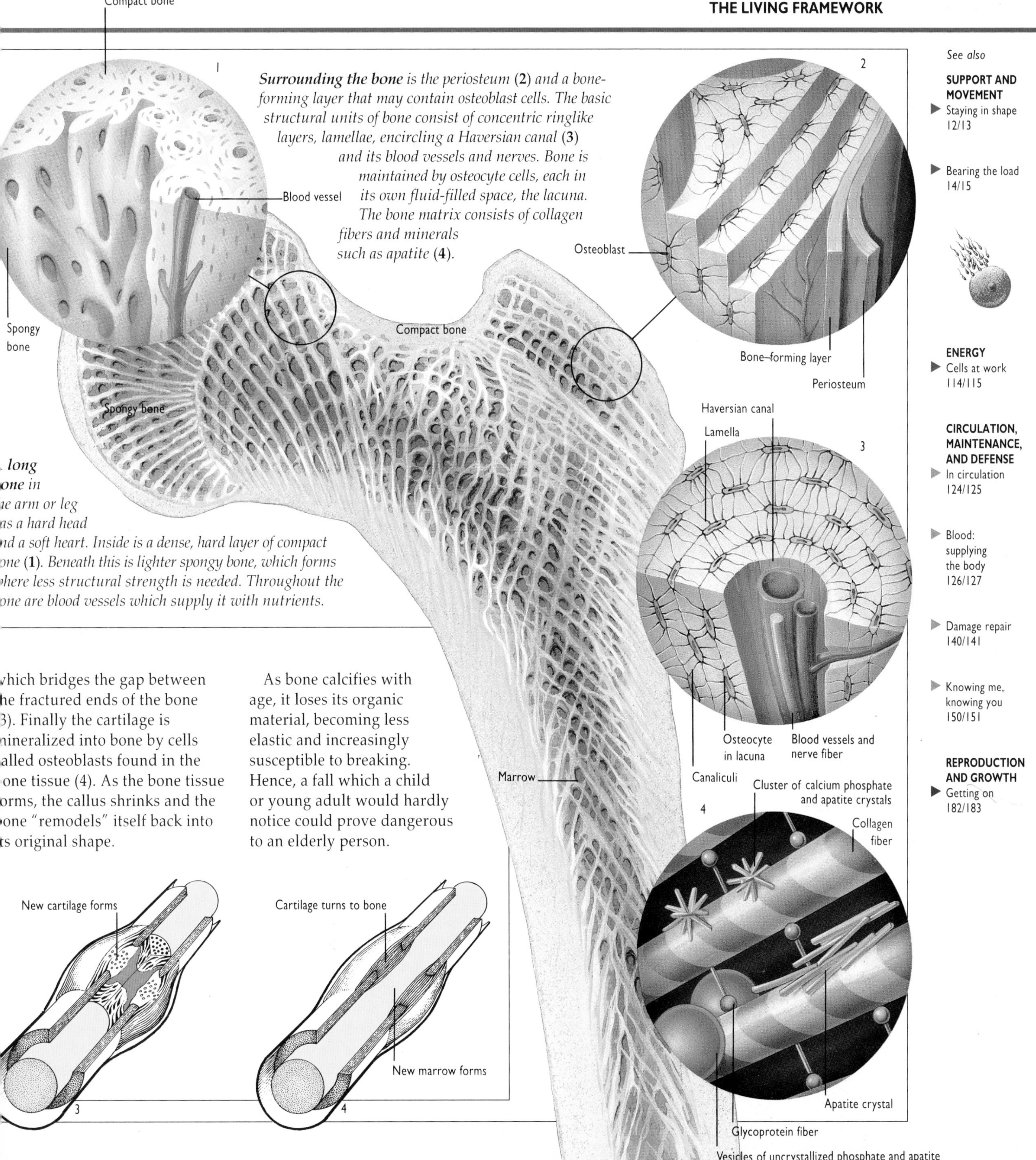

Surrounding the bone is the periosteum (**2**) and a bone-forming layer that may contain osteoblast cells. The basic structural units of bone consist of concentric ringlike layers, lamellae, encircling a Haversian canal (**3**) and its blood vessels and nerves. Bone is maintained by osteocyte cells, each in its own fluid-filled space, the lacuna. The bone matrix consists of collagen fibers and minerals such as apatite (**4**).

. long ***one*** *in*
ıe arm or leg
ıas a hard head
ıd a soft heart. Inside is a dense, hard layer of compact
one (**1**). *Beneath this is lighter spongy bone, which forms*
here less structural strength is needed. Throughout the
one are blood vessels which supply it with nutrients.

which bridges the gap between he fractured ends of the bone 3). Finally the cartilage is nineralized into bone by cells alled osteoblasts found in the one tissue (4). As the bone tissue orms, the callus shrinks and the one "remodels" itself back into ts original shape.

As bone calcifies with age, it loses its organic material, becoming less elastic and increasingly susceptible to breaking. Hence, a fall which a child or young adult would hardly notice could prove dangerous to an elderly person.

See also

SUPPORT AND MOVEMENT
- ▶ Staying in shape 12/13
- ▶ Bearing the load 14/15

ENERGY
- ▶ Cells at work 114/115

CIRCULATION, MAINTENANCE, AND DEFENSE
- ▶ In circulation 124/125
- ▶ Blood: supplying the body 126/127
- ▶ Damage repair 140/141
- ▶ Knowing me, knowing you 150/151

REPRODUCTION AND GROWTH
- ▶ Getting on 182/183

Bone junction

Virtually every move you make depends on one of the body's engineering marvels – the joints between bones.

Nature has solved the problem of how to protect the ends of adjoining bones in a remarkably effective way. Movable bones do not actually touch, but instead have a long-lasting, wear-resistant bearing between them – a joint. The main type, which usually allows great freedom of movement, is the synovial joint. Its basic design is similar in all mammals, including humans. Bone is rough and hard, so its surfaces need a hardy, smooth coating. Inside a synovial joint, bone is covered by a pearly-smooth elastic substance – the articular cartilage. This is similar in structure to bone, being collagen (protein fibers) in a stiff matrix, but it does not have bone's hardening minerals. The cartilage-covered parts of bone can thus pivot or rotate on each other with little friction and wear. The cartilage also acts as a shock-absorber, so if the joint is jarred, the bones do not shatter.

Synovial joints have lubrication, too. The cartilage-covered bone ends are encased in a flexible bag – the articular capsule – which links the two bones together. This bag encloses a space, the synovial cavity. It is lined by a thin, shiny membrane, the synovial membrane, which produces synovial fluid. This thick, slippery substance fills the synovial capsule and is the "oil" that lubricates the joint. The articular capsule is thickened and strengthened by collagen at various sites, to form ligaments. These elastic strips link the two bones, to stop them from pulling apart or moving too far and dislocating the joint.

In the body, there is great variety in the numbers of bones in a joint, the shapes of their cartilage-covered ends, and the type and amount of movement they allow. The similarities to mechanical joints give them various names, such as pivots and hinges. There are also several other types of joints in the body, including stationary joints that allow no movement at all.

Ball-and-socket joints, *such as those at the shoulder and hip, have a spherical head (ball) that fits into a bowl-shaped cavity (socket). They allow movement in two planes plus some twisting. The socket in the shoulder blade (scapula) is shallower than that in the hip (pelvis), allowing more freedom of movement, but reducing stability.*

Hinge joints, *as at the elbow and knee, move in only one plane, but they trade limited movement for great stability. Other hinge-type joints are found in the smaller knuckles of the fingers and toes and between some bones in the ankle.*

When you throw a ball, *the shallow ball-and-socket shoulder joint lets the upper arm swing from behind the body, right around to the front. The hinged elbow joint snaps the arm straight, to flick the hand. It is aided by more than 20 hinge, gliding, and other joints between the 8 wrist bones and their neighbors. The fingers push and straighten at their shallow hinge joints, giving the final impetus to send the ball on its way.*

Washer joints *are a special feature of the spinal column. Between each pair of vertebrae is a pad of spongy fibrous cartilage, the inter-vertebral disk, which lets them both rock slightly. These movements, over all its bon[es], enable the whole spine to bend double.*

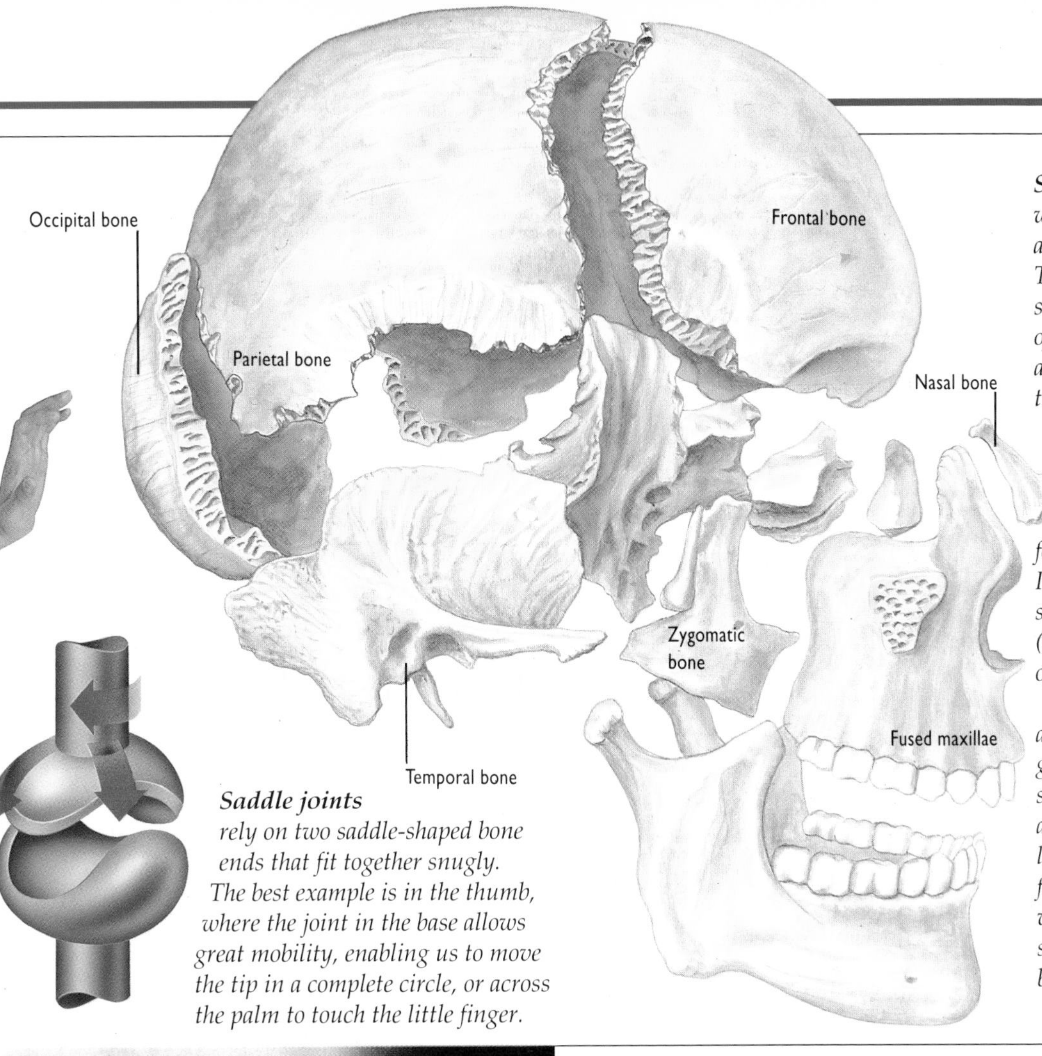

Saddle joints rely on two saddle-shaped bone ends that fit together snugly. The best example is in the thumb, where the joint in the base allows great mobility, enabling us to move the tip in a complete circle, or across the palm to touch the little finger.

Some bones are joined *in a secure way so that they move little, if at all, and are unlikely to come apart. The most solid joints are fibrous sutures, found between the bones of the adult skull. The various bones are butted and united by fibrous tissue and function as a solid unit.*

In fibrous syndesmoses, the bones are linked firmly by fibers, but are not touching and can move slightly, as with the two forearm bones, the radius and ulna. In cartilaginous joints, the bones, such as the ribs and the breastbone (sternum), are cemented to a bridge of flexible cartilage between them.

Perhaps the simplest joint that allows significant movement is the gliding joint, where two almost flat surfaces slide across or twist against each other, but only by a limited amount. Gliding joints are found between various bones in the wrist and ankle, and between the shoulder blade (scapula) and collar bone (clavicle).

See also

SUPPORT AND MOVEMENT

▶ Staying in shape 12/13

▶ Bearing the load 14/15

▶ The living framework 16/17

▶ Holding it together 20/21

▶ Making a move 22/23

CONTROL AND SENSATION

▶ Into action 50/51

▶ In the balance 70/71

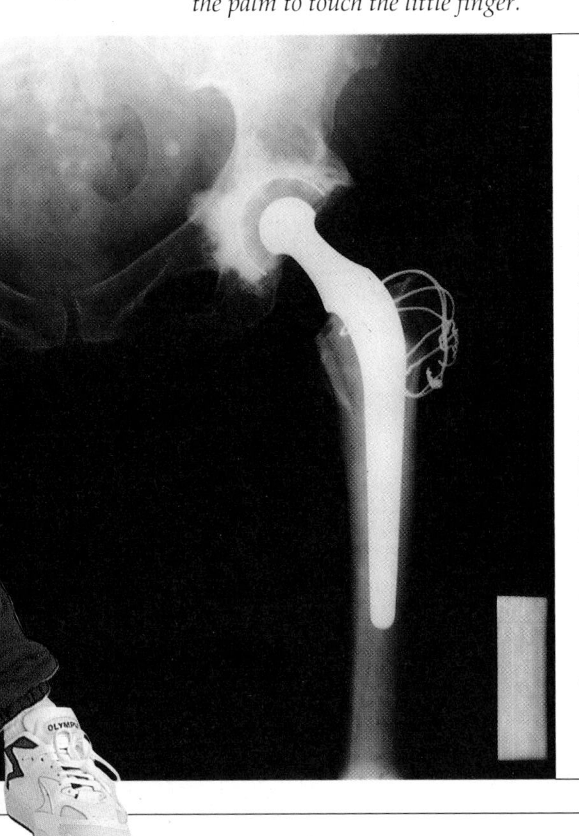

REPLACING THE NATURAL JOINT

Since Roman times, physicians have tried to treat joint problems, such as osteoarthritis where the cartilage flakes and degenerates, with prostheses (artificial parts). Arthritic hips, in particular, can severely limit mobility. In medieval times the prosthesis was external, strapped onto the skin over the joint to help take the weight. In the 1960s, British surgeon John Charnley devised a successful internal prosthesis, the artificial hip joint. Its ball-and-socket design mimicked the natural version. The ball is of highly polished stainless steel, mounted on a long spike cemented into the marrow cavity of the thigh bone (femur). The socket, made of high-density plastic, is cemented onto the hip bone. Artificial elbows, knuckles, and even knees are now available. The hunt continues for tougher, longer-lasting joint materials that can withstand the tremendous stresses imposed by body movements – stresses that can be observed in photographs of plastic models taken in polarized light (**right**).

Holding it together

Connective tissues, the body's "living glues and fillers," hold our tissues and organs together.

The body is made up of cells, grouped into tissues, grouped into organs. Connective tissues are found between organs, between various types of tissues and even in the spaces between cells. They fill, pack, support, protect, wrap, restrict, cushion, contain, restrain, insulate and, of course, connect.

Connective tissue has three major components. The first is cells – mainly fibroblasts, which make all-important fibers. Other connective tissue cells include adipose cells, swollen with globs of fat; macrophages, which eat germs and debris to keep the tissue clean; and reticular cells, plasma cells, and others involved in body defense and immunity.

The second and distinctive component is fibers such as collagen, elastin, and reticulin. The main type is the whitish protein collagen – resembling a microscopic length of rope – found in almost all connective tissues. Collagen is tough and can bend but hardly stretch; its fibers can be packed densely or scattered loosely, and can be neatly woven and well aligned or irregular and random. The highly flexible and elastic yellowish elastin gives young skin its rubbery stretchiness. Thinner, branching reticulin fibers are embedded in connective tissue's third component, known as the matrix or ground substance. This ranges from a runny syrup to a thick, stiff gel and is made mainly of large molecules such as glycoproteins.

Almost every body part has some type of connective tissue. In loose, or areolar, irregular connective tissue, the fibers are arranged at random and far apart in their jellylike matrix, to form spongy, flexible wrappings for body organs, blood vessels and nerves. In dense irregular connective tissue, the fibers are close together; in regular connective tissue, they are aligned. Adipose, or fatty, connective tissue has many adipose cells and is soft and squishy, for insulation and cushioning. More specialized types of connective tissue are bone and cartilage – and blood. Of course, blood does not connect anything. But it has the same embryonic origins as other connective tissues, as well as the three components: cells, fibers (for blood clotting) and a liquid matrix – the plasma.

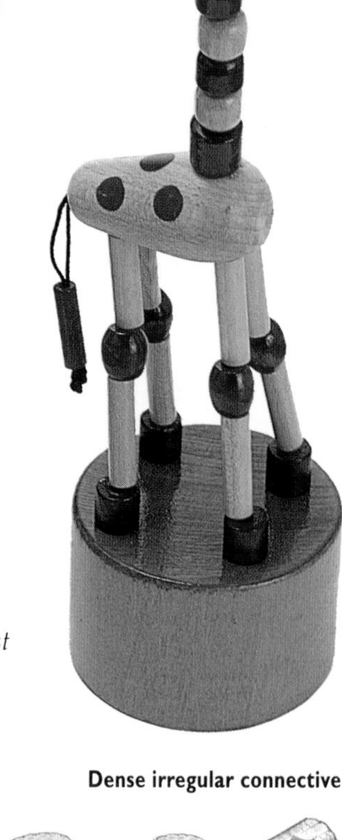

This model giraffe collapses *in a heap* ***(below right)*** *when a button on its base is pressed. The button releases the tension in a set of cords that run through its separate parts; without the tension the parts flop.*

The human body would do the same if its connective tissue disappeared. Bone, the stiffening structural core of the body, is a type of connective tissue, and so is cartilage, which forms the framework of parts such as the nose, ears, and larynx (voice box). Cartilage also plays an important role in another body part, the trachea, or windpipe. Bands of cartilage hold it in shape and prevent it from being squashed flat, so that air can pass in and out of the lungs.

Dense irregular connective tissue *is flexible and tough, able to stretch in almost any direction. It is widespread inside and around organs, forming sheaths and capsules and linking parts together. It forms the middle layer of skin, encases nerves, muscles, and blood vessels, forms the outer capsules of glands and the eyeball, and makes up the outer "skin" of bones and cartilages.*

Dense irregular connective tissu

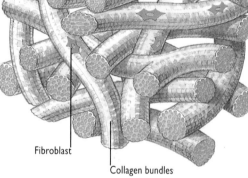

Fibroblast

Collagen bundles

Ligament

Ligaments*, the stretchy straps that hold bones together in and around joints, are made of ligamentous regular connective tissue. The bundles of collagen, sometimes interwoven with elastin, curve gently and are packed in layers or sheets. Their curves can be straightened under tension, but spring back when released.*

Beneath the outer layer of the skin, *and just above the underlying muscles and tendons, is the adipose connective tissue, or fat. This type of connective tissue is unusual in that it does not have a supportive role; it does, however, provide insulation against heat loss and both cushions and protects.*

Tendon

Tendons, *which connect muscles to what they pull, consist of tendinous regular connective tissue, or sinew, that is made of bundles of collagen packed together and twisted like rope fibers. Sheets of tendinous connective tissue on and within muscles are known as fascia and help to bind and align the tissues in the muscle. Sheets of tendinous and ligamentous tissue combined form aponeuroses, which act as anchor points for tendons and muscles.*

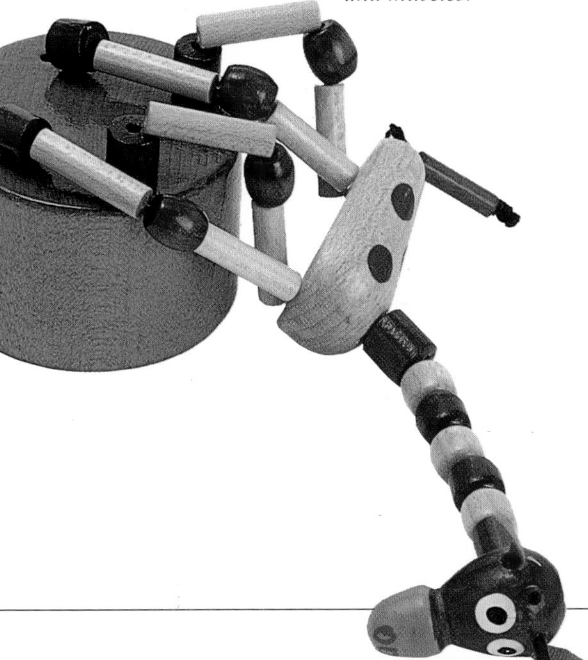

Connecting *the fingers to the muscles in the upper forearm that make them move are tendons. In fact, 21 long, cordlike tendons run through the wrist. They are "glued" to the finger bones at one end and to the forearm bones near the elbow at the other end.*

Wrapped around the tendons at the wrist, like a watchband under the skin, are two bands of fibrous tissue called retinacula. They keep the tendons aligned and following the correct line of pull as they slide beneath it. For instance, the extensor retinaculum on the back of the wrist restrains the extensor muscle tendons which straighten the fingers.

At sites where the tendons might rub and chafe adjacent tissues, they are encased in slippery bags known as synovial sheaths. These have the same membranes, smooth surfaces and lubricating fluid as bone-to-bone synovial joints.

See also

SUPPORT AND MOVEMENT
- Staying in shape 12/13
- The living framework 16/17
- Bone junction 18/19
- Muscles at work 26/27

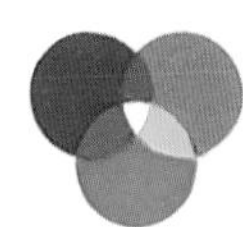

CONTROL AND SENSATION
- The chemicals of control 38/39
- The biological camera 56/57

CIRCULATION, MAINTENANCE, AND DEFENSE
- In circulation 124/125
- Blood: supplying the body 126/127
- Outer defenses 142/143
- Helpers and killers 148/149

Making a move

Every body movement is powered by muscles. But muscles can only pull, so how do you push?

Our muscles – what we call the "red meat" on other animals – make up almost half the body's weight and produce all the body's movements and postures. The muscles that move you around are called skeletal, or voluntary, muscles and are under direct control of your will.

A muscle is a biological puller. It works by getting shorter, or contracting, and moving the bones that are attached to it. The body has opposing, or antagonistic, pairs of muscles – one to lift the leg, for example, the other to lower it. The "lifting" muscle pulls its bone one way and its opposing partner, or antagonist – the "lowering" muscle – pulls the same bone in the opposite direction. In each case, as one muscle contracts and exerts a force, its opposite number automatically relaxes and is passively stretched.

However, this two-way partnership is a considerable simplification. In reality, muscles work in teams of 20, 30, or even more. They move, lift, and rotate bones so that you can pull, push, squeeze, stretch, walk, jump, and run. As muscle teams in one region of the body make the primary movements, others elsewhere adjust and compensate so that you are always poised and balanced.

MUSCLES FACT FILE

The most variable muscle is probably the platysma in the side of the neck: it covers the whole region in some people, is straplike in others, and completely missing in a few.

No. of skeletal, or voluntary, muscles	640 approx.	
Bulkiest	gluteus maximus	2¼ pounds (1 kg) or more
Smallest	stapedius	⅕ inch (0.5 cm)
Longest	sartorius	19½ inches (50 cm)
Longest group	erector spinae	35½ inches (90 cm)
Widest	external oblique	17¾ inches (45 cm)
Percentage of body weight as muscle	male	40–50
	female	30–40
Muscle fiber length	ave.	1⅕ inches (3 cm)
	max.	12 inches (30 cm)
	min.	1/25 inch (0.1 cm)

The typical skeletal muscle *has a bulging body, or belly, that tapers at each end into a cordlike tendon, which is firmly anchored into a bone of the skeleton. Some muscles are triangular, others are sheetlike, according to the job they do and how they are attached. For instance, the upper back is dominated by the trapezius, a large triangular muscle that extends up into the neck and from the spine to the shoulder blade (scapula). By pulling or lifting this bone, it can alter the position of the whole arm. The deltoid, one of six main muscles that stabilize the shoulder joint, lifts and twists the arm. The shoulder joint requires support since it is inherently unstable – the upper-arm bone is set in the shallowest of sockets.*

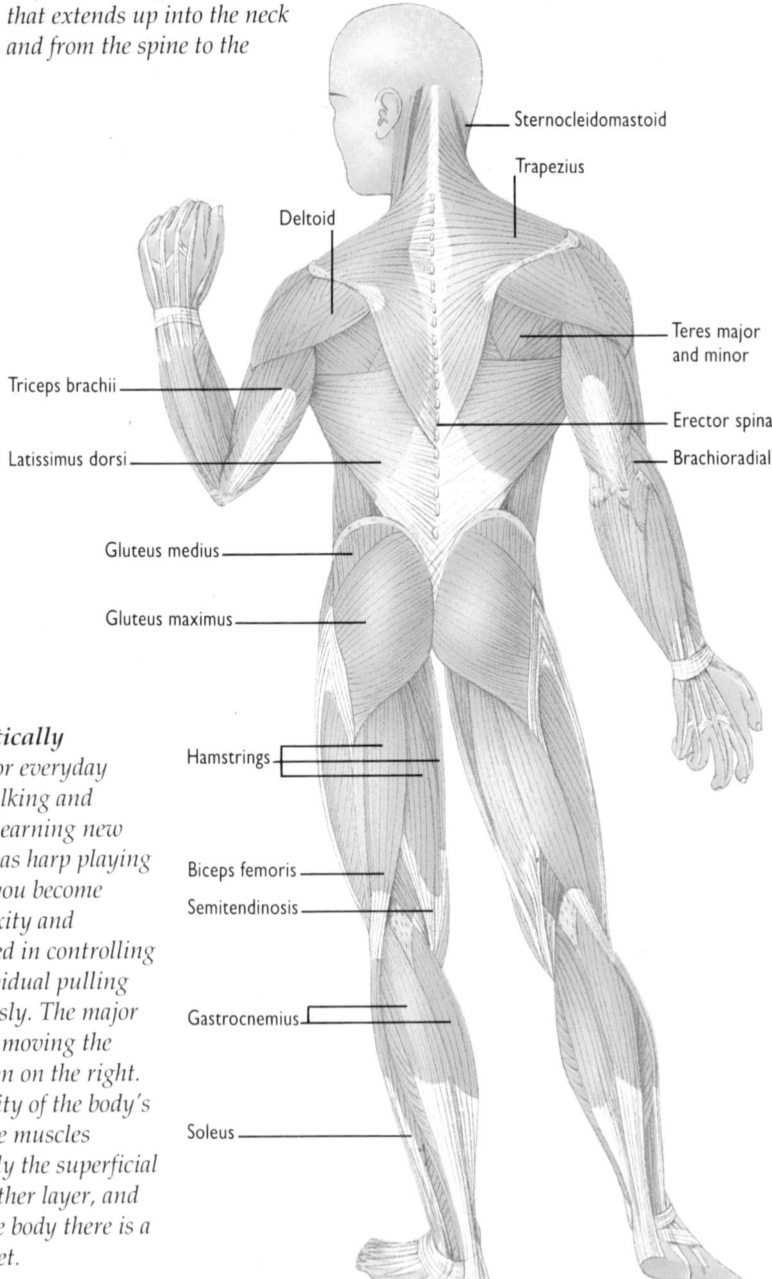

The brain automatically controls muscles *for everyday activities such as walking and sitting. Only when learning new physical skills, such as harp playing or windsurfing, do you become aware of the complexity and coordination involved in controlling more than 600 individual pulling devices simultaneously. The major muscles involved in moving the body about are shown on the right. Such is the complexity of the body's musculature that the muscles depicted here are only the superficial ones; beneath is another layer, and in some places in the body there is a third, deep muscle set.*

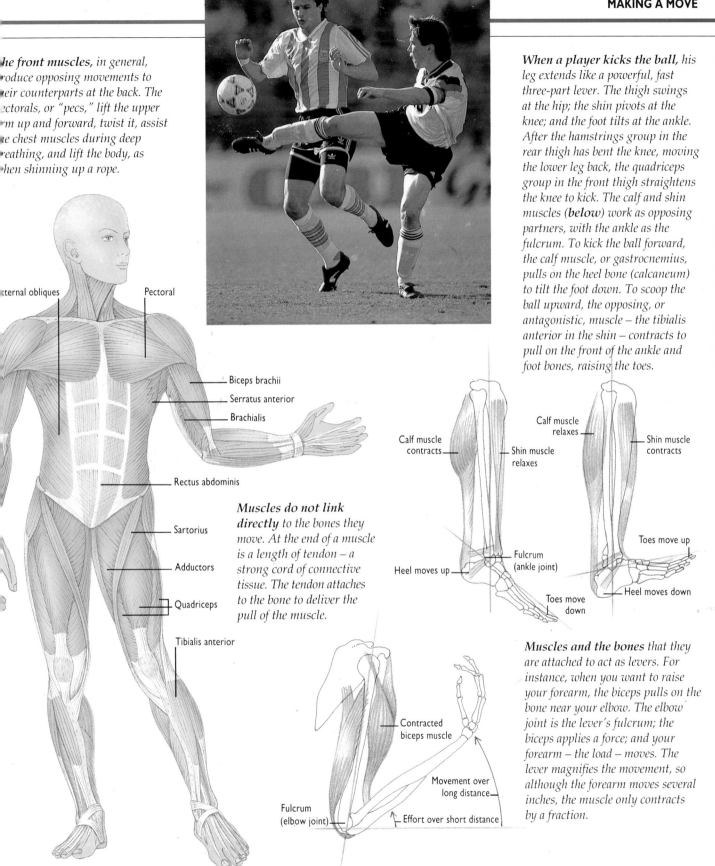

***he front muscles,** in general, roduce opposing movements to eir counterparts at the back. The ctorals, or "pecs," lift the upper m up and forward, twist it, assist e chest muscles during deep eathing, and lift the body, as hen shinning up a rope.*

*When a player kicks the ball, his leg extends like a powerful, fast three-part lever. The thigh swings at the hip; the shin pivots at the knee; and the foot tilts at the ankle. After the hamstrings group in the rear thigh has bent the knee, moving the lower leg back, the quadriceps group in the front thigh straightens the knee to kick. The calf and shin muscles (**below**) work as opposing partners, with the ankle as the fulcrum. To kick the ball forward, the calf muscle, or gastrocnemius, pulls on the heel bone (calcaneum) to tilt the foot down. To scoop the ball upward, the opposing, or antagonistic, muscle – the tibialis anterior in the shin – contracts to pull on the front of the ankle and foot bones, raising the toes.*

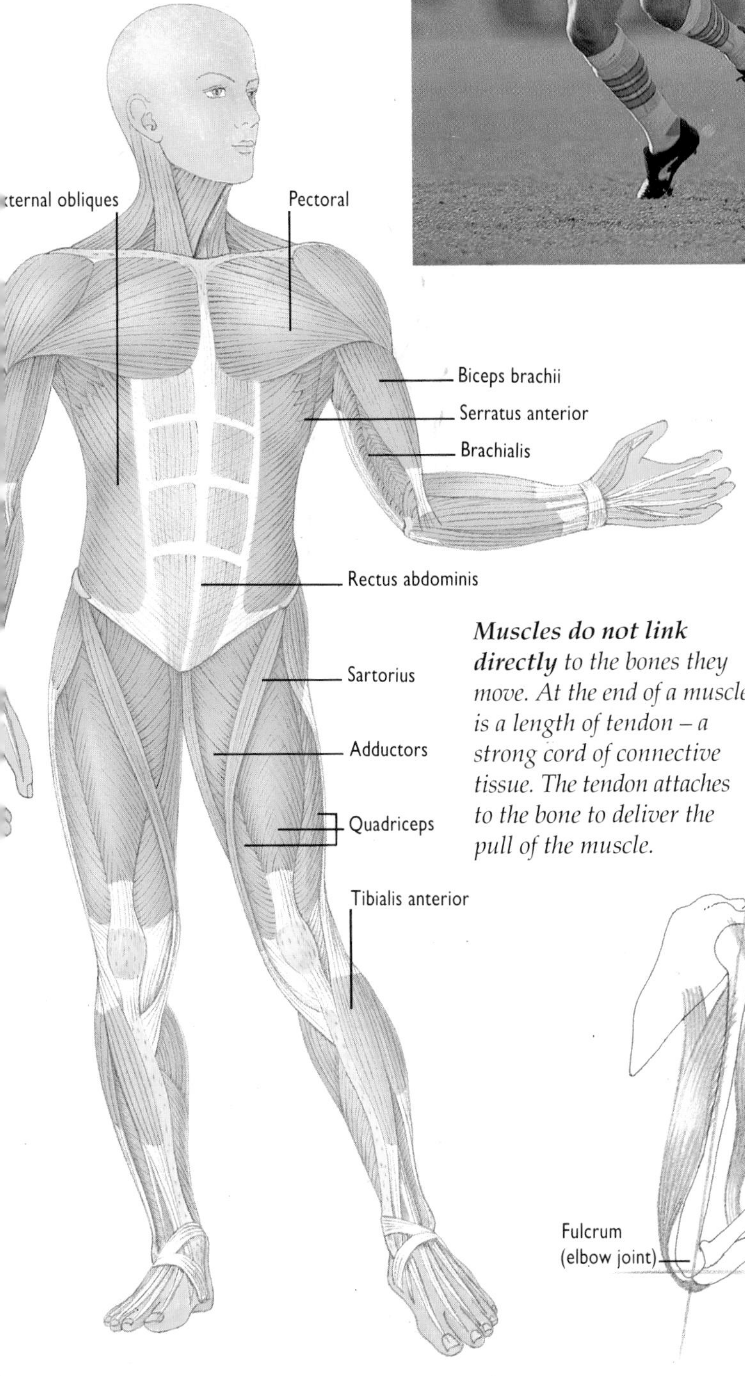

xternal obliques
Pectoral
Biceps brachii
Serratus anterior
Brachialis
Rectus abdominis
Sartorius
Adductors
Quadriceps
Tibialis anterior

***Muscles do not link directly** to the bones they move. At the end of a muscle is a length of tendon – a strong cord of connective tissue. The tendon attaches to the bone to deliver the pull of the muscle.*

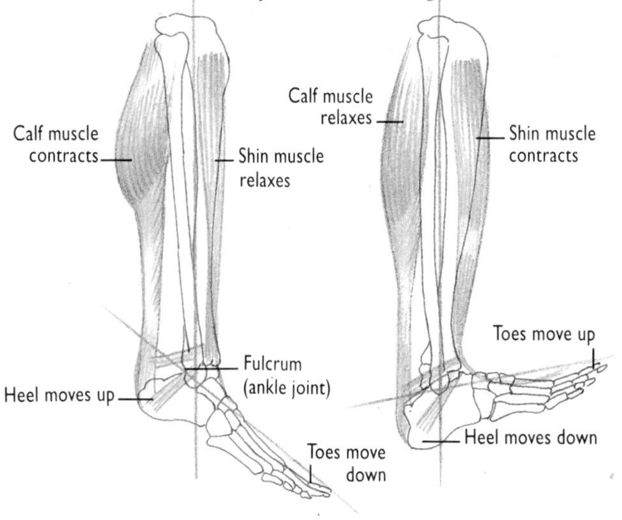

Calf muscle contracts
Shin muscle relaxes
Heel moves up
Fulcrum (ankle joint)
Toes move down
Calf muscle relaxes
Shin muscle contracts
Toes move up
Heel moves down

Contracted biceps muscle
Movement over long distance
Fulcrum (elbow joint)
Effort over short distance

***Muscles and the bones** that they are attached to act as levers. For instance, when you want to raise your forearm, the biceps pulls on the bone near your elbow. The elbow joint is the lever's fulcrum; the biceps applies a force; and your forearm – the load – moves. The lever magnifies the movement, so although the forearm moves several inches, the muscle only contracts by a fraction.*

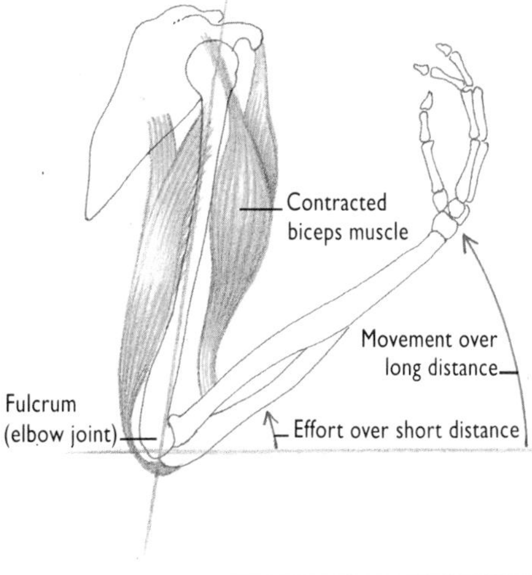

See also

SUPPORT AND MOVEMENT
▶ Staying in shape 12/13
▶ Bearing the load 14/15
▶ The living framework 16/17
▶ Bone junction 18/19
▶ Holding it together 20/21
▶ Fine movement 24/25
▶ Muscles at work 26/27

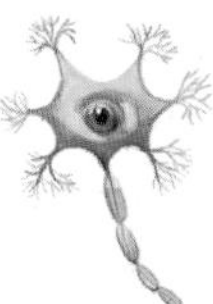

CONTROL AND SENSATION
▶ Down the wire 46/47
▶ Into action 50/51

ENERGY
▶ Fueling the body 94/95

REPRODUCTION AND GROWTH
▶ Building bodies 160/161

Fine movement

Some carefully controlled muscles create extremely delicate movements.

Look in the mirror, and raise your eyebrows slowly. As they lift, your expression changes from slight surprise to total astonishment. Along with body movements, facial expressions are a major part of our everyday life, as we flash a smile or frown in anger. Like the bigger movements, such as running and kicking, they depend on muscles – but thinner, shorter muscles for tiny, delicate movements. For instance, a complex network of neck muscles adjusts the larynx, or voice box, to control speaking volume and pitch. If the combined effect of these muscles stretches your vocal cords by just 1/10 inch (2 mm), the pitch of your voice will rise sharply, making it into a shrill shriek.

Dozens of the body's muscles, especially those around the face and in the hands and feet, are as thin as string. They cannot pull with any great strength, but they can make delicate, well-controlled movements. This is due to their small size and to the way in which the nerves are wired up to the hair-thin muscle fibers that are the basic units of a muscle. The biggest muscles have hundreds of thousands of muscle fibers; the smallest ones possess only a few hundred. In a large muscle, such as the gluteus in the hip, one nerve fiber (motor neuron) stimulates the contraction of thousands of muscle fibers. In a small muscle, like the straplike muscles that swivel an eye, each nerve fiber stimulates as few as 12 muscle fibers. This gives amazing precision of control, allowing some small muscles to make their own delicate movements, especially around the eyes, nose, and lips. It also allows some small muscles to fine-tune large-scale moves, as when you balance on one foot and small muscles in your toes constantly adjust their pressure to keep you upright.

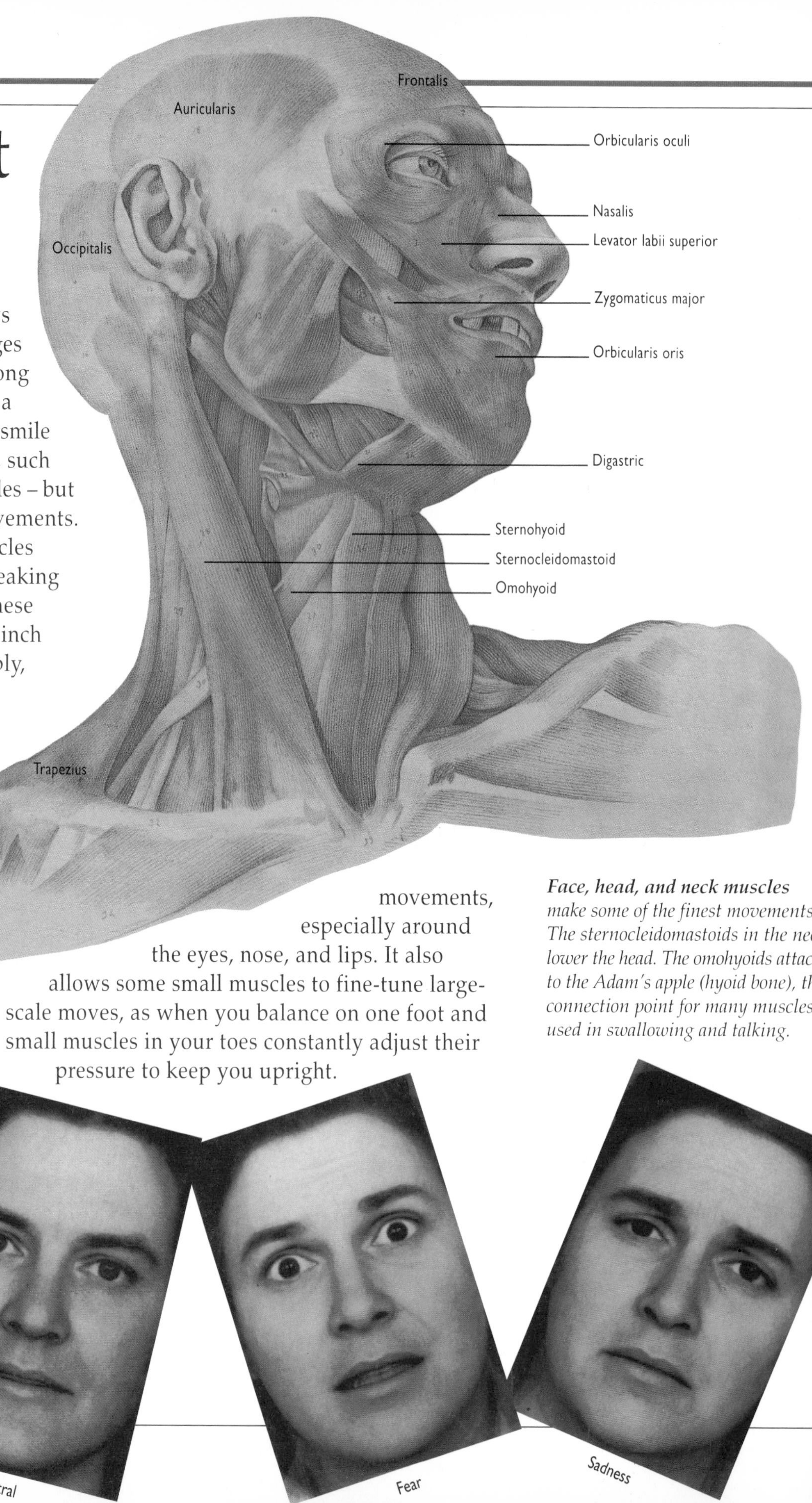

Face, head, and neck muscles *make some of the finest movements. The sternocleidomastoids in the nec[k] lower the head. The omohyoids attac[h] to the Adam's apple (hyoid bone), th[e] connection point for many muscles used in swallowing and talking.*

LOOKING AROUND

When you shift your gaze from a tree in the distance to another tree next to it, your eyeball swivels in an arc of less than one degree. This calls for extremely small and accurate movements by the muscles that move the eye. In fact, they work to tolerances of hundredths of an inch, and their stretch sensors tell the brain about the eye's angle of view for distance and depth perception.

Each eyeball has six muscles, attached by their front ends to different areas around its sides. Most of the muscles extend backward to anchor in a ring of fibrous tissue at the rear of the eye socket (orbit). The muscles partially contract or relax with great precision to swivel the eye from side to side, up and down, twist it clockwise and counterclockwise, and any combination thereof. The angle of pull of one eye muscle – the superior oblique – is altered where its tendon passes around a trochlea, or bony "pulley." So this muscle rotates or twists one eye clockwise, the other counterclockwise.

Happiness

Anger

Surprise

Disgust

More than 50 small muscles *crisscross the face and head, just under the skin. Some are attached to the skull bones at both ends. Others have a bone anchorage at only one end and are attached either to each other or to a common anchor point by fibrous straps or sheets. In closely coordinated action, they produce a galaxy of facial expressions. Each zygomaticus major runs from the cheekbone down to the corner of the mouth, and less than* $\frac{1}{5}$ *inch (5 mm) contraction in them produces the flicker of a smile. The orbicularis oris purses lips and has an important role in speech; each nasalis flares a nostril; each frontalis raises an eyebrow; the occipitalis draws the scalp back; the levator labii superior raises the top lip; and in a few people, the auricularis wiggles an ear.*

After intensive research, psychologists have found that there are seven universally recognized facial expressions. Computer-created composites can be made using photographs of actors whose features are accurately measured while assuming each of these expressions. Such images are now being used in research into areas such as autism and Parkinson's disease, whose sufferers may have problems controlling facial expressions.

See also

SUPPORT AND MOVEMENT
- Bearing the load 14/15
- Making a move 22/23
- Muscles at work 26/27

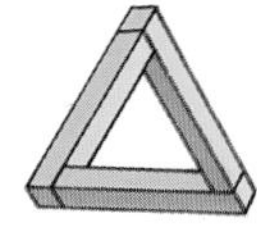

CONTROL AND SENSATION
- Down the wire 46/47
- Into action 50/51
- Seeing the world 54/55
- The biological camera 56/57
- In the balance 70/71
- A sense of self 74/75

ENERGY
- Chewing it over 98/99

Muscles at work

Powered molecular hinges are the basis of all muscle movement.

With your palms down and fingertips facing each other, slide the straight fingers of one hand between those of the other, so that the fingers interlock. On a scale millions of times smaller, at the level of molecules, this is a simple model of how your arm and hand muscles are producing the very movements you are now making.

Your fingers correspond to two body proteins, actin and myosin, the essential ingredients of muscle tissue. They are extremely small – about 80,000 myosins laid side by side would form a flat multistrand ribbon 1/25 inch (1 mm) wide, and actins are only half as thick. Actin is a long ropelike molecule; myosin consists of a main backbone, with armlike cross bridges protruding at regular intervals. Each cross bridge supports a myosin head, which is instrumental in muscular contraction.

The myosin heads attach to a neighboring actin and bend, pulling the actin along. The heads then detach from the actin, straighten, and repeat the process. It is like pulling on a rope with your hands. Billions of these incredibly small movements happen in a fraction of a second throughout the muscle, to make it contract. The process is driven by high-energy molecules of ATP (adenosine triphosphate), which the body uses to power many of its chemical processes. Muscle contraction is controlled by nerve signals from the brain, arriving along motor nerves.

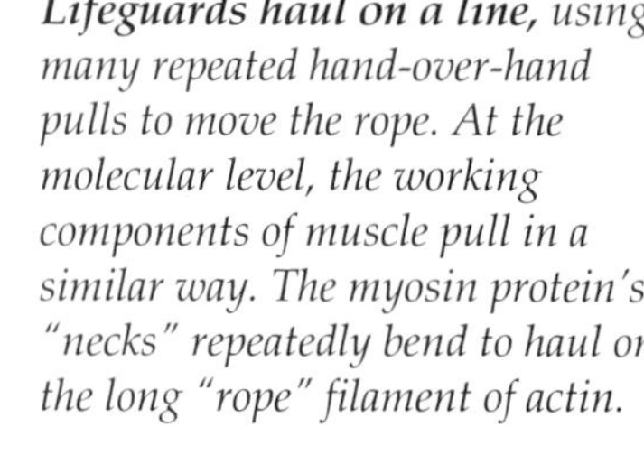

***Lifeguards haul on a line,** using many repeated hand-over-hand pulls to move the rope. At the molecular level, the working components of muscle pull in a similar way. The myosin protein's "necks" repeatedly bend to haul on the long "rope" filament of actin.*

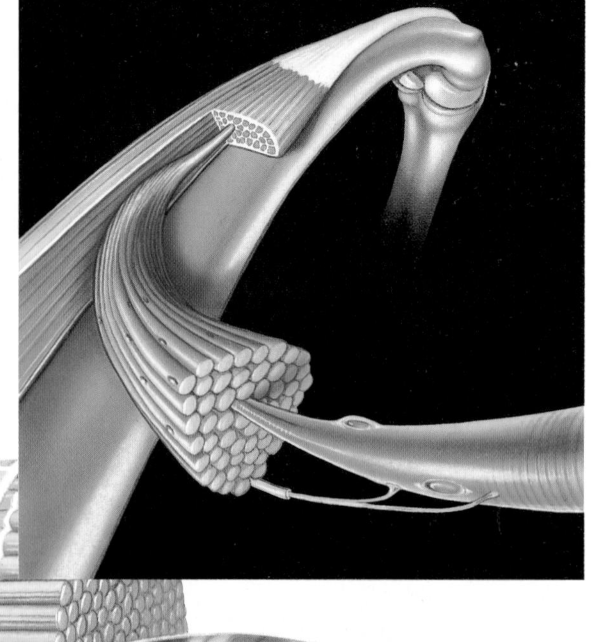

***The muscles that move us** are made of bundled fibers. These fibers in turn consist of hair-fine "threads" known as muscle cells, or myofibers. These rod-shaped cells have many nuclei and can be up to 1 foot (30 cm) long. Each myofiber contains a bundle of hundreds of even smaller fibers, the myofibrils, which are made of two types of muscle filaments (myofilaments) – actin and myosin. Seen through a microscope, the myofibrils give the myofibers a regular pattern of stripes and bands.*

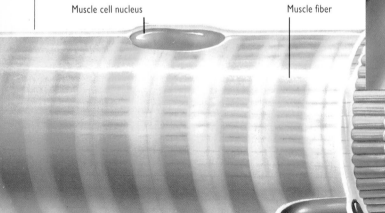

Muscle cell nucleus

Muscle fiber

Motor nerve endings

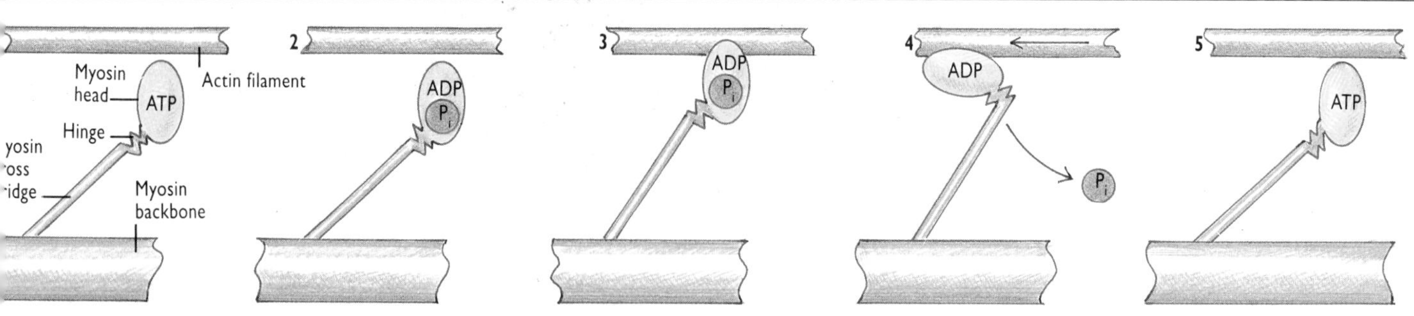

*he **chemical mechanism** that pulls individual muscle filaments past each her (**above**), making muscles contract, is known as the cross-bridge cycle. nergy-rich ATP (adenosine triphosphate) attaches to the myosin head (**1**). then breaks down into ADP (adenosine diphosphate) and P_i, an inorganic hosphate group (**2**). This split changes the shape of the myosin head, and it taches, or binds, to the actin myofilament (**3**). As the P_i leaves the myosin ead, the "neck," or hinge, kinks and drags the actin along slightly (**4**). Finally, e ADP separates, to be replaced by another high-energy ATP molecule, and e myosin neck straightens (**5**). The whole cycle repeats hundreds of times a cond, on thousands of actin–myosin myofibrils in each myofiber, to produce e powerful pull of the whole muscle.*

***In a fully relaxed,** or stretched out, muscle, the ends of the actin filaments project only slightly into the bundles of myosins (**1**), and the I-band is at its widest. During contraction, thousands of myosin heads pull the actin myofilaments past, "hand over hand." The result is that the adjacent contractile units (sarcomeres) of the myofiber shorten (**2**). The I-bands become shorter while the A-bands stay the same width. Fully contracted, the I-bands narrow to nearly nothing, and the H-zone (within the A-band) can disappear (**3**). At full contraction the whole muscle is almost half of its relaxed length.*

Troponin protein

Actin filament

***ctin** is made of b-units twisted gether. Troponin oteins switch yosin head binding n and off under the fluence of calcium.*

Myosin cross bridge

Myosin head

***Myosin,** the thick myofilament, has a long backbone with many armlike cross bridges. These end in heads, which tilt to and fro producing muscular contraction.*

osin filament

Myosin backbone

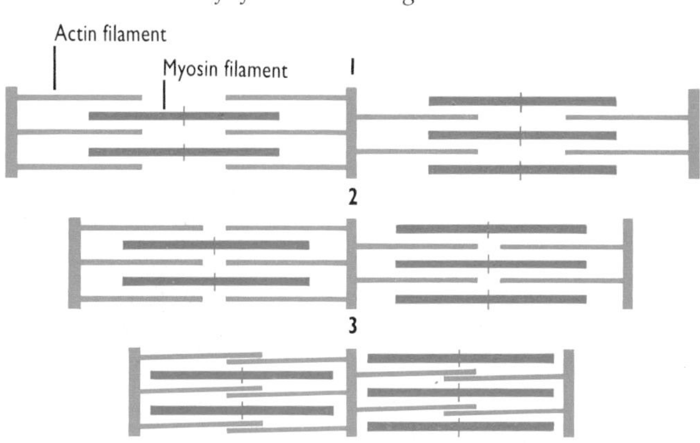

***Skeletal muscle is striped** because of the way actin and myosin interlock and overlap. The A-band is a stack of myosins. The I-band is actins where they do not overlap myosins. The actins are joined at their ends to their neighbors, forming the Z-lines.*

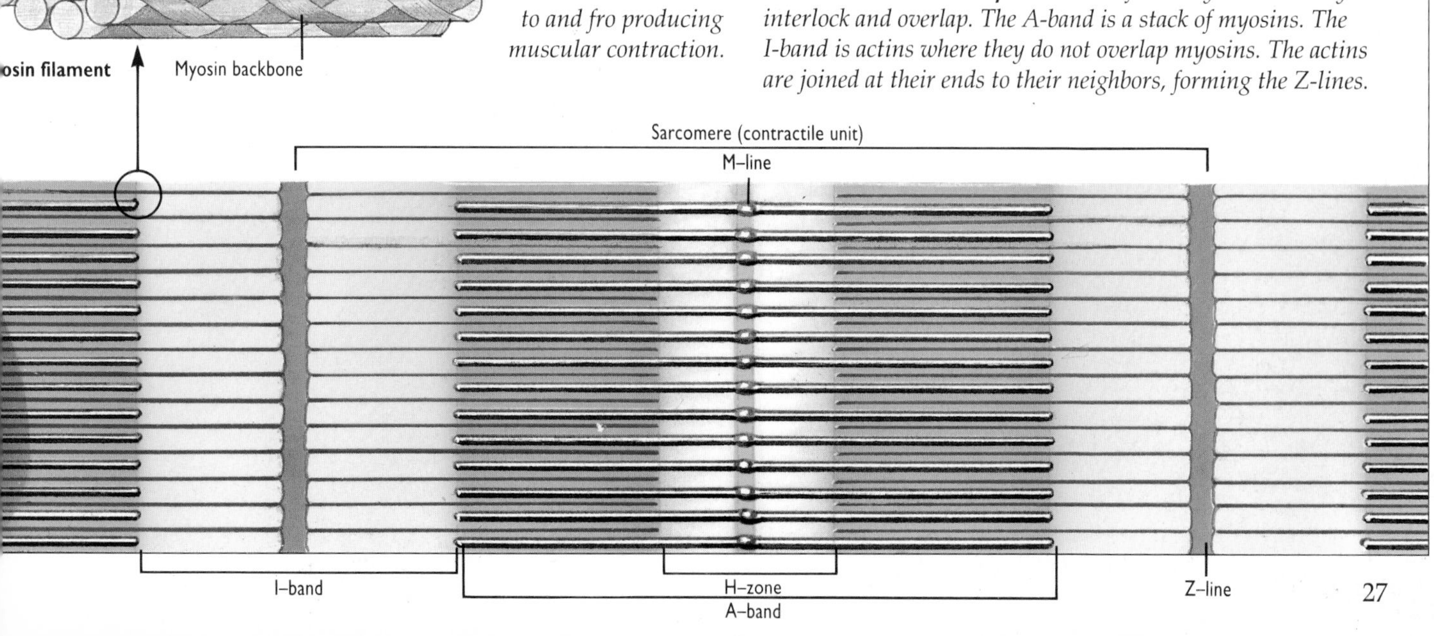

See also

SUPPORT AND MOVEMENT
- Making a move 22/23
- Fine movement 24/25
- The heart's special muscle 28/29
- Smooth operators 30/31

CONTROL AND SENSATION
- Down the wire 46/47
- Into action 50/51

ENERGY
- Fueling the body 94/95
- The cell and energy 112/113
- Cells at work 114/115

CIRCULATION, MAINTENANCE, AND DEFENSE
- Blood: supplying the body 126/127

REPRODUCTION AND GROWTH
- Building bodies 160/161

The heart's special muscle

Human life depends on the heart muscle's untiring and automatic contractions, continuously pumping blood around the body.

Unlike the muscles in, say, your arm, heart, or cardiac, muscle never tires. One reason for this is that the blood supply to heart muscle is copious. It has to be to supply the cardiac muscle with the oxygen and high-energy glucose it needs, and to flush away wastes, such as lactic acid, that might otherwise cause fatigue and cramps.

Heart muscle is like a combination of skeletal and visceral, or smooth, muscle. Looked at through a microscope, it has the regularly patterned bundles of fibers of skeletal muscle, but they branch, recalling the random arrangement of smooth fibers. Cardiac muscle also has its own built-in rhythm of about 100 beats each minute. Nerve signals from the brain usually slow this to about 60–80 in resting adults.

The inherent beating rhythm starts in a small patch of cells – the sinoatrial node – in the wall of the upper right atrium. This is the heart's pacemaker. It generates electrical signals that pass along nervelike tracts to a relay station at the base of the right atrium – the atrioventricular node. The signals continue along a thick conducting tract known as the bundle of His, then through its left and right branches, which subdivide further into a network of modified muscle fibers – the Purkinje fibers – in the walls of the ventricles. As each burst of electricity from the sinoatrial node reaches the muscle fibers, they contract, and so the heart pumps. The design of the wiring system means that the wave of contraction starts at the pointed base of the heart and works its way upward. Blood is thus ejected up into the main vessels, rather than being trapped in the bottom of the heart.

The muscles of a fully fit rowing eight *on their starting burst of strokes can produce almost as much power as a small car. But if the crew members do not row in a controlled, coordinated fashion, they lose their timing. The boat lurches and jolts, and much of the power is wasted. This is where the cox comes in. The cox times the strokes, sets the rhythm, and shouts instructions. The different sections of heart muscle are similarly synchronized in their rhythmic contractions by the heart's own natural pacemaker, the sinoatrial node.*

SHOCKED BACK TO LIFE

Cardiac arrhythmias are medical conditions in which the heart loses its coordinated, rhythmic beating. Its various sets of muscles "do their own thing" and contract incompletely or out of sequence, upsetting the regular pumping of blood. Cardiac arrhythmias can be due to problems with the electrical coordinating system of the heart, with heart muscle itself, or with its blood supply.

In ventricular fibrillation, for example, the lower main pumping chambers – the ventricles – contract very quickly, almost in a trembling fashion, yet pump hardly any blood. This can be a life-threatening emergency. One treatment is defibrillation: giving the heart a brief electric shock to jerk it out of its arrhythmia. The electric current, produced by a machine termed a defibrillator, is dispensed via two large metal plates (electrodes) held against the chest. The heart stops beating momentarily, in many cases allowing the sinoatrial node to regain control.

Defibrillator electrodes, *pressed onto the chest skin on each side of the heart, deliver a burst of electricity that overwhelms the heart's own electrical conducting system. Several bursts may be needed, and the electricity makes other muscles contract, too, so the patient may convulse and jerk around. Other people must stand well clear.*

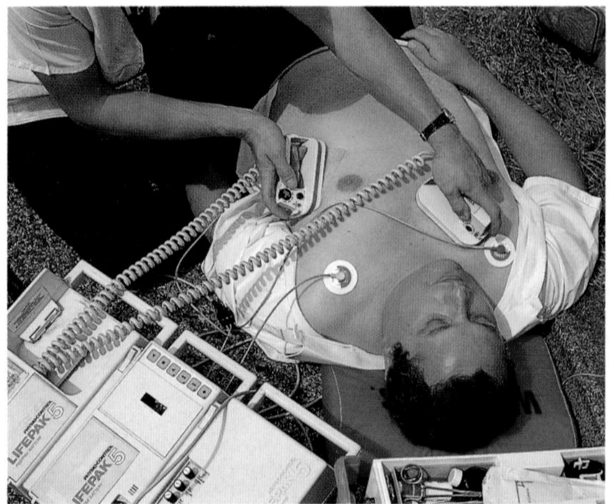

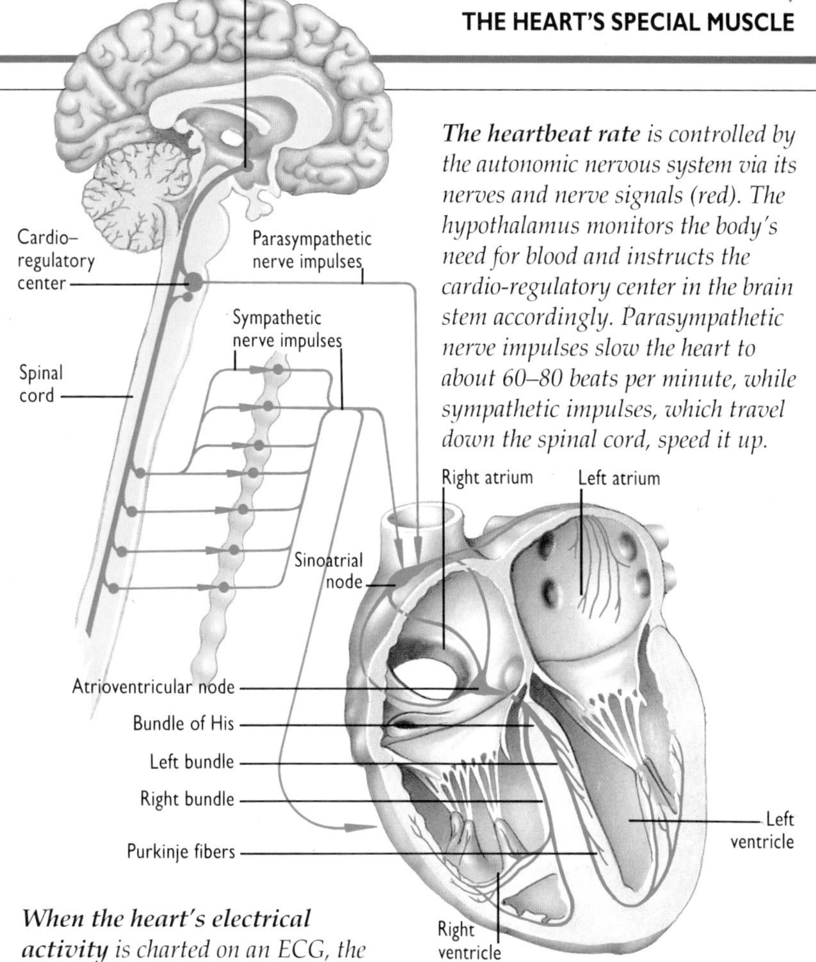

***The heartbeat rate** is controlled by the autonomic nervous system via its nerves and nerve signals (red). The hypothalamus monitors the body's need for blood and instructs the cardio-regulatory center in the brain stem accordingly. Parasympathetic nerve impulses slow the heart to about 60–80 beats per minute, while sympathetic impulses, which travel down the spinal cord, speed it up.*

***When the heart's electrical activity** is charted on an ECG, the P-wave represents signals passing to the atrioventricular from the sinoatrial node. This is followed by atrial systole, when the atria pump blood into the ventricles. The dip at Q records signals passing along the bundle of His, and R and S mark their passage along the bundle branches and Purkinje fibers. These signals trigger ventricular systole, the contraction of the main chambers. At T the ventricles are relaxing during diastole.*

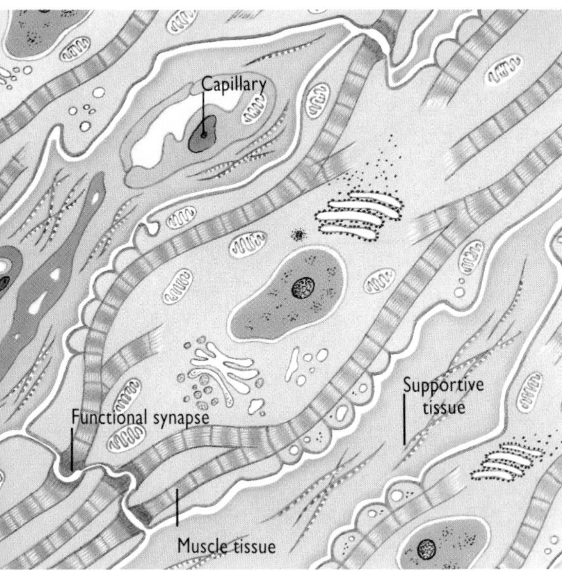

***Purkinje fibers (above)** are the heart's "nerve cells." They are cardiac muscle cells that also have impulse-transmitting membranes typical of true nerve cells. The impulses, which they conduct at 5¼ feet/sec (1.6 m/sec), reach every minute nook and cranny of the ventricle walls to stimulate the heart muscle to contract. Signals cross from fiber to fiber at gaps that are similar to nerve-cell synapses.*

The first heart sound – heard using a stethoscope – is the two valves between the ventricles and atria closing. The second is the two exit valves from the ventricles slamming shut.

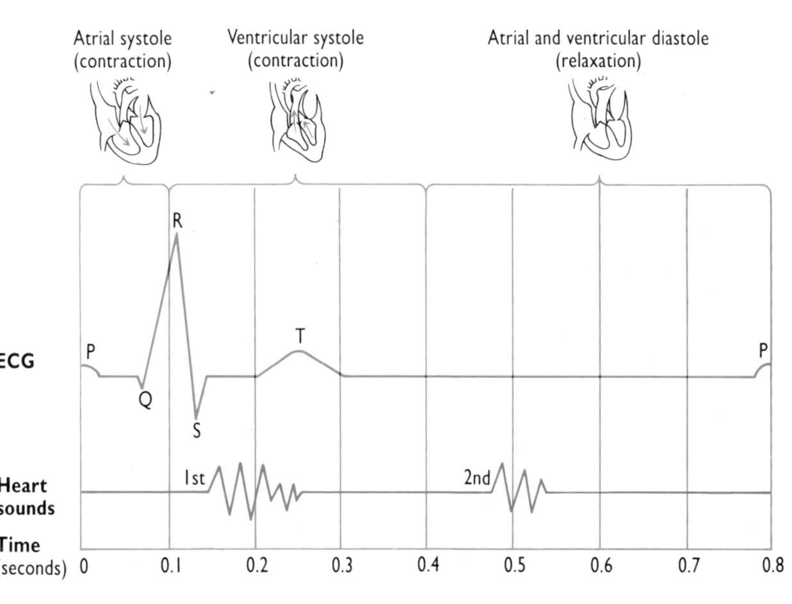

See also

SUPPORT AND MOVEMENT
▶ Muscles at work 26/27

▶ Smooth operators 30/31

CONTROL AND SENSATION
▶ Control systems 36/37

▶ The chemicals of control 38/39

▶ Down the wire 46/47

▶ The automatic pilot 48/49

ENERGY
▶ Cells at work 114/115

CIRCULATION, MAINTENANCE, AND DEFENSE
▶ The living pump 120/121

▶ Growing heart 122/123

▶ In circulation 124/125

▶ Blood: supplying the body 126/127

Smooth operators

The next time you hear your stomach gurgle, the chances are that you are listening to muscles in action.

When a person is asleep, most of the skeletal (voluntary, or striped) muscles show little activity. But the breathing and heart muscles continue their rhythmic contractions, under autopilot control from the brain. There is also activity of muscles in the walls of internal body organs, such as the esophagus (gullet), stomach, intestines, and bladder, and in the small airways in the lungs and small arteries all over the body. These never-resting muscles are made of a type of tissue which – to contrast it with skeletal muscle – is known as visceral, involuntary, or smooth muscle. This is because it makes up much of the bulk of the visceral (abdominal) organs, is largely under involuntary, or unconscious, control from the brain, and has a smooth, unstriped appearance under a microscope.

Smooth muscle is not connected to bones. Generally, it forms tubes or sacs and contracts to lessen the space it encloses. The fibers of smooth muscle are also usually arranged in layers, each contracting in a different direction. In a tube such as the gullet, fibers arranged in a longitudinal layer (lengthwise) contract to shorten and widen the tube, while the circular layer (fibers arranged around the circumference) contracts to narrow and lengthen it. This so-called peristalsis pushes its contents along.

***A snake or worm** essentially consists of two long tubes of muscle one inside the other. The inner one is the digestive tract, which pushes food along by waves of contraction known as peristalsis. The egg-eating snake can force an egg bigger than its head down its peristalsing gullet where sharp spines on the vertebral neck bones crack and pierce the egg to collapse the shell and release the contents. The outer tube of muscle is the body wall. In worms, this also writhes in peristaltic waves to propel the creature forward. Snakes, by contrast, tend to move by bending the body from side to side in S-shaped curves, pushing against any small irregularity to slither forward.*

ASTHMA AND SMOOTH MUSCLE

The lungs' airways, the bronchi and bronchioles, have smooth muscle in their walls. In susceptible people, this muscle may go into spasm (contract) due to an allergic or hypersensitive reaction. The spasm is associated with inflammation of the bronchial lining and excess production of mucus.

Each of these three factors – muscle spasm, swollen lining, and excess mucus – makes the airway narrower, leading to shortness of breath and also to breathing problems known as asthma. This condition tends to occur in episodes and can be triggered by a number of things, from exposure to house dust, pollen, animal fur, or feathers to emotional upset or sudden inhalation of cold air.

***Many asthma sufferers** gain relief from symptoms – which in a bad attack can cause the sufferer literally to gasp for breath – using a bronchodilator drug. Because the smooth muscles in the airways have contracted, air is held in the lungs, and the sufferer cannot breathe out properly. The drug is squirted from an inhaler and is breathed in as a fine mist, deep into the lungs, to act directly on the narrowed airways and dilate (widen) them.*

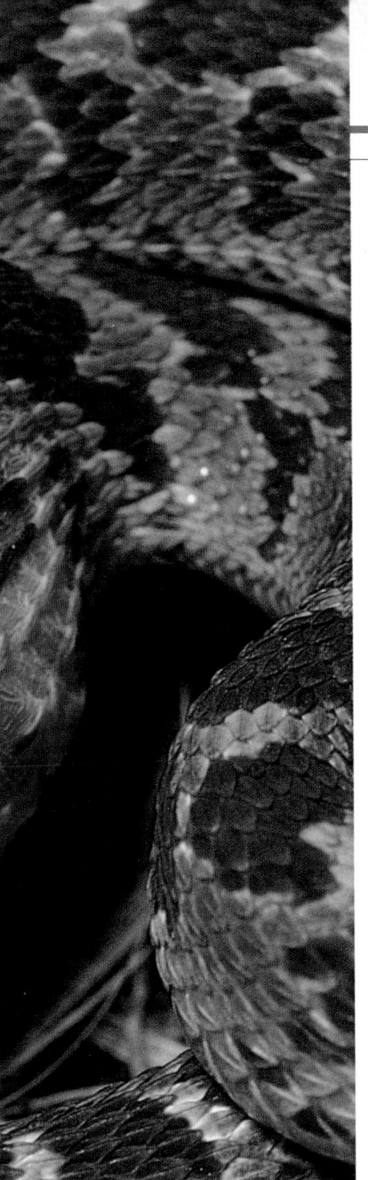

One of the clearest examples of peristalsis *is found in the esophagus (gullet), which propels swallowed lumps (boluses) of food down to the stomach. The gullet's circular muscle layer contracts behind the bolus in a traveling wave, massaging it on its way. Meanwhile, the longitudinal layer ahead of the bolus also contracts, making space for it by widening the tube. Gullet peristalsis is so strong that you can swallow even while upside-down. It can also go into reverse and throw the stomach contents up and out of the mouth. In the intestines, where the semi-digested food has a more souplike consistency, peristalsis does not deal with individual boluses. It has a general massaging effect on the liquid contents.*

In car tires *the strengthening materials – cords or plies – are arranged in bands of fibers lying in different orientations, rather like layers of smooth muscle in body organs. Each layer gives strength and stability in a certain direction, so that the whole design is firm yet supple.*

Smooth muscles in certain arterial walls can contract or relax to vary the diameter of the vessel, controlling blood pressure and distribution.

Sphincters – ring-shaped structures with a central hole – are also made of smooth muscle. They stay contracted involuntarily, keeping the hole shut; the muscles are relaxed voluntarily to open the hole and let out the contents. One such sphincter is in the anus at the end of the digestive tract.

At the molecular level, the fibers (myofibers) of smooth muscle contract in much the same way as skeletal myofibers, although they are shorter and more spindle-shaped. They also work a "roster system": while some contract, others relax and recover so that overall contraction in smooth muscle is maintained without fatigue.

See also

SUPPORT AND MOVEMENT
▶ Muscles at work 26/27

▶ The heart's special muscle 28/29

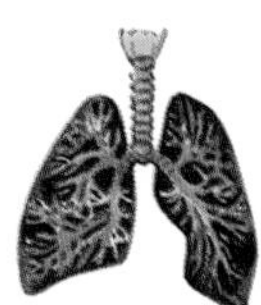

CONTROL AND SENSATION
▶ Control systems 36/37

▶ The chemicals of control 38/39

▶ The automatic pilot 48/49

ENERGY
▶ Chewing it over 98/99

▶ The food processor 100/101

▶ Absorbing stuff 104/105

▶ A deep breath 108/109

CIRCULATION, MAINTENANCE, AND DEFENSE
▶ In circulation 124/125

▶ The water balance 132/133

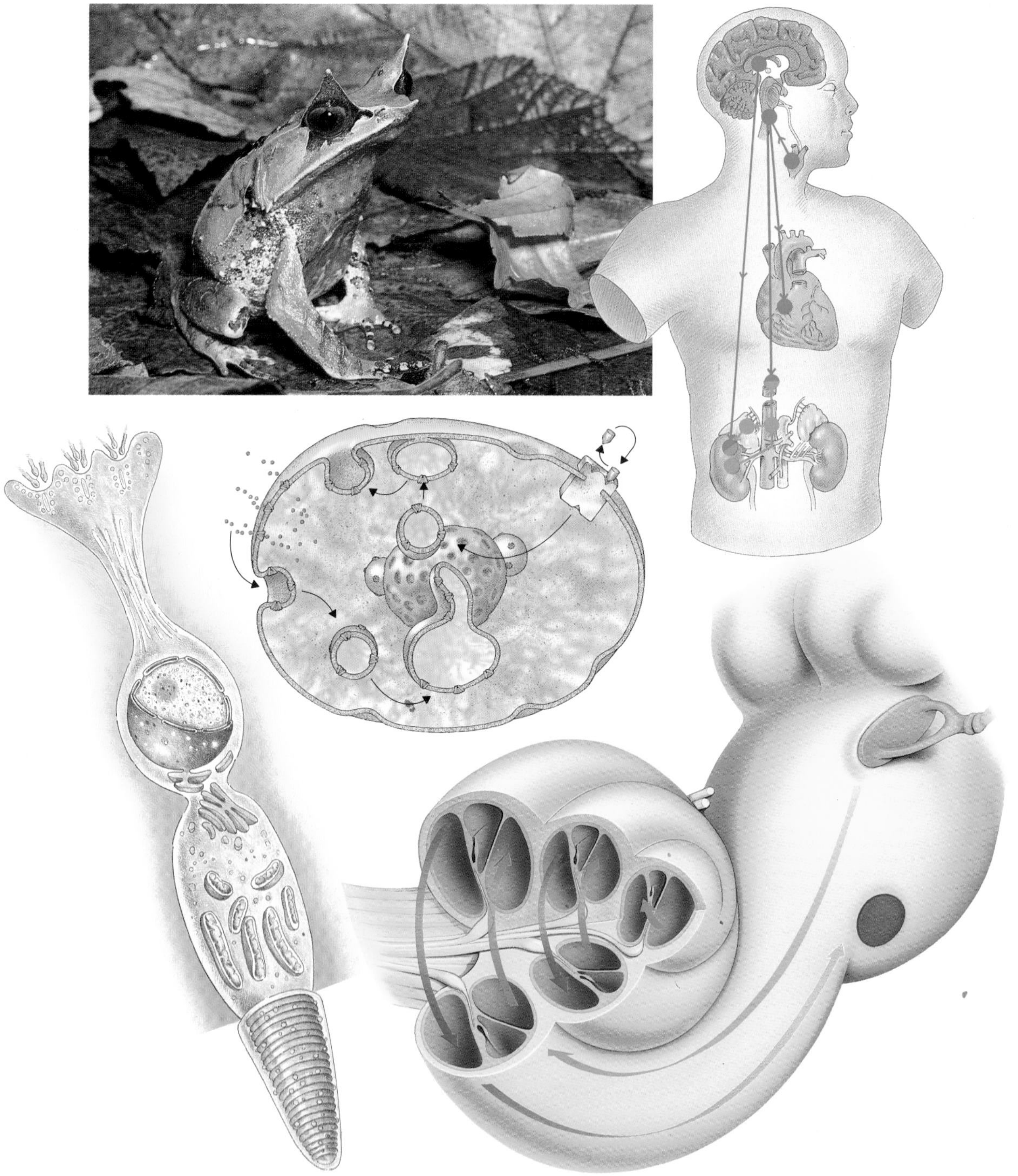

Control and Sensation

To walk down a street, you simply will the action to happen and off you go. You do not tie up your conscious awareness in controlling the complex muscle movements involved. Neither do you consciously adjust heartbeat and breathing rates to supply the muscles with the blood they need. And you certainly do not monitor and control the many other functions that maintain your physiological equilibrium. These duties are taken care of by parts of the brain and nervous system and by complex chemical control systems working behind the scenes. This allows you to focus on the myriad impressions from your senses and frees your mind to roam in the realm of pure thought where problems are solved, memories recalled, and dreams dreamed. And all these higher functions happen in a biological computer the size of a small melon – your brain.

***Left (clockwise from top):** camouflage in action; nerves and hormones in control; from sound to sensation in the ear; a seeing cell; the glucose story. **This page (top):** a call to arms; (**right**) matching up the puzzle.*

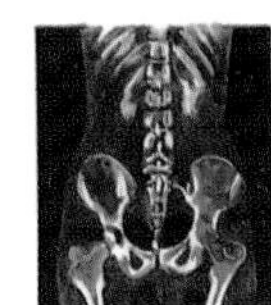

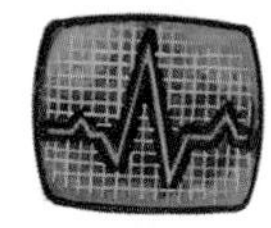

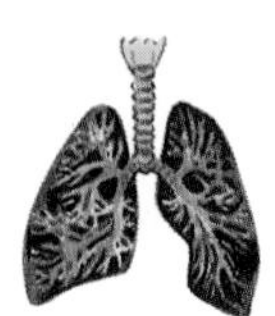

Steady as you go

Body cells are fussy about their environment – they function well only in constant conditions.

When winter approaches, with its weak sun and long, cold nights, have you ever wished that you could curl up and hibernate until spring, like a brown bear or hedgehog? This would involve great changes in your body. Its temperature would fall from a steady 98.6°F (37°C) to nearer 41°F (5°C), and the functions of the heart, lungs, digestive tract, kidneys, muscles, and other body parts would slow almost to a standstill. But the human being is a non-hibernating mammal – we are active and "warm-blooded" all year round. Indeed, conditions within the body – including body temperature, the regular pulsing of blood within the circulatory system, the quantities of fluids, and the concentrations of hundreds of chemicals, minerals, and other substances – remain remarkably constant. They are kept within close limits, day in, day out, whether we are active or inactive, awake or asleep, and whether we have just eaten or not eaten for 24 hours.

Why does the body keep its internal conditions within such narrow variations? The reason is rooted in our mammalian ancestry. Our cells, like those of most mammals, are extremely fussy and delicate. They have evolved over millions of years to work best in certain conditions: at a temperature of about 98.6°F (37°C), and with specific concentrations of sodium, potassium, and dozens of other substances. If conditions alter even slightly, the fine-tuning of the hundreds of complex chemical pathways inside each cell is disrupted. And since the cells and tissues of the body are interdependent – each cell has its own task, and they all work together for the good of the whole – breakdown in one area rapidly spreads to others. So the body must be homeostatic – that is, it must stay the same inside.

This constancy of the internal environment is achieved by numerous systems that continuously monitor, regulate, and adjust the conditions to provide continuing optimum surroundings for the cells. There are many hundreds of "body variables," and even a short list of factors affecting the blood alone includes more than 100 minerals, nutrients, waste products, and other chemicals, which, if they vary beyond certain set limits, can bring disruption and poor health.

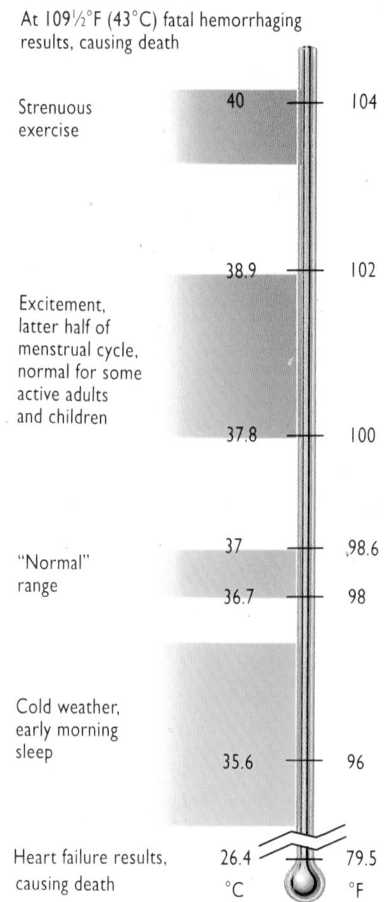

The temperature in the core of the body *is normally about one degree either side of 98.6°F (37°C). There is a slight natural daily variation, reaching a high point at about 11 a.m. and a low point at around 4 a.m. Superimposed on this are variations due to bodily activity, chiefly muscle contraction.*

Exercise or infection can raise temperature to about 104°F (40°C). If it goes much higher, death can occur due to malfunction of the cardiovascular system. At the other end of the scale, people can survive fairly cool core temperatures, but below a certain point the heart stops.

*Skin temperature, as shown on a thermograph (**right**), where white and red are hot and blue and black are cold, is a good guide to internal temperature.*

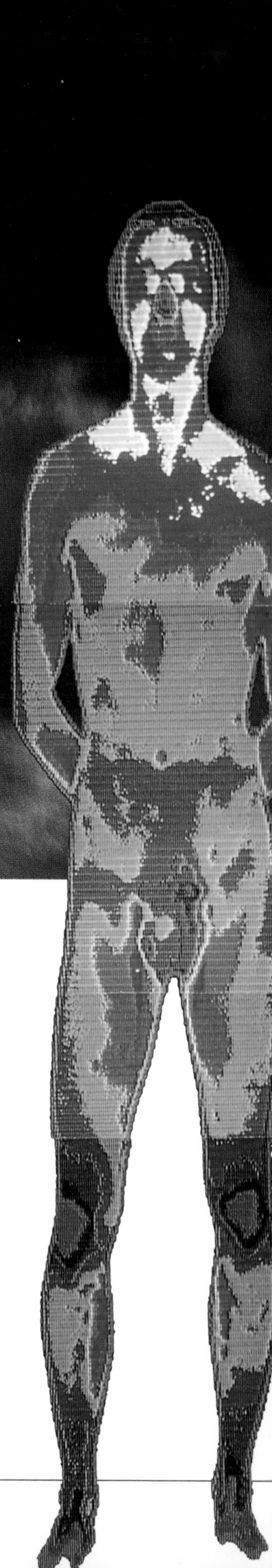

THE BODY'S BALANCING ACT

Many body systems that are regulated within narrow bands involve several control methods. For instance, blood acidity is determined partially by carbon dioxide levels. Carbon dioxide is a potentially poisonous by-product of energy production in the body's cells (cellular respiration). After it has been made in a cell, it initially dissolves in the water that makes up most of the blood plasma, forming a weak acid known as carbonic acid.

This would normally make the blood more acidic, with a lower pH. But blood pH is tightly controlled between 7.45 and 7.35 (almost neutral), so instead, the breathing rate is raised and the excess carbon dioxide is exhaled through the lungs. However, this is not the only way blood pH is adjusted. It is also controlled by excretion of acid-type chemicals through the urinary system; by the presence of chemicals called buffers, which mediate acid–alkali balance; and by the level of sodium in blood.

Exercise makes you hot and sweaty. *Vigorous activity has various effects that can be perceived externally, such as a flushed, perspiring face, heavy breathing, and a racing pulse. These processes are designed to return the body to its resting state, by reversing the changes brought on by the busy, active muscles. These use up oxygen and energy-rich glucose (blood sugar) and produce lots of heat. So the skin sweats and flushes to lower the raised temperature; the lungs work faster to raise the lowered blood oxygen level; and the heartbeat speeds up to distribute the oxygenated blood around the body. In addition to these outward signs, many other restabilizing processes are going on within the body.*

The body's "thermostat" is centered in the hypothalamus in the front lower part of the brain. Groups of cells here are sensitive to raised temperature. When they become too warm, they send signals via the nervous system to trigger the body's cooling processes. Sweat is released onto the skin and evaporates, drawing warmth from within; blood vessels in the skin widen and also allow heat to be lost. In addition, warming processes such as general cellular chemistry and involuntary muscle activity are inhibited. If the body becomes too cool, cold sensors in the spinal cord, hypothalamus, skin, and other sites reverse the situation. For instance, the blood flow to the surface of the skin is lowered to reduce heat loss, and shivering generates heat in the muscles.

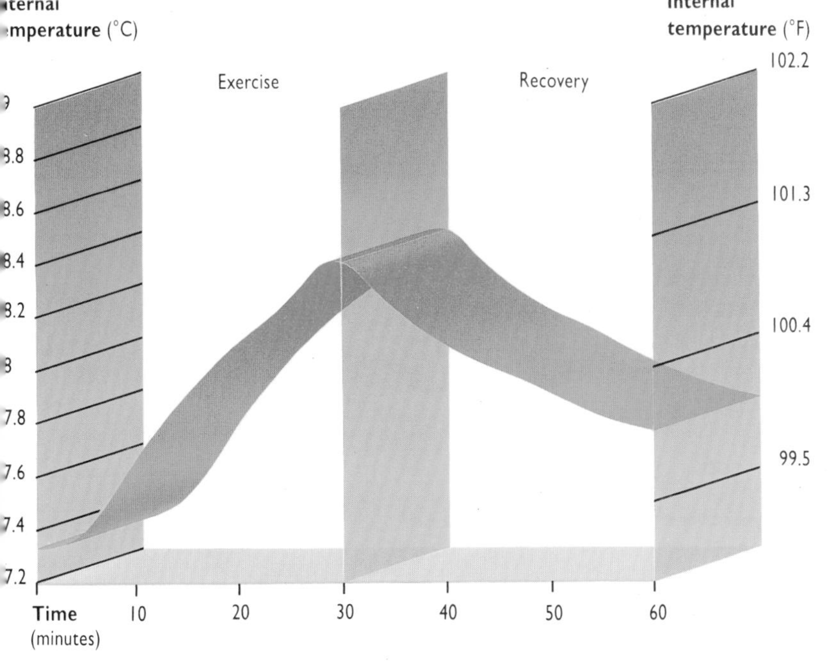

See also

CONTROL AND SENSATION

▶ Control systems 36/37

▶ The chemicals of control 38/39

▶ Key chemicals 42/43

▶ The automatic pilot 48/49

ENERGY

▶ The speed of life 92/93

▶ Fueling the body 94/95

▶ The cell and energy 112/113

▶ Cells at work 114/115

CIRCULATION, MAINTENANCE, AND DEFENSE

▶ Blood: supplying the body 126/127

▶ Waste disposal 130/131

▶ Inside the kidney 134/135

Control systems

How does the body govern the vast number of processes that make it work?

If you leave faucets running, after a few minutes water pours across the floor. This is a failure of a control system. The purpose of the system is to monitor events, obtain information, and set in motion the appropriate responses. Your brain and eyes failed to monitor the bath water level, so you missed the information of water lapping around the rim and did not respond by turning off the faucets or letting the water out.

Mopping up after a spillage of water is a tedious chore, but imagine a similar failure in the workings of a huge and complex machine such as a jet plane or an ocean liner. These have thousands of processes occurring every second and are packed full of control and monitoring systems to maintain the smooth running of the whole. The body is unimaginably more complex than even these large machines – it has even more control systems, sensors, monitoring equipment, feedback loops, and responding devices. Many of these systems are tied into each other.

In the body, most of the control systems work using either the tiny electrical signals of the brain and nervous system, which flash along

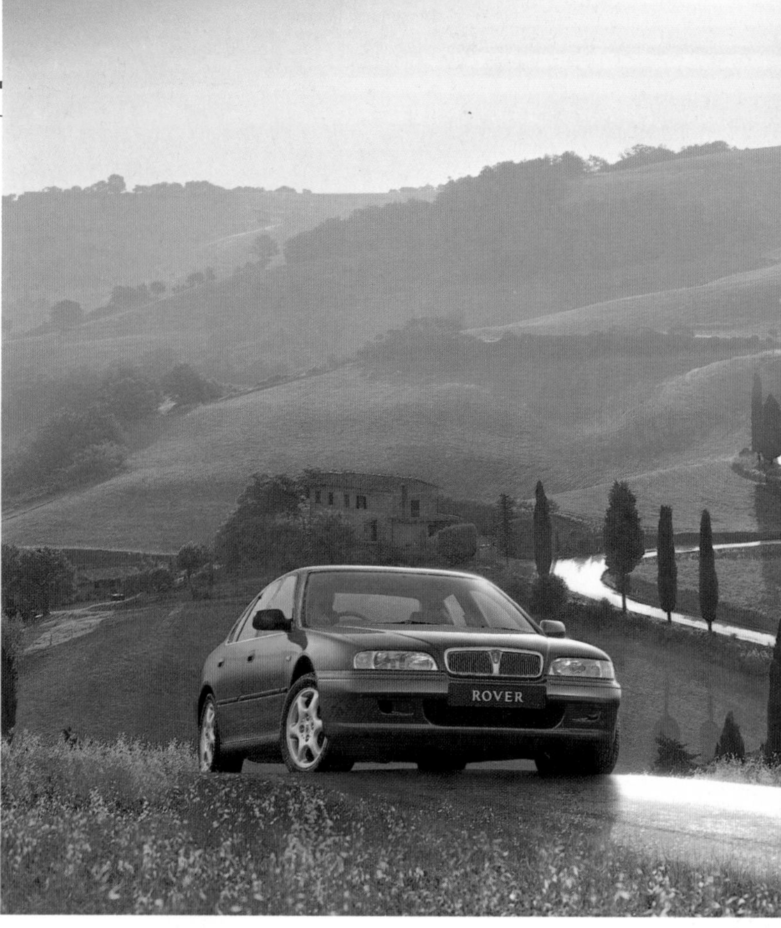

***When traveling** along a winding road, the driver of a car can keep it off the verge only by constantly observing the progress of the car along the desired route and by compensating for and correcting any over- or understeering. But feedback about what is happening outside the car is only part of the story. The driver also has to monitor the warning lights and gauges for engine temperature, fuel level, and oil pressure among others. The body, too, has a multitude of sensors to detect variables such as temperature, blood pressure, the levels of oxygen, nutrients, minerals, and other substances in the blood and body fluids. In fact, feedback is vital for keeping the body on course in its daily life of moving, eating, resting, and sleeping.*

SENDING OUT CONTROL SIGNALS

To communicate with one specific person, you might make a telephone call or address a letter, thus targeting the information. To communicate with lots of people, you could send out a mass mailing or broadcast on the radio. Many people would receive the information, but only those who were interested or susceptible would respond.

The body's information and control systems are similarly targeted or broadcast. Nerves carry targeted signals direct from source to destination, for example from your brain to your finger, to make it twitch. No other nerve circuits are involved. But a hormone released from its endocrine gland circulates in the blood and spreads all around the body. Like people reading a mass mailing, every cell has the opportunity to respond, but only those programmed to be sensitive to the hormone do so.

RAA hormone system

In the renin-angiotensin II-aldosterone (RAA) system, as blood pressure falls, kidney sensors trigger the release of renin. Angiotensin II is then made, which raises pressure by constricting arteries and stimulating the adrenals to release aldosterone, which makes the kidneys filter out less salt and water.

ADH hormone system

As blood pressure falls, sensors in the hypothalamus trigger it to release ADH (antidiuretic hormone). Like aldosterone, this makes the kidneys produce less urine, thus filtering out a smaller amount of salts and water, increasing the level of body fluids and, as a consequence, raising blood pressure.

VMC nervous system

The brain's vasomotor center (VMC) monitors thoughts and emotions, as well as input from pressure sensors in main arteries and carbon dioxide and oxygen sensors in the brain stem. It sends signals through the autonomic nerve system to control heart rate, blood-vessel constriction, and thus blood pressure.

***Blood pressure** is controlled by both the nervous and hormonal systems. For good health, it must continuously be monitored and adjusted toward normal. Too much pressure can damage and even burst blood vessels, while too little means the brain lacks oxygen and nourishment, causing dizziness or fainting. Several hormones and nerve links work together to keep blood pressure within acceptable limits, the nerves acting within seconds and the hormones on a longer time scale.*

erves, or the messenger chemicals called hormones, hich circulate in the bloodstream. Nerves carry nessages very fast, and the effects of nerve ontrol can be felt quickly. Hormones, made by ndocrine glands, work on a longer time basis, om several seconds through to hours, days, nd even years.

Most control systems involve negative, or orrective, feedback. The term "feedback" is used or the process by which a given bodily function is ontinually monitored. Information detected by a eceptor is "fed back" to a control center which can then nake any necessary adjustments to that function, keeping working smoothly. It is like the thermostat in an oven that nonitors the oven's temperature, turning the heat on when it ets cooler and then turning it off when it gets too hot. Thus, egative feedback keeps functions under control, within et limits, by turning something off. In positive feedback, owever, a rise in the function being monitored causes the ontrol center to reinforce the function. The body has few uch systems. One is birth, when labor stimulates the release f a hormone that increases the strength of contractions.

See also

CONTROL AND SENSATION
- Steady as you go 34/35
- The chemicals of control 38/39
- Hormones of change 40/41
- Key chemicals 42/43
- The automatic pilot 48/49

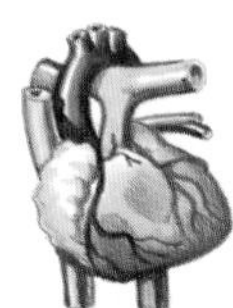

SUPPORT AND MOVEMENT
- The heart's special muscle 28/29

CIRCULATION, MAINTENANCE, AND DEFENSE
- The water balance 132/133
- Inside the kidney 134/135

The chemicals of control

Chemicals secreted directly into the blood by special glands have the power to bring about dramatic changes.

Our body's chemical control system consists of about 10 or so important glands (plus many minor ones) and the chemicals they make, known as hormones. This name comes from the Greek *hormao*, "I excite," and refers to the fact that each hormone excites or stimulates specific parts – known as its target cells or tissues – into different levels of activity. Hormones are made in endocrine glands, which pass their secretions from their cells straight into the blood flowing through the gland rather than emptying their products down a duct or tube. In general, the higher the amount of a hormone in the blood, the greater its effects on its targets. The level of a hormone or the degree of its effect is monitored by sensors and controlled by feedback loops.

The hypothalamus, part of the lower brain, is the main nerve–hormone link. It regulates much of the system by its action on the "chief" hormone gland, the pituitary, which is just below it. This navy-bean-sized gland makes 10 main hormones, 4 of which control other endocrine glands. One of the pituitary's target glands is the thyroid, producing its own hormones, two of which regulate the body's basal metabolic rate – the overall speed of the chemical reactions in cells. Also under direct pituitary control are the adrenals, whose hormones are concerned with stress responses, fluid

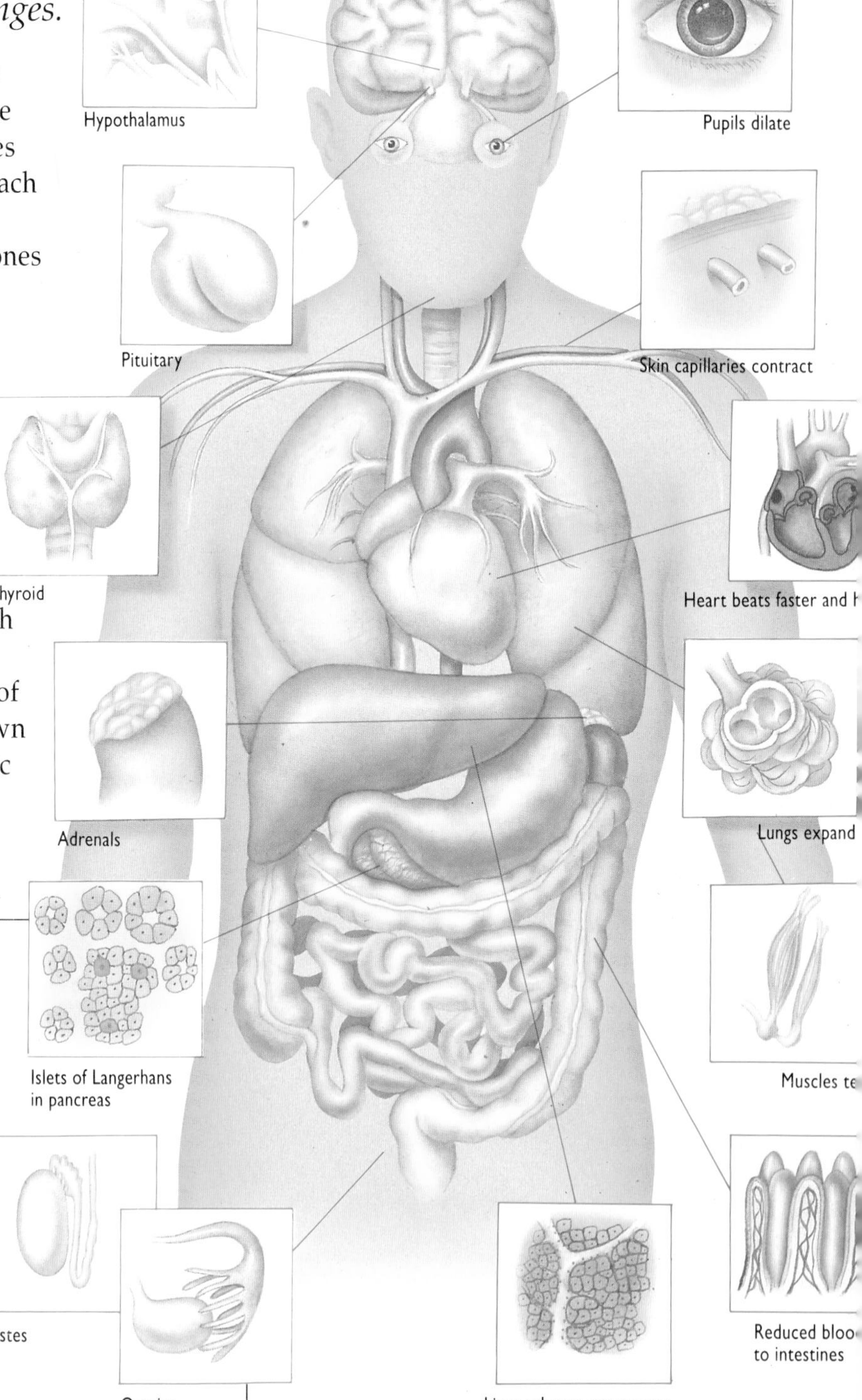

GLANDS FACT FILE

Gland	Size	Weight	Hormones produced include
Pituitary (1)	1/3 x 2/5 inch (0.8 x 1 cm)	1/50 ounce (0.5 g)	Prolactin, growth hormone, vasopressin, oxytocin, TSH, ACTH, FSH, LH, ICSH, MSH
Thyroid (1)	4¾ x 2½ inches (12 x 5 cm)	1 ounce (25 g)	Thyroxine (T4), tri-iodothyronine (T3), calcitonin
Parathyroids (4)	1/5 x 1/6 inch (0.6 x 0.4 cm)	1/500 ounce (0.05 g)	Parathormone
Adrenals (2)	2 x ¾ inch (5 x 2 cm)	2 ounces (50 g)	Steroid hormones, epinephrine, and norepinephrine
Pancreas (1)	5½ x 1½ inches (14 x 4 cm)	4¼ ounces (120 g)	Insulin, glucagon
Ovaries (2)	1 x 3/5 inch (2.5 x 1.5 cm)	1/5–1/3 ounce (5–10 g)	Estrogen, progesterone
Testes (2)	1¾ x 1 inch (4.5 x 2.5 cm)	½ ounce (12 g)	Testosterone

Fear or a sudden fright *can trigger the epinephrine-aided "fight or flight" response, which prepares the body to stay to do battle or to run from danger. In our evolutionary past, when confronted by a saber-toothed tiger or giant cave bear, only these two options applied. There was no third choice of sitting down for discussions, or the fourth option of simple inaction. Today, the last two choices predominate, and the body's inability to "work off" the effects of epinephrine by physical action can lead to stress-related disorders.*

Working in tandem *with the nervous system, the hormone epinephrine brings about body changes to produce the stress response (**left**) that prepares the body for action. The pupils of the eyes enlarge to let in more light. Small blood vessels contract in the skin, turning it paler, and in the digestive and other internal organs, causing "butterflies in the stomach." Meanwhile, the blood vessels that feed the muscles compensate by dilating, allowing more blood to be supplied as the muscle fibers tense for sudden movement.*

The blood receives more oxygen from expanded lungs in a puffed-up chest and more high-energy glucose from supplies liberated by the liver. And blood is circulated more rapidly by a heart that pumps harder and faster.

balance, and kidney function. Tiny clumps of cells in the pancreas, a dual digestive–endocrine gland, make the hormones insulin and glucagon, which regulate the blood sugar that provides body cells with energy. The hormones of the gonads (sex organs) are involved in sexual development and the production of eggs and sperm.

Trying to obtain a complete view of the body's endocrine system is akin to memorizing the timetables of the trains, buses, and planes serving a big city. There is far too much detail. But certain general principles apply. More than 200 hormones or hormonelike substances have now been discovered. And every few months medical researchers discover another new hormone or similar chemical to add to the vast jigsaw.

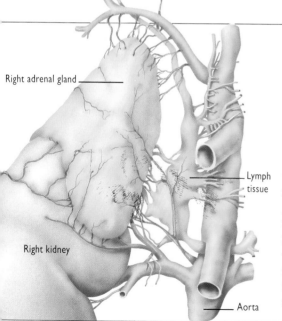

Right adrenal gland

Lymph tissue

Right kidney

Aorta

THE ADRENAL GLAND – THE "KIDNEY'S BONNET"

One of the best-known hormones is epinephrine (adrenaline), from the adrenal gland just above the kidney. It is also one of the fastest-acting hormones, producing results within seconds. It pours into the bloodstream in response to stress or a fright and readies the whole body for action.

Early anatomists called the kidney the renis and described the adrenal (ad-renal) or supra-renal gland as "perched atop it like a bonnet." Like all endocrine glands, the adrenal receives a plentiful blood supply, in this case through three sets of branches from three major arteries – the phrenic and renal arteries, and the aorta (the main artery of the body). It is a triangular gland with two main parts, an outer cortex and inner medulla. The cortex makes steroid hormones, numbering about 40 at a recent count, all derived from the cholesterol molecule. The main ones are aldosterone, which affects the balance of the minerals sodium and potassium by selective absorption in the kidneys, and cortisol, which promotes glucose production and helps the body to resist stress, injuries, traumas, inflammation, and allergic reactions. The medulla produces epinephrine and norepinephrine.

See also

CONTROL AND SENSATION

▶ Steady as you go 34/35

▶ Control systems 36/37

▶ Hormones of change 40/41

▶ Key chemicals 42/43

▶ The automatic pilot 48/49

▶ Moods and emotions 84/85

ENERGY

▶ The food processor 100/101

▶ Absorbing stuff 104/105

CIRCULATION, MAINTENANCE, AND DEFENSE

▶ The living pump 120/121

▶ Inside the kidney 134/135

REPRODUCTION AND GROWTH

▶ Cycle of life 164/165

▶ Sexual maturity 180/181

Hormones of change

Three glands – one of which triggers the other two – have dramatic and widespread effects in the body.

Many of the body's glands – including the ovaries (in women) and the testes (in men) – are controlled by the pituitary. Its two lobes have different functions. The posterior (rear) lobe makes two main hormones: oxytocin and ADH, or vasopressin. Oxytocin affects uterine contractions in pregnancy and birth, and the subsequent release of breast milk. ADH helps to regulate blood fluids and mineral levels in the body.

The anterior (front) lobe makes six main hormones: growth hormone regulates cell growth and multiplication, and thereby body growth; TSH controls the thyroid gland; ACTH controls the adrenals; prolactin promotes breast development and milk production in new mothers; follicle-stimulating hormone (FSH) and luteinizing hormone (LH) control the ovaries and testes. In women, LH and FSH relate to menstruation and pregnancy. FSH also encourages the ovaries to produce estrogen, which brings about changes such as the development of the sex organs and breasts, laying down of fat which accentuates body curves, and hair growth in the armpits and genital region. In men LH, or ICSH (interstitial cell-stimulating hormone), encourages cells in the testes to make testosterone, producing puberty changes including development of muscle and physique, growth of facial and body hair, and deepening of the voice.

The two female sex glands *in which estrogen is made are the ovaries. These are found in the lower abdomen on either side of the uterus (womb). Each ovary contains many egg cells, one of which will usually ripen and be released during every other menstrual cycle. The egg is gathered by the funnel-like fimbria and passes along the Fallopian tube, or oviduct, to the uterus.*

During puberty, more estrogen is manufactured by cells, known as theca cells, that surround each egg cell, bringing about the changes from girl to woman. Overlaid on this general increase in levels of estrogen is its fluctuation during the menstrual cycle.

Estrogen levels, *detected in urine, are raised during the reproductive years. They rise and fall in each menstrual cycle.*

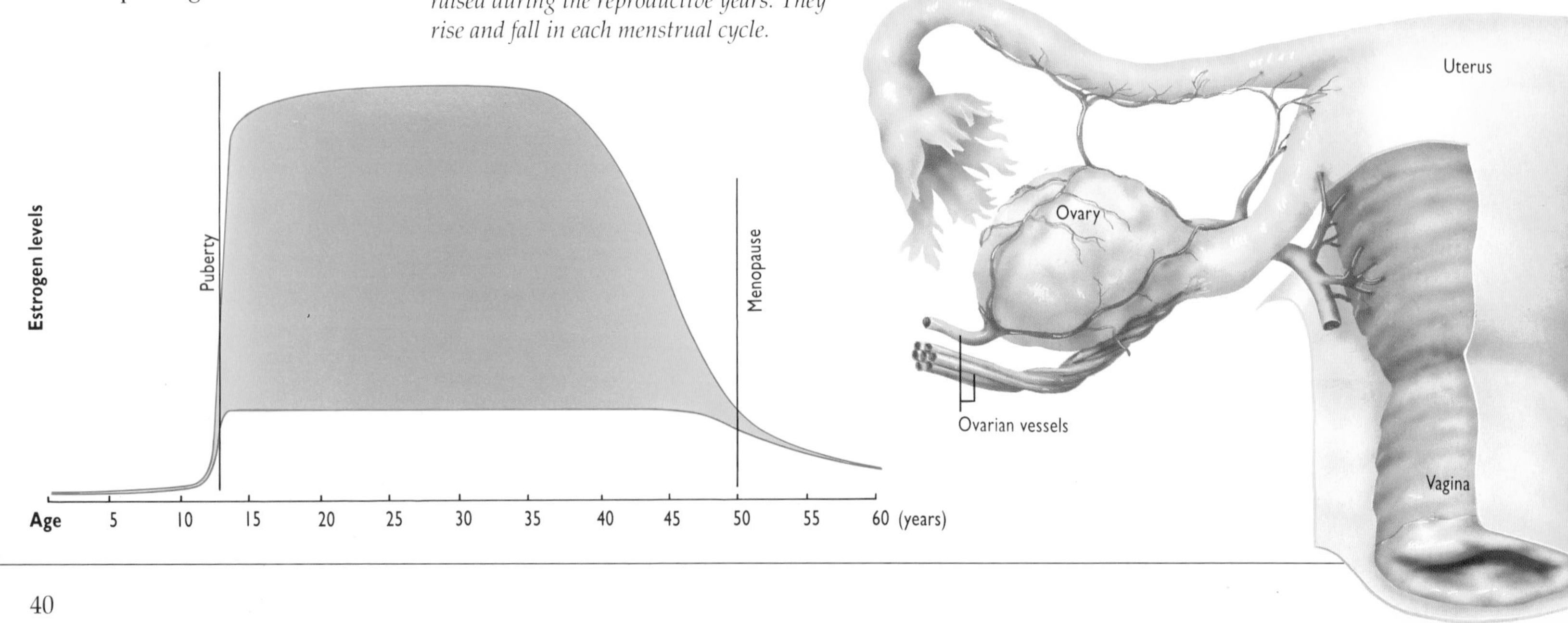

ı any large group of mixed ages ıd genders, several people will be ıdergoing relatively rapid physical ıange. Children will be making ·eir growth spurts. Adolescents ill be going through puberty, and ıder women may be experiencing ·e menopause. These changes are ıe to hormones produced in the tuitary and sex glands.

Optic chiasma

Optic nerve

Pituitary

Pons

THE PITUITARY AND GROWTH HORMONE

After a post-birth growth spurt, *height gain steadies in childhood. Then we "shoot up" during puberty, peaking at a height gain of 3–3½ inches (8–9 cm) a year around the age of 12 for girls and 14 for boys.*

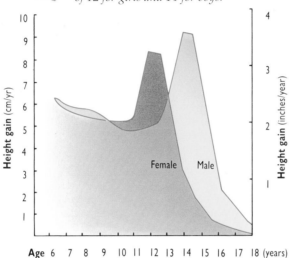

Height gain (cm/yr)

10 9 8 7 6 5 4 3 2 1

Height gain (inches/year)

4 3 2 1

Female

Male

Age 6 7 8 9 10 11 12 13 14 15 16 17 18 (years)

Sited almost in the center of the head, above the bulge in the brain stem called the pons, is the center of the endocrine system – the pituitary. Growth hormone from its anterior lobe stimulates increase in cell size and rate of cell multiplication in all tissues, making the body grow. It works by increasing the manufacture of proteins, the body's major structural molecules. With the sex hormones, growth hormone produces a spurt in size and development during puberty. Lack of growth hormone causes dwarfism. This can be treated by injections of the hormone, which was once obtained from the pituitaries of corpses, but is now made using genetically altered microbes.

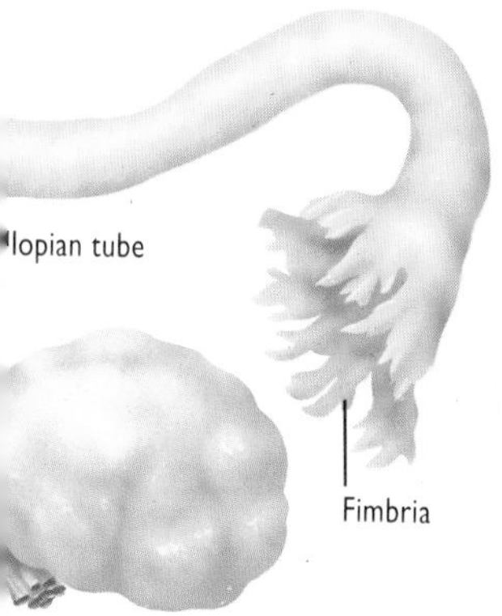

espite the gradual drop *in ·stosterone levels, many men ·main sexually active and ·oduce viable sperm into their)s or even 70s and later.*

Levels of the chief male hormone, *testosterone, rise dramatically in boys during their early teens, causing the main changes of puberty, and peak in the early 20s. Thereafter, levels fall steadily.*

The male glands, *the testes, make and store sperm, which pass along the vas deferens for ejaculation. Testosterone is made by interstitial, or Leydig, cells between the sperm-making tubes inside each testis.*

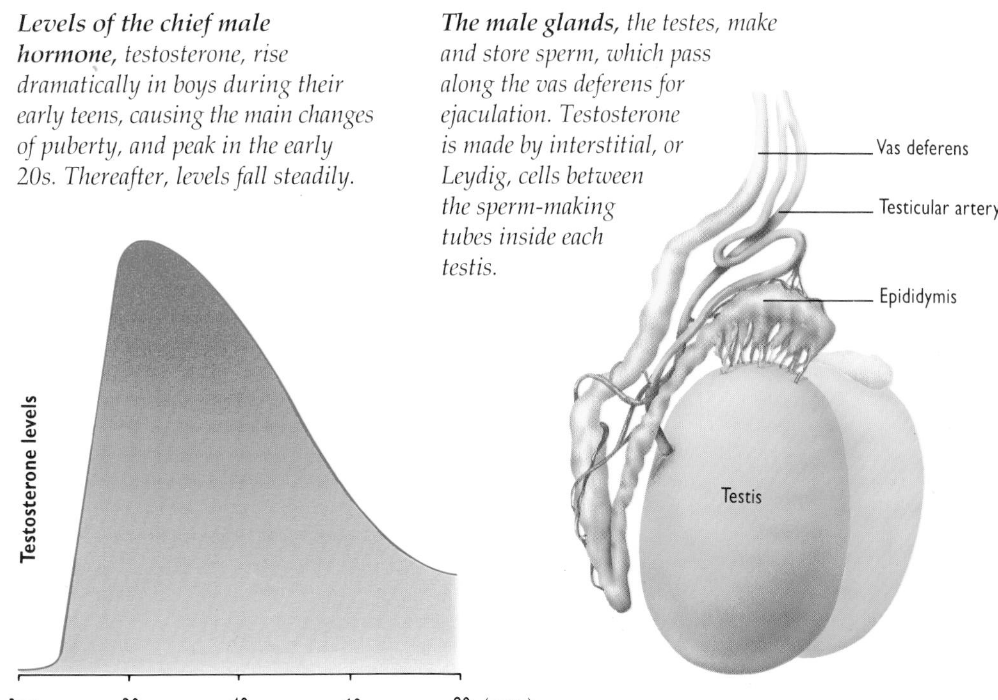

See also

Key chemicals

A vital group of hormones deals with the regular running of the body, hour by hour, day by day.

Probably the best-known chemical involved in running the body is the hormone insulin – a lack of it causes the condition diabetes. But there are many others, including those made by the thyroid gland, the parathyroids, and the kidneys.

Insulin, made in the pancreas, helps control levels of the sugar glucose – the energy source for many chemical processes in cells. Constant supplies of glucose are needed all over the body. Too little glucose in the blood (hypoglycemia) starves the brain of energy; too much (hyperglycemia) upsets the body's fluid balance. Either way, the result may be coma or even death.

Two hormones regulate blood glucose levels. As food is digested, sugars are absorbed into the blood, raising its level of glucose. Cells in the pancreas sense this and release insulin, which allows all cells to take up more glucose. Insulin also encourages the conversion of glucose into glycogen, which is stored in places such as the liver and muscles. Both processes remove glucose from the blood. To stop glucose levels from falling too far is the job of insulin's partner, glucagon. It reverses the conversion to glycogen, so glucose pours back into the blood.

GLUCOSE TESTING

Some people with diabetes keep track of the amount of glucose in their blood; if there is too much, they can give themselves an injection of insulin to correct the level. In the past, physicians would sip the urine of suspected diabetics. If it tasted sweet, the diagnosis was confirmed, since glucose in diabetics builds up in the blood and spills over into the urinary system. A more recent test is to put a chemical-coated "dipstick" into a urine sample. Its color change shows how much glucose is present. Modern tests use chemical-coated strips, color charts, or small electric meters that check blood glucose levels directly.

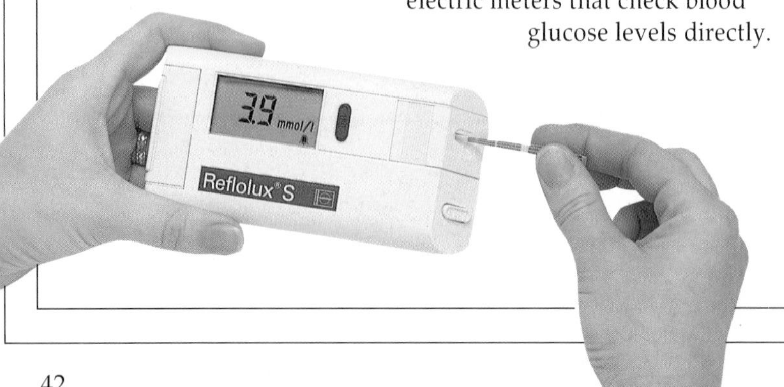

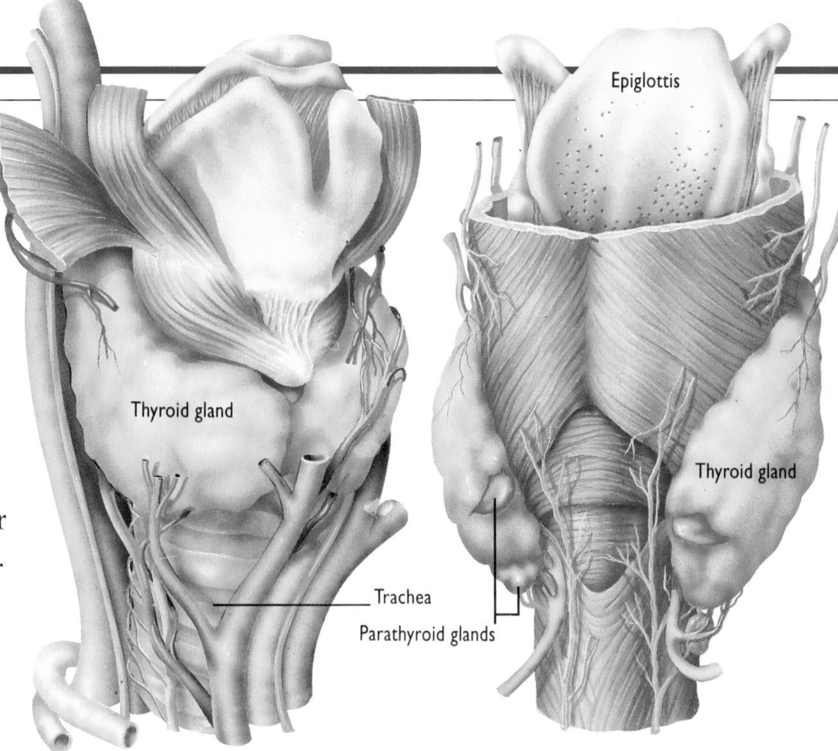

The thyroid *would be just under a bow tie (if you wore one), in the same place and roughly the same size and shape. It makes three main hormones: calcitonin, which affects calcium levels; thyroxine (T4); and tri-iodothyronine (T3). T3 and T4 affect metabolic rate – the speed at which cells perform chemical reactions. Higher levels speed metabolism.*

The parathyroids *are four pea-size glands embedded in the posterior (rear) parts of the thyroid. They make parathormone (parathyroid hormone) which, with calcitonin, regulates calcium levels. More parathormone means more calcium is released from bones to satisfy the constant, small demands of muscles and nerves for this vital mineral.*

The kidneys' *main roles are to filter blood and remove wastes and excess water. But they, too, have hormone-producing functions. They make renin, which takes part in the daily control of blood pressure. Renin is made by the juxta-glomerular cells which are between the tiny arterial blood vessels and urine-carrying tubules of each kidney filtering unit. Another kidney hormone is erythropoietin, which stimulates the development of new red blood cells in bone marrow. The kidneys also make the hormone 1,25-dihydroxy-cholecalciferol. This works with parathormone from the parathyroid glands and calcitonin from the thyroid to influence calcium levels.*

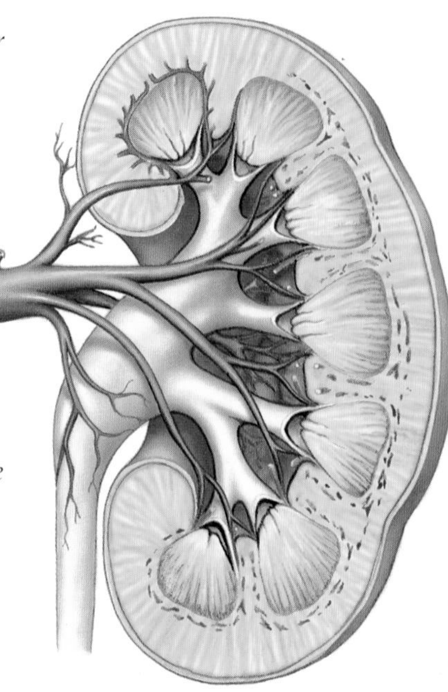

Kidney

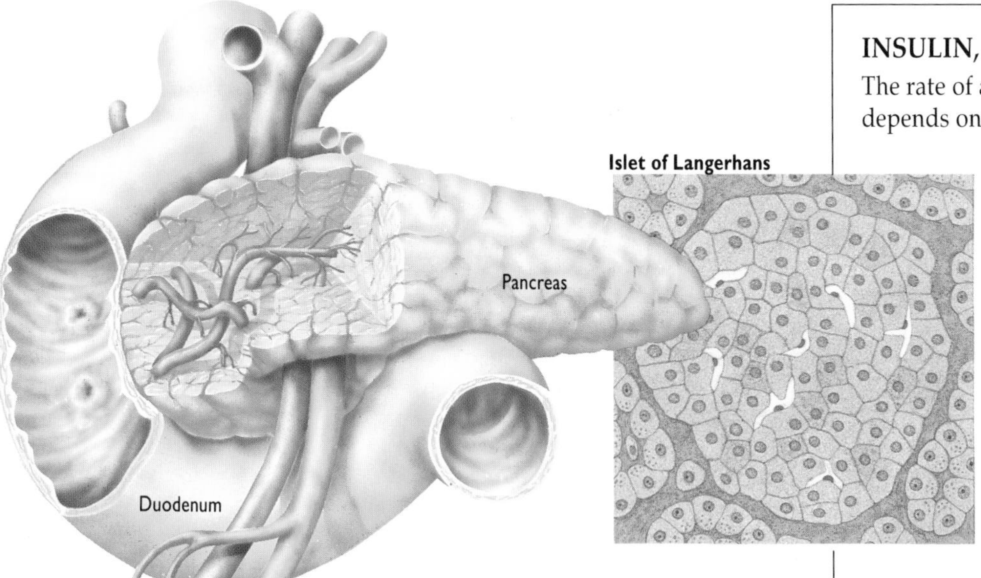

The pancreas is really two glands in one. About 98 percent of its bulk makes secretions that flow along a network of collecting ducts and out through a tube. The secretions are a cocktail of powerful digestive enzymes and chemicals, which flow along the pancreatic duct into the duodenum. The other 2 percent makes hormones which it releases directly into the blood vessels flowing through it. It consists of about a million tiny clumps of cells (islets of Langerhans) scattered through the gland. Cells in the islets are of two main types, A and B. A cells make glucagon; B cells produce insulin – both are hormones intimately involved in the control of blood glucose.

INSULIN, GLUCOSE, AND THE CELL

The rate of absorption of glucose by a cell depends on insulin levels. When an insulin molecule (**1**) attaches to its receptor, a specially shaped part of a cell's membrane, a protein molecule (**2**) inside the cell is activated – the signal for glucose transporters to move. Some transporters are already embedded in the membrane, pulling glucose molecules into the cell from outside. Others are "in storage" in vesicles (**3**) – small blobs of membrane – or in endosomes (**4**) – main depot-like larger blobs. The signal makes vesicles move to the main cell membrane and join with it (**5**), exposing their glucose transporters to the outside, so they can begin importation (**6**). With more transporters on cell membranes hauling glucose into cells, glucose levels in the blood reduce and less insulin is made. When this happens, parts of the transporter-bearing cell membrane bud inward (**7**), forming vesicles (**8**). These wander to the cell's interior; coalesce into endosomes (**9**); and await the next signal when vesicles re-form (**10**) and the process repeats itself.

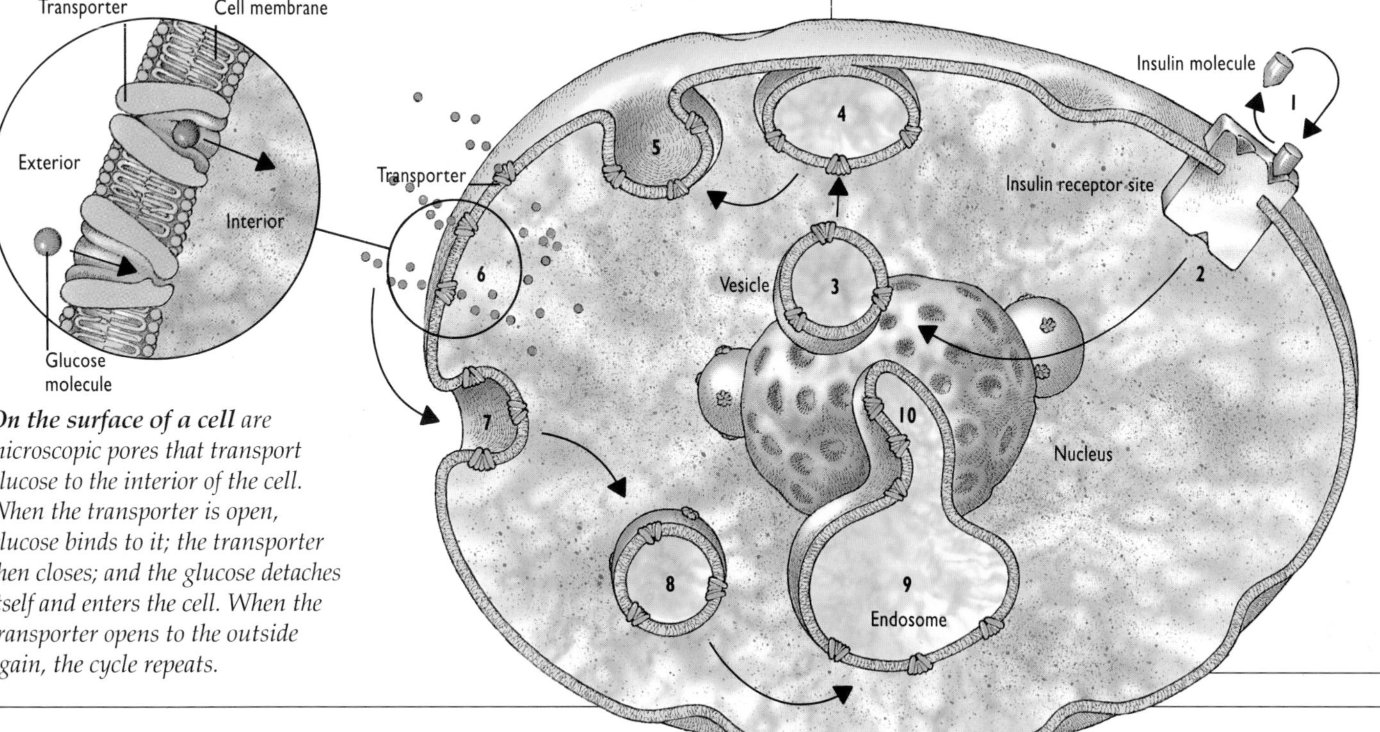

On the surface of a cell are microscopic pores that transport glucose to the interior of the cell. When the transporter is open, glucose binds to it; the transporter then closes; and the glucose detaches itself and enters the cell. When the transporter opens to the outside again, the cycle repeats.

See also

CONTROL AND SENSATION
- Steady as you go 34/35
- Control systems 36/37
- The chemicals of control 38/39
- Hormones of change 40/41
- The automatic pilot 48/49

ENERGY
- Fueling the body 94/95
- The cell and energy 112/113
- Cells at work 114/115

CIRCULATION, MAINTENANCE, AND DEFENSE
- Inside the kidney 134/135

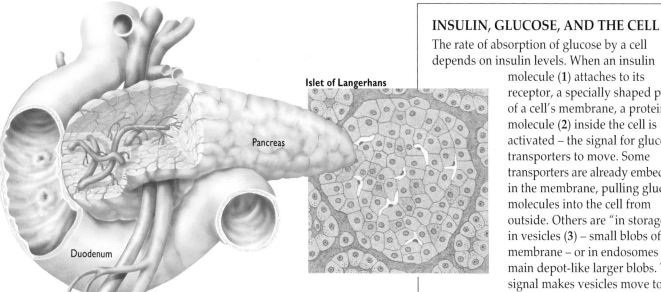

The nerve net

From scalp to toes, an extraordinarily sophisticated network of nerves controls and monitors the body.

Dozens of pale, cordlike objects snake through the body, reaching every part. These are the body's nerves. They are made of bundles of nerve cells, or neurons, which have long, thin, wirelike parts – the fibers or axons. A nerve as thin as a piece of thread is actually made up of many hundreds of fibers. Nerve cells are specialized to carry tiny electrical nerve signals, whose number and timing represent information traveling from one part of the body to another. In this way, through the nerve net, the body controls and coordinates many of its thousands of processes and actions.

Here and there, the nerves expand into lumplike ganglia made mainly of nerve cell bodies with many interconnections. The ganglia are like sorting and relay stations. There are also sites where larger nerves branch and join, like the classification yard of a rail network. These are nerve plexuses, and they are major re-routing sites for nerve fibers.

But at two places in the body, nerve cells form much larger concentrations. One is the spinal cord, the body's biggest nerve, containing many millions of nerve fibers. It is well protected inside a tunnel of bone formed by the lined-up holes inside the vertebrae (spinal bones or backbones). At its top, the spinal cord joins the second nerve concentration, the brain; together they form the central nervous system (CNS). The other major division of the nerve net, the peripheral nervous system (PNS), is the bodywide network of nerves.

Some nerves carry two-way traffic. Sensory nerve messages come in from the skin and other sense organs, along the nerve to the brain, while motor nerve messages go from the brain along the nerve to the muscles, instructing them to move. The facial nerve is an example, bringing information about taste from tongue to brain and taking information from the brain to facial muscles to "make faces." Other nerves are only one-way. The olfactory nerve is only sensory, bringing messages about smell from nose to brain. Individual nerve fibers – a bundle of which make up a nerve – carry one-way traffic only.

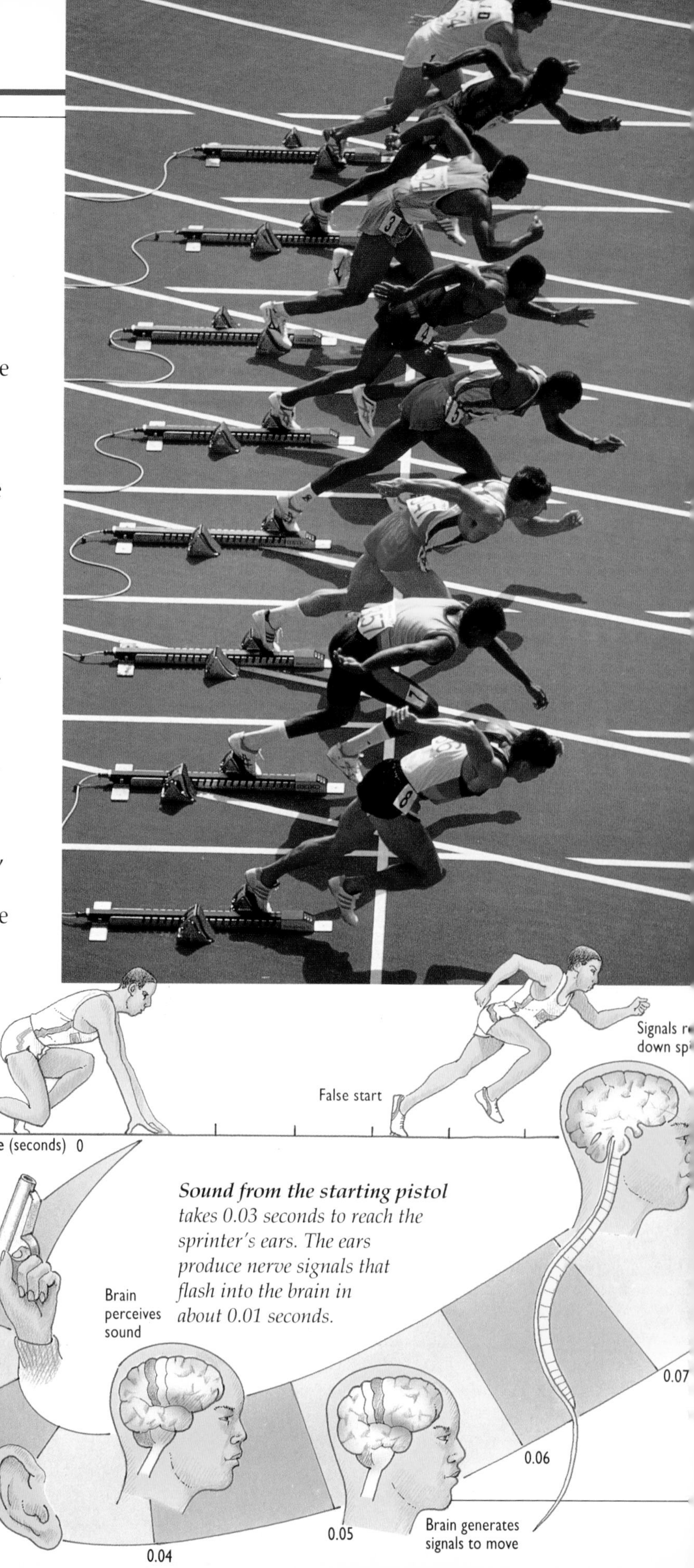

Sound from the starting pistol *takes 0.03 seconds to reach the sprinter's ears. The ears produce nerve signals that flash into the brain in about 0.01 seconds.*

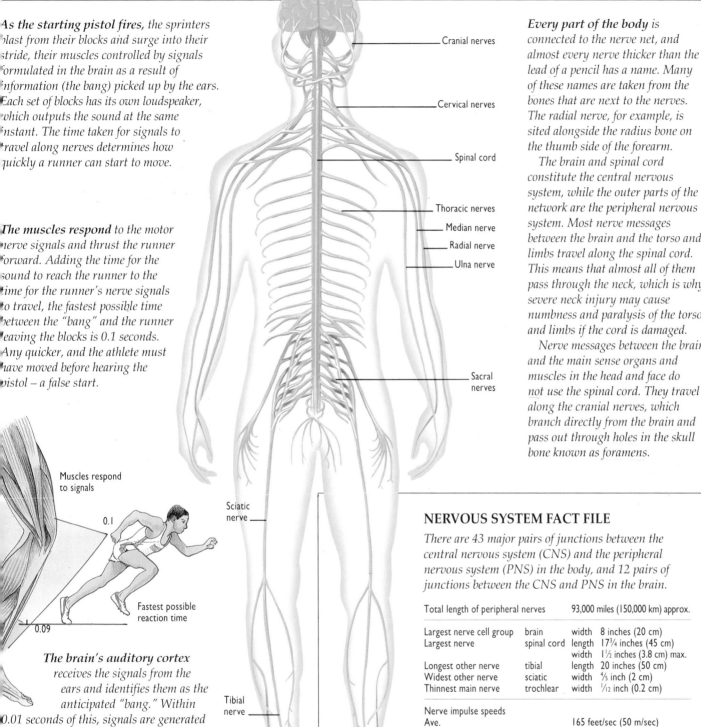

As the starting pistol fires, the sprinters blast from their blocks and surge into their stride, their muscles controlled by signals formulated in the brain as a result of information (the bang) picked up by the ears. Each set of blocks has its own loudspeaker, which outputs the sound at the same instant. The time taken for signals to travel along nerves determines how quickly a runner can start to move.

The muscles respond *to the motor nerve signals and thrust the runner forward. Adding the time for the sound to reach the runner to the time for the runner's nerve signals to travel, the fastest possible time between the "bang" and the runner leaving the blocks is 0.1 seconds. Any quicker, and the athlete must have moved before hearing the pistol – a false start.*

Muscles respond to signals

0.1

Fastest possible reaction time

0.09

The brain's auditory cortex *receives the signals from the ears and identifies them as the anticipated "bang." Within 0.01 seconds of this, signals are generated in the motor cortex, which controls voluntary body movements. These signals travel down the spinal cord and out along the peripheral nerves to muscles all over the body, but chiefly to those in the legs.*

Brain

Cranial nerves

Cervical nerves

Spinal cord

Thoracic nerves

Median nerve

Radial nerve

Ulna nerve

Sacral nerves

Sciatic nerve

Tibial nerve

Every part of the body *is connected to the nerve net, and almost every nerve thicker than the lead of a pencil has a name. Many of these names are taken from the bones that are next to the nerves. The radial nerve, for example, is sited alongside the radius bone on the thumb side of the forearm.*

The brain and spinal cord constitute the central nervous system, while the outer parts of the network are the peripheral nervous system. Most nerve messages between the brain and the torso and limbs travel along the spinal cord. This means that almost all of them pass through the neck, which is why severe neck injury may cause numbness and paralysis of the torso and limbs if the cord is damaged.

Nerve messages between the brain and the main sense organs and muscles in the head and face do not use the spinal cord. They travel along the cranial nerves, which branch directly from the brain and pass out through holes in the skull bone known as foramens.

NERVOUS SYSTEM FACT FILE

There are 43 major pairs of junctions between the central nervous system (CNS) and the peripheral nervous system (PNS) in the body, and 12 pairs of junctions between the CNS and PNS in the brain.

Total length of peripheral nerves			93,000 miles (150,000 km) approx.
Largest nerve cell group	brain	width	8 inches (20 cm)
Largest nerve	spinal cord	length	17¾ inches (45 cm)
		width	1½ inches (3.8 cm) max.
Longest other nerve	tibial	length	20 inches (50 cm)
Widest other nerve	sciatic	width	⅘ inch (2 cm)
Thinnest main nerve	trochlear	width	1/12 inch (0.2 cm)
Nerve impulse speeds			
Ave.			165 feet/sec (50 m/sec)
Slowest – small uninsulated fibers			2½ feet/sec (0.7 m/sec)
Fastest – large insulated fibers			395 feet/sec (120 m/sec) or more
Max. voltage of nerve impulses			100 millivolts
No. of impulses			
Max. – large insulated fibers			300/sec
Min. – small uninsulated fibers			50/sec

See also

CONTROL AND SENSATION

▶ Control systems 36/37

▶ Down the wire 46/47

▶ The automatic pilot 48/49

▶ Into action 50/51

▶ The range of senses 52/53

▶ Making sense 72/73

SUPPORT AND MOVEMENT

▶ Making a move 22/23

ENERGY

▶ Cells at work 114/115

Down the wire

Every moment, countless numbers of signals are whizzing through all parts of your body as the nerve cells go about their business.

A nerve cell's body is much like that of any other typical cell, but what makes it different are its long, thin, branching projections, called neurites, and its special membrane. There are two main kinds of neurites – axons and dendrites. Each nerve cell, or neuron, has only one axon projecting from it, though that axon may have several branches, known as teleodendria. The axon is the real "wire" of the nervous system. Some nerve cells in the limbs have whole axons more than 3 feet (90 cm) long, making neurons the body's longest cells, though each one is vanishingly thin. Dendrites are even thinner – and shorter – than axons, and many project from the cell body; they nearly always branch.

The nerve cell's membrane, which extends around and along all the neurites, is special because it is excitable – it carries electrochemical nerve impulses (which do not go through the cell's core). Although nerve cells pass nerve impulses to each other, they do not touch. The nerve impulses are gathered by the dendrites, travel toward and around the cell body, and then away along the axon membrane. Between the membranes of an axon and the dendrite of a neighboring nerve cell, the signal is carried over a gap – the synapse – by chemicals known as transmitter or neurotransmitter molecules. These cross the gap and when they reach the dendrite, they stimulate it to make its own nerve impulse, which flashes away along its membrane. It takes about one-thousandth of a second for an impulse to jump the gap.

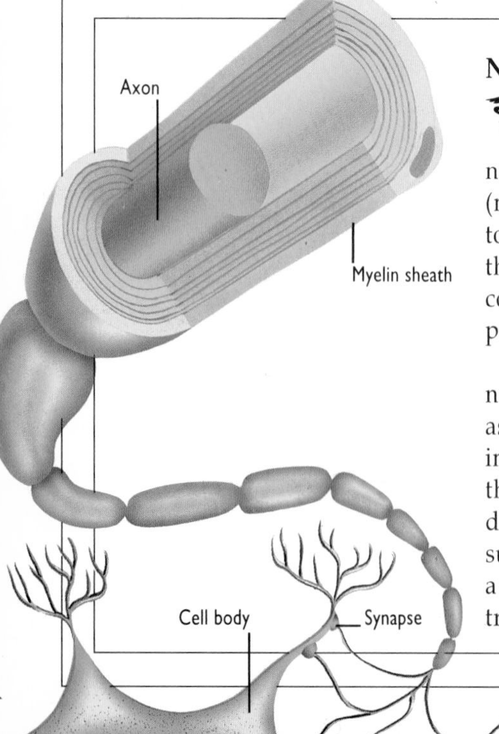

NERVE CELL PROBLEMS

A nerve cell's intricate workings can be upset if a component fails, resulting in diseases of the nervous system. In Lou Gehrig's disease, motor (muscle-controlling) nerve cells, from the brain stem to the muscles, shrink and die, and the muscles waste through lack of use. If motor nerve cells in the spinal cord are destroyed, by the polio virus for example, paralysis, wasted muscles, and deformation can result.

When the myelin sheaths around a number of nerve cells' axons degenerate, damaged areas known as plaques can cause multiple sclerosis. This leads to impairment of function, depending on the sites of the plaques. Parkinson's disease develops when dopamine-producing cells in a brain area called the substantia nigra degenerate. The lack of dopamine – a vital movement-control neurotransmitter – causes tremors and loss of coordination.

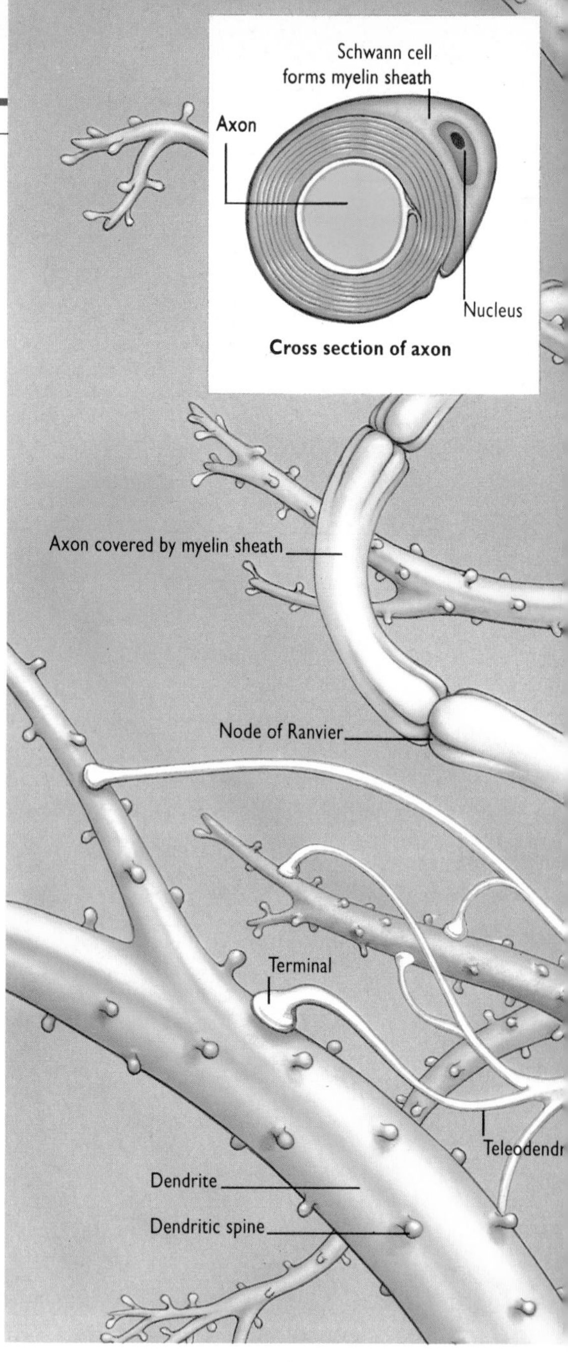

The body of a nerve cell (above) *appears to dwarf its slender projections, or neurites – the axons and dendrites. In fact, dendrites are so numerous that they represent more than 80 percent of the cell's total surface area. Nerve cell bodies and neurites form a dense, seemingly tangled web of links with other nerve cells. But the web is highly organized, developing partly from genetic programming and partly in response to movements, physical skills, memories, and other mental processes. In the brain, a single nerve cell may have 50,000 dendrite branches and be able to communicate with 250,000 other nerve cells. This provides incredible numbers of possible pathways for nerve impulses, which are the basis of thinking, learning, and memory.*

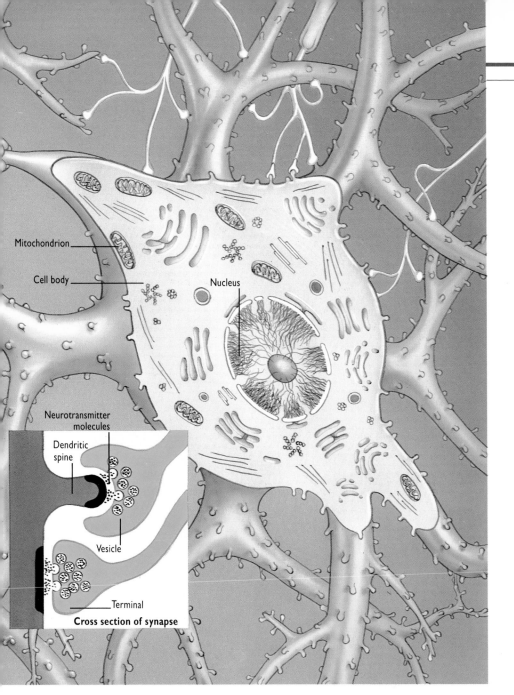

Mitochondrion

Cell body

Nucleus

Neurotransmitter molecules

Dendritic spine

Vesicle

Terminal

Cross section of synapse

Some nerve cells' axons (inset far left) *are covered by a layer or sheath of a fatty substance called myelin; this helps to insulate them electrically from other axons and speeds the passage of nerve impulses. The sheath is made of a Schwann cell which, during the body's early development, coils around the axon, leaving a spiral of myelin. There is also a bulge to house the Schwann cell's relatively large nucleus.*

A single Schwann cell does not stretch the length of the axon. There are many of them, like sausages in a string, separated by "pinched" areas called nodes of Ranvier. The nerve impulse "leaps" along the axon from one node to the next.

At the synapse (inset left)*, an arriving nerve impulse causes small bags, or vesicles, in the axon's terminals to release neurotransmitter molecules. These seep across the gap to stimulate receptors on a dendrite, setting off another nerve impulse.*

There are many neurotransmitters in different parts of the nervous system, including acetylcholine, epinephrine, norepinephrine, dopamine, and serotonin, and amino acids such as glutamate, aspartate, gamma aminobutyric acid (GABA), and glycine. Synapses exist between an axon's terminal and a dendritic spine, and between terminals and flat parts of the dendrites.

See also

CONTROL AND SENSATION

▶ Control systems 36/37

▶ The nerve net 44/45

▶ The automatic pilot 48/49

▶ Into action 50/51

▶ The biological camera 56/57

▶ Making sense 72/73

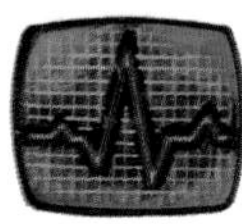

SUPPORT AND MOVEMENT

▶ The heart's special muscle 28/29

THE SIGNAL DOWN THE "WIRE"

A nerve impulse is a "traveling wave" of electrochemical activity that passes along a nerve cell's membrane. The impulse is due to movements of electrically charged atoms – ions – of sodium (Na) and potassium (K) through the membrane. At rest there is a potential difference, or negative voltage, across the membrane of -80 millivolts: the resting potential. The impulse, or action potential, is achieved by pumping positive potassium ions outside so the cell's potential becomes +40 millivolts. Sodium ions are then pumped back in to restore the resting potential. The impulse passes a given spot in 2 to 5 milliseconds; then there is a recovery period before the next one.

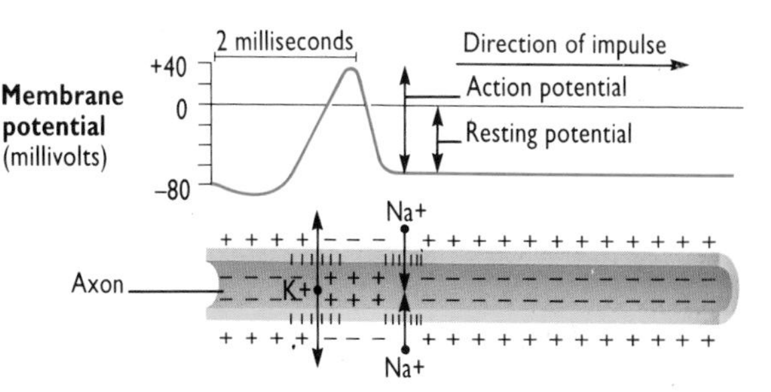

The average speed *of a nerve impulse traveling along the membrane of a nerve cell is 165 feet/sec (50 m/sec).*

The automatic pilot

You are hardly aware of your brain's complex auto-functions, but every second they keep you alive.

When you daydream about nothing in particular, it may seem that your brain has little to do. But in many parts of the brain, dozens of activities carry on constantly, running your body systems like an automatic pilot flying a jet plane without the human pilot having to touch the controls. "Behind the scenes" activities include control of heartbeat rate and the force of contraction, breathing rate and the amount of air breathed in and out, blood pressure, intestinal writhing, and the many other vital processes that keep you alive.

The main region of control is in the lower part of the brain – the brain stem. Each vital function has its own control center (some have more than one), such as the respiratory center which controls the breathing muscles and the cardiovascular center which regulates the heart. These centers receive nerve signals from sensors about the state of the systems they control and send out nerve signals that control the activity of their target systems, whether they are glands or smooth muscles such as those found in the intestine.

The networks for incoming and outgoing signals, together with the control centers, make up the autonomic, or "automatic," nervous system (ANS). Some sections of the ANS – in the brain and spinal cord – make up the central nervous system (CNS). Other parts form some of the nerves and other structures of the peripheral nervous system (PNS), which is outside the CNS. The part of the ANS that conveys outgoing control signals outside the CNS is split into two divisions, the sympathetic and parasympathetic. These generally counterbalance each other in the effects they have on the various body systems.

A small patch of brain tissue – the hypothalamus – is the overall controller of the ANS. Essentially, the hypothalamus maintains stable internal body conditions. In effect, the ANS allows the body to be regulated by the automatic pilot parts of the brain.

Like a movie camera *focused on a film's action, you only notice the ongoing narrative of your life. Ou of shot of the camera is complex behind-the-scenes activity. Similarly, the ANS runs your bod leaving your conscious mind free t concentrate on thoughts and action*

Both divisions of the autonomic nervous system *(ANS) contain motor nerves that carry signals telling an organ or muscle to do something. In the parasympathetic division, motor nerves (green) branching off the central nervous system (CNS) go to remote ganglia (groups of nerve cell bodies) close to the organs or systems they control. There they pass their signals to nerves that finally deliver the instructions.*

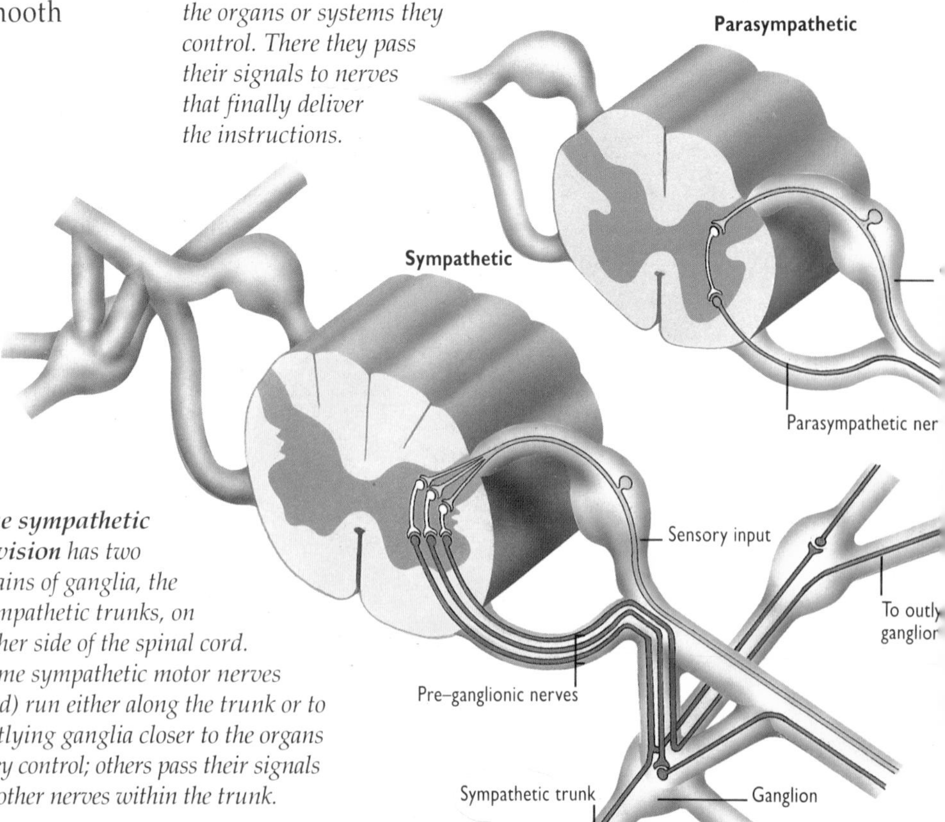

The sympathetic division *has two chains of ganglia, the sympathetic trunks, on either side of the spinal cord. Some sympathetic motor nerves (red) run either along the trunk or to outlying ganglia closer to the organs they control; others pass their signals to other nerves within the trunk.*

Brain

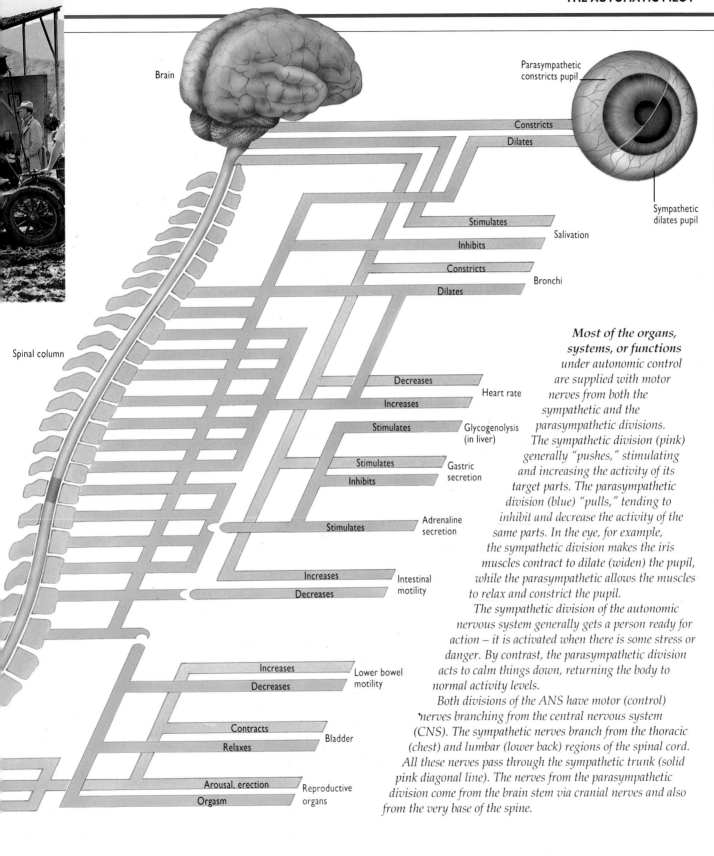

Parasympathetic constricts pupil

Constricts
Dilates

Sympathetic dilates pupil

Stimulates
Inhibits
Salivation

Constricts
Dilates
Bronchi

Spinal column

Decreases
Increases
Heart rate

Stimulates
Glycogenolysis (in liver)

Stimulates
Inhibits
Gastric secretion

Stimulates
Adrenaline secretion

Increases
Decreases
Intestinal motility

Increases
Decreases
Lower bowel motility

Contracts
Relaxes
Bladder

Arousal, erection
Orgasm
Reproductive organs

Most of the organs, systems, or functions *under autonomic control are supplied with motor nerves from both the sympathetic and the parasympathetic divisions.*

The sympathetic division (pink) generally "pushes," stimulating and increasing the activity of its target parts. The parasympathetic division (blue) "pulls," tending to inhibit and decrease the activity of the same parts. In the eye, for example, the sympathetic division makes the iris muscles contract to dilate (widen) the pupil, while the parasympathetic allows the muscles to relax and constrict the pupil.

The sympathetic division of the autonomic nervous system generally gets a person ready for action – it is activated when there is some stress or danger. By contrast, the parasympathetic division acts to calm things down, returning the body to normal activity levels.

Both divisions of the ANS have motor (control) nerves branching from the central nervous system (CNS). The sympathetic nerves branch from the thoracic (chest) and lumbar (lower back) regions of the spinal cord. All these nerves pass through the sympathetic trunk (solid pink diagonal line). The nerves from the parasympathetic division come from the brain stem via cranial nerves and also from the very base of the spine.

See also

CONTROL AND SENSATION
- ▶ Steady as you go 34/35
- ▶ Control systems 36/37
- ▶ The nerve net 44/45
- ▶ Down the wire 46/47

SUPPORT AND MOVEMENT
- ▶ The heart's special muscle 28/29
- ▶ Smooth operators 30/31
- ▶ The chemicals of control 38/39

CIRCULATION, MAINTENANCE, AND DEFENSE
- ▶ In circulation 124/125

Into action

Just pointing a finger needs a complex internal command and control system.

When you make any movement, whether chewing gum or turning a page, the action starts as a burst of electrical nerve activity in the wrinkled cerebral cortex – the uppermost "thinking" part of your brain. The intention to move crystallizes as a plan, known as the central motor program, in the brain's motor and association areas.

The chief site for carrying out the plan is the motor cortex. But it is aided by associated sites, especially the premotor cortex, which receives feedback from the body about movements already in progress. It even monitors non-movements, since simply holding your posture involves minimal and slowly shifting contractions, or muscle tone, which stop you from flopping over.

Each patch of the motor cortex sends out nerve signals destined for a certain part of the body. Parts which have very fine control, such as the fingers and lips, have bigger patches of motor cortex. The signals emerge from the base of the brain and flash down the spinal cord, out along peripheral nerves to the muscles. Nerves that carry messages from brain to muscles are motor nerves.

In muscles that need close and delicate control, like those that swivel the eye, a single nerve fiber (an axon) may control 10 or fewer muscle fibers (cells). In a large limb muscle, a nerve fiber may control several thousand muscle fibers. The outward journey for nerve messages is only part of the story, since the cerebellum, basal ganglia, premotor cortex, and motor cortex are continually assessing the results and amending their orders. This is why you do not chew your gum too hard or tear the page out of the book.

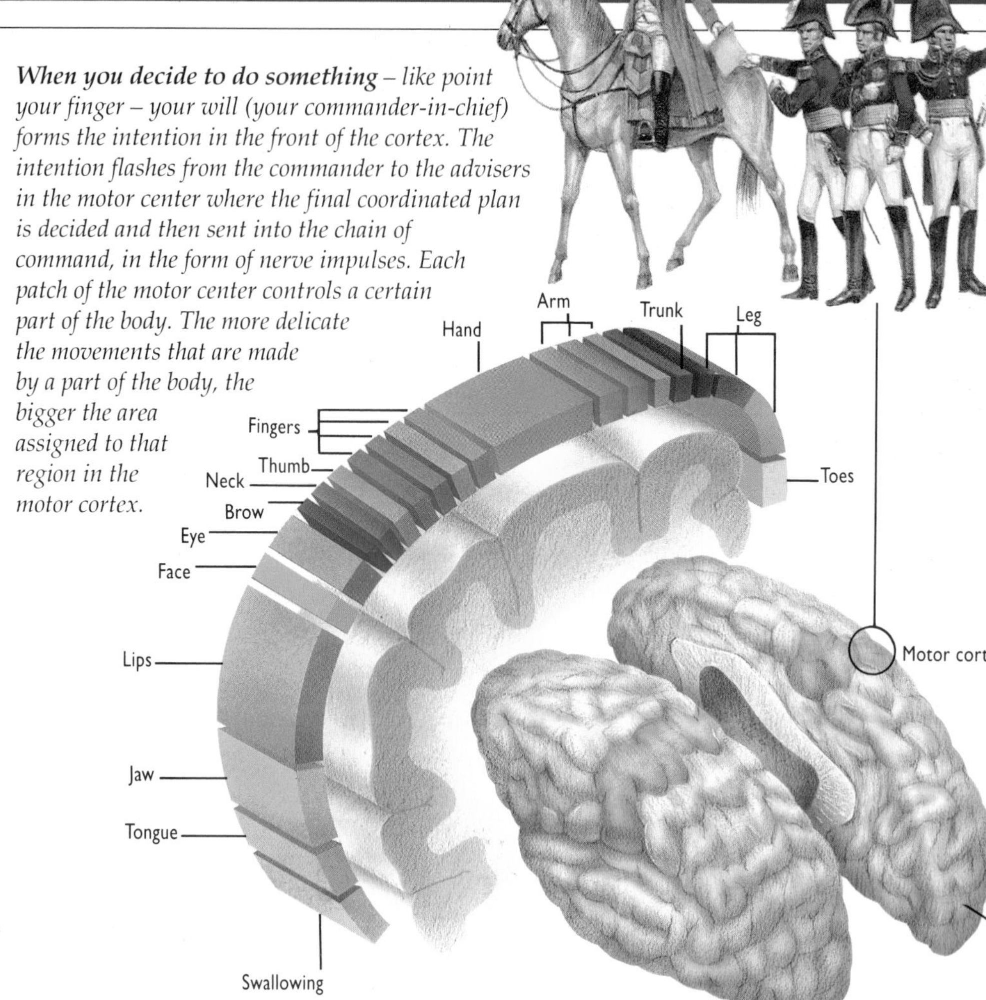

***When you decide to do something** – like point your finger – your will (your commander-in-chief) forms the intention in the front of the cortex. The intention flashes from the commander to the advisers in the motor center where the final coordinated plan is decided and then sent into the chain of command, in the form of nerve impulses. Each patch of the motor center controls a certain part of the body. The more delicate the movements that are made by a part of the body, the bigger the area assigned to that region in the motor cortex.*

THE MINI BRAIN

Nestling at the lower rear of the brain, like a mini version of the large wrinkled cerebral cortex above, is the cerebellum. It coordinates movements, especially the fine, rapid, accurate movements of skilled actions such as writing or playing a sport.

The cerebellum receives motor signals from the motor cortex, as well as sensory signals from the muscles, joints, and skin about how a movement is progressing. It feeds signals back through the brain stem and spinal cord to the muscles to fine-tune their contractions, and it returns signals to the motor center, telling it of the latest developments. This happens with each movement, even turning this page.

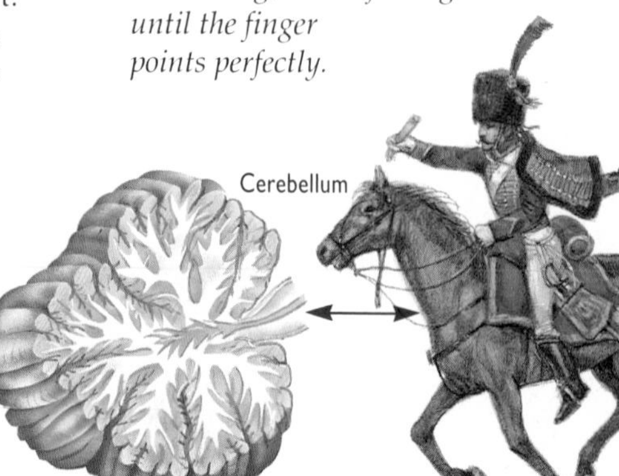

***In the cerebellum** the initial orders are quickly assessed, fleshed out with fine detail, and relayed to the motor cortex and back, via the brain stem and spinal cord, to the arm and hand. The to and fro process of moving, monitoring, and adjusting continues until the finger points perfectly.*

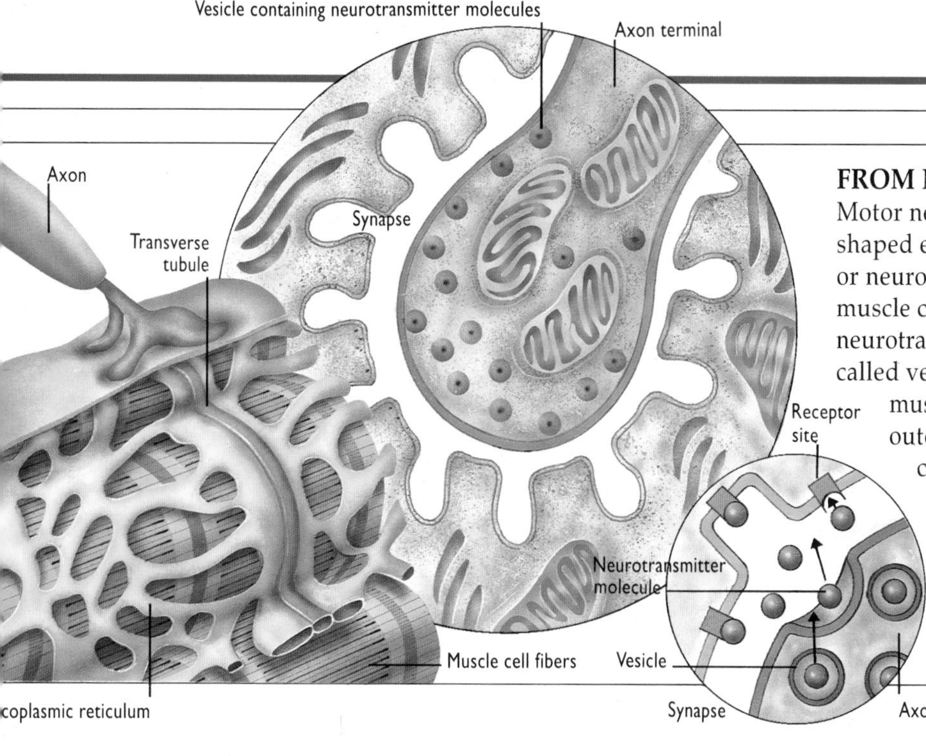

FROM NERVE TO MUSCLE

Motor nerve cell axons terminate in hand-shaped endings – known as nerve-muscle, or neuromuscular, junctions – that sit close to muscle cells (fibers). When an impulse arrives, neurotransmitter molecules are released from sacs called vesicles. These dash across the gap to the muscle cell, where they stimulate the cell's outer membrane. Deep infoldings in the cell membranes (transverse tubules) let impulses reach all of the cell in a few thousandths of a second. Impulses trigger the release of calcium from storage sites (sarcoplasmic reticula), starting contraction.

See also

CONTROL AND SENSATION
- Control systems 36/37
- The nerve net 44/45
- Down the wire 46/47
- Learning skills 80/81

SUPPORT AND MOVEMENT
- Muscles at work 26/27

ENERGY
- Cells at work 114/115

REPRODUCTION AND GROWTH
- Building bodies 160/161

Like a dispatch officer, *the brain stem passes on initial orders to the cerebellum, sending some of the clearer ones via the spinal cord to field commanders. Orders go from the spinal cord at nerve roots along peripheral nerves to the arm and hand muscles – the troops who do the work. Field observers (the senses) soon bring news – when they sense or observe the finger starting to point – and help to coordinate the action.*

Nerve pathways cross over in medulla

Mid–brain and pons

Spinal cord

Nerve root

Skeletal muscles

Signals from position sensors *go back up the spinal cord, with news of the finger's movement. News about the hand's position comes via a different pathway, from the eyes. Monitoring and feedback have begun.*

A model of a human being (right)*, with each part of the body sized in proportion to the area of motor cortex in the brain that controls it, creates an odd being known as a motor homunculus (little man). The areas with greatest motor control, such as the fingers, tongue, lips, and mouth, are largest.*

The range of senses

Our senses are incredibly acute, but they detect only part of the big picture.

Some people add balance as a sixth sense to the "big five" – sight, hearing, smell, taste, and touch. But the list also extends beyond the familiar senses that provide information about the external world to those that are concerned mainly with conditions inside the body. The proprioceptive sense, for instance, tells you what position your limbs are in without your having to look. Then there is a whole battery of internal senses that monitor levels of vital substances such as sugar, oxygen, and carbon dioxide in the blood, as well as physical conditions such as blood pressure, abdominal distension, and body temperature. We are not generally aware of the workings of these "internal" senses.

One thing that all senses have in common is that when they detect something, they register its presence in a common form – the nerve signal, which goes along sensory nerves to the brain. But our brain is only able to perceive a part of what is going on in the world outside, because each of our senses has limitations. For instance, there are thousands of molecules in the air and chemicals in our food that we cannot smell or taste. Creatures such as insects and reptiles see ultraviolet or infrared light, and we cannot. And our ears are deaf to ultrasounds (high frequencies) or infrasounds (low ones), which birds, dogs, bats, and many other animals hear.

10 nm

To X-rays

Ultraviolet

We can extend *the range of our senses by using detectors that turn a phenomenon we cannot detect into one that we can. A device that picks up radiation from Jupiter produces signals that can be turned into an image visible to us (**left**).*

The sense of smell *relies on the detection of chemicals floating in the air. When a molecule has the right shape to fit into a receptor in the nose, it triggers nerve signals which, after interpretation by the brain, we perceive as a smell. We call these molecules odor, smell, or scent chemicals. A person can detect and recognize 10,000 or more shapes of molecules and thus the same number of different smells. We are most sensitive to chemicals such as those in musk scents and skunk sprays, which can be detected at a concentration of 1 part in 20 billion.*

Humans have about 16 million smell, or olfactory, cells; rabbits 100 million; and eels only some 800,000.

Taste, *much complemented by smell, is probably our least sophisticated sense. Nevertheless, we have about 9,000 taste buds (the organs of taste) in the mouth, many more than a chicken or a cat. Humans have fewer than pigs and cows.*

All flavors are mixes of four basic taste sensations: sweet, sour, salty, and bitter. The tongue can detect concentrations of molecules such as quinine (a bitter chemical) at levels as low as about 1 part in 2 million.

Number of taste buds

Touch sensitivity *is measured by the minimum gap needed for two sharp points touched to the skin to be perceived separately. The smaller the gap, the more sensitive the skin.*

Back 1⅗ inch (4 cm)

Palm of hand ⅖ inch (1 cm)

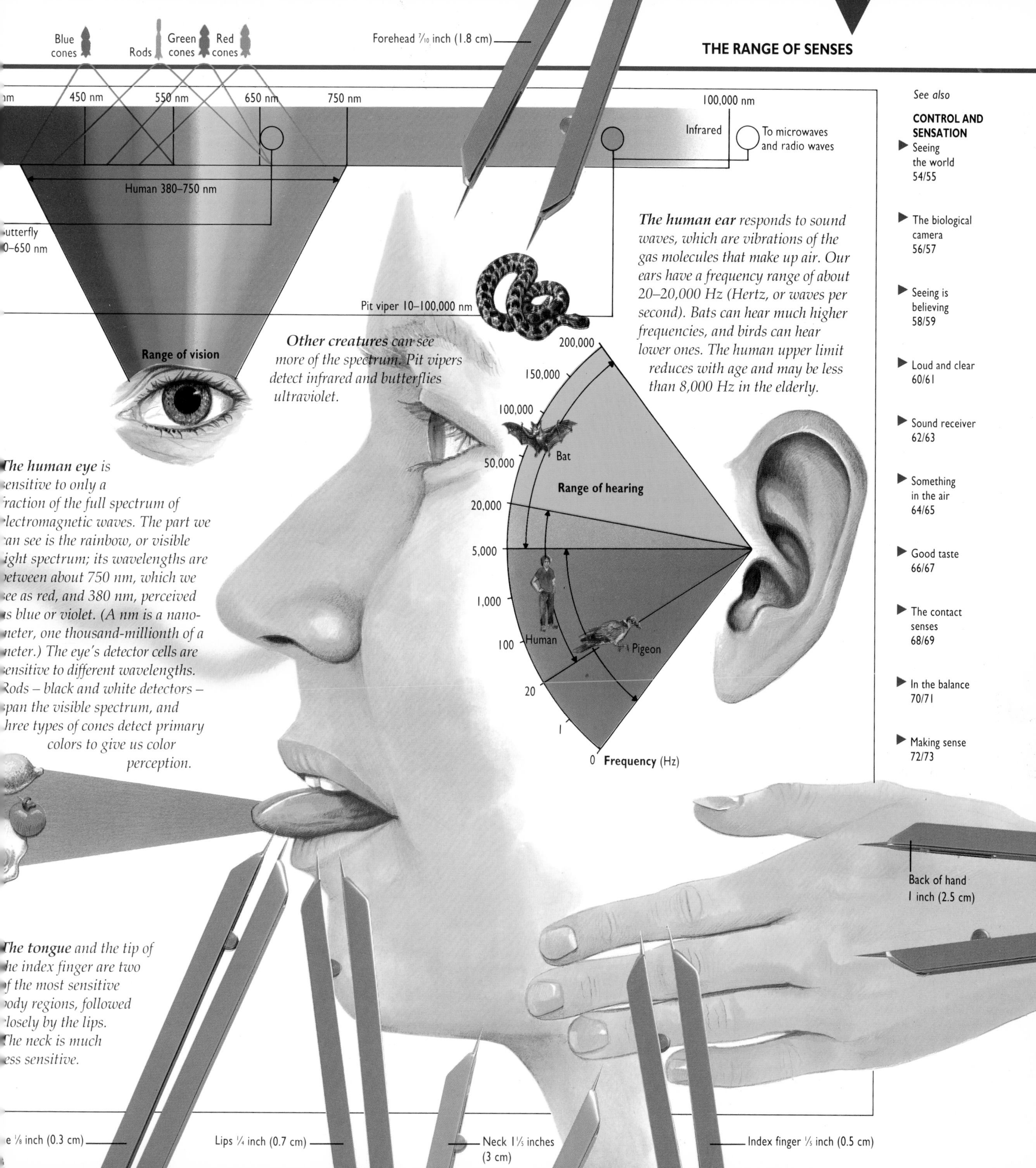

The human ear responds to sound waves, which are vibrations of the gas molecules that make up air. Our ears have a frequency range of about 20–20,000 Hz (Hertz, or waves per second). Bats can hear much higher frequencies, and birds can hear lower ones. The human upper limit reduces with age and may be less than 8,000 Hz in the elderly.

Other creatures can see more of the spectrum. Pit vipers detect infrared and butterflies ultraviolet.

The human eye is ɛensitive to only a ·raction of the full spectrum of ·lectromagnetic waves. The part we ·an see is the rainbow, or visible ·ight spectrum; its wavelengths are ɔetween about 750 nm, which we ·ee as red, and 380 nm, perceived ıs blue or violet. (A nm is a nano- ·neter, one thousand-millionth of a ·neter.) The eye's detector cells are ·ensitive to different wavelengths. Rods – black and white detectors – ·pan the visible spectrum, and hree types of cones detect primary colors to give us color perception.

The tongue and the tip of he index finger are two ·f the most sensitive ·ody regions, followed ·losely by the lips. The neck is much ·ess sensitive.

See also

CONTROL AND SENSATION

▶ Seeing the world 54/55

▶ The biological camera 56/57

▶ Seeing is believing 58/59

▶ Loud and clear 60/61

▶ Sound receiver 62/63

▶ Something in the air 64/65

▶ Good taste 66/67

▶ The contact senses 68/69

▶ In the balance 70/71

▶ Making sense 72/73

Seeing the world

Open your eyes and the world springs into life – objects in your visual field have color, movement, and depth.

Almost since the first films, special-effects experts have been trying to fool audiences. With today's computer technology, models, and animation, they can almost always create wholly convincing images. Yet we may still notice slight shimmers or blurs around shapes; places where the separate pictures do not quite join; or jerky, unnatural movements. We know instinctively when something is not quite right, since the eyes are able to spot even the tiniest flaw or defect.

Sight is our primary and dominant sense, but seeing does not happen in the eyes. These delicate and sensitive devices convert light energy into patterns of nerve signals which pass to the brain where analysis and scene re-creation occur. Vision is estimated to take up more than one-third of our total sensory awareness, and is the input route (as words, pictures, and other visuals) for more than half of the information in the brain.

Whenever there is an image to see, *the eyes pick out lines and shapes as we as colors of every imaginable hue. They also detect movements of all kinds – from large and slow to fast and small. They can pick out shadows, contours, an perspective, giving clues to size and distance, as well as many other features. True? Not exactly. The eyes are chiefly biological movie cameras, sensors of light rays with some preliminary processing power, though still extraordinaril sensitive and adjustable. The business of analyzing the signals to create the world view takes place in the "mind's eye" – the brain.*

Judging distance *depends partly on the interpretation of visual clues. For instance, experience tells us that an object that seems to be in front of another is likely to be the nearer of the two. Move your head sideways, and nearer objects seem to pass in front of farther ones, an effect known as parallax.*

If an object is readily identifiable, its distance can be gauged from its size in the context of the scene. Thus, if you can discern an image on a postage stamp, even though of a huge statue, it must be close by; if the real-life statue seems small, it must be far away. Also, colors fade with distance, so duller-looking objects must be farther away. Parallel lines converge in the distance, an effect of perspective, and lines or shapes become hazier and less distinct with distance because of dust, smoke, or water vapor in the air, giving additional clues.

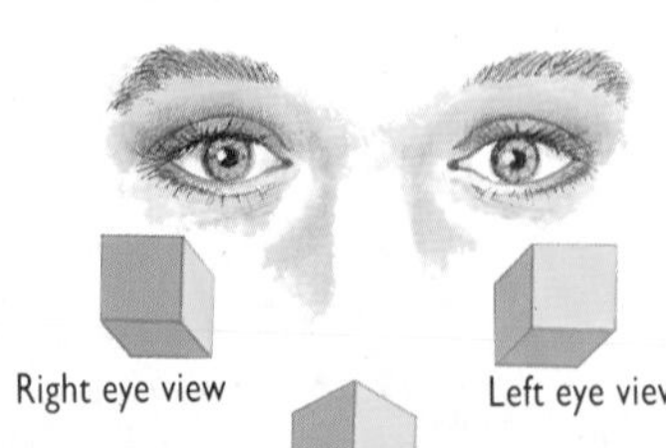

How depth perception works *for nearby objects is demonstrated by looking at something close first with one eye, then the other – each view is slightly different. Judging distance in this way, from two views, is called binocular or stereoscopic vision. The brain uses the overlapping left and right views in several ways to build a three-dimensional image – one with depth. First, the more dissimilar the views, the nearer the object. Second, the brain knows exactly where the eyes are looking, from stretch sensors in the eye-moving muscles. The farther the eyes point inward, the nearer the object. Third, the brain has feedback from the eyes' focusing mechanism.*

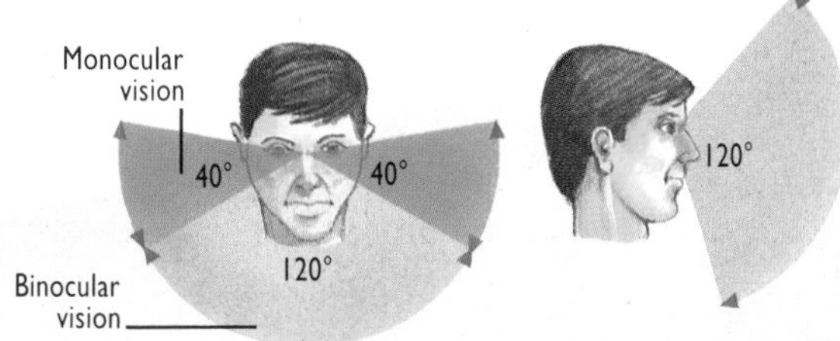

Unlike some animals, *such as the zebra and the rabbit, we cannot see behind us. In order to get two good overlapping views, for the depth perception of stereoscopic vision, we have sacrificed an all-round panorama of the world. In humans the field of view – what you can see to the edge of your vision without moving the eyes or the head – is about 120 degrees up and down, and 200 degrees from side to side (of which about 120 degrees is the overlapping field seen by both eyes).*

To perceive color, *the eye has three different types of light-sensitive cone cells. One type responds best to red light, another to green, and the third to blue. The color perceived depends on the combination of nerve signals from the three sets of cone cells. Recent research shows that the brain itself "adds" to the perception of color, according to the brightness and colors of surrounding objects, and the contrasts between them. Color is thus deduced as well as sensed.*

Aovement attracts our attention, *and ve can track moving objects because our eyes re swiveled by their own muscles as well as eing moved when the head turns. But some novements are simply too fast for our eyes to etect: we can see the wingbeats of a duck, ut a hummingbird's wings move so fast that hey are just a blur. Our brains come to the escue and we understand, even if we cannot see.*

The eye's light-sensitive interior *is extremely delicate and has a self-protecting mechanism that prevents burnout in ultrabright conditions. The pupil – the hole at the front of the eye which lets in light – becomes narrower in brightness and wider in dimness. So despite varying conditions, the interior of the eye receives a fairly constant amount of light, which is within its working range for seeing clearly and comfortably and measured in units of luminance. When it gets too dim, the cone cells, which detect colors and detail, finally lose their sensitivity and can no longer adapt and see. So the scene becomes grayer and less distinct. In time, the rod cells continue to adapt to increasing darkness, and see in grays on all but the blackest nights.*

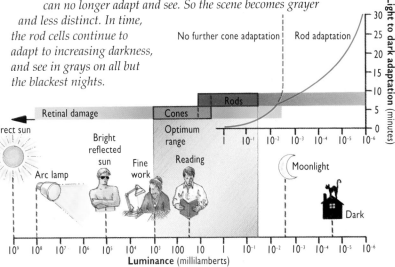

No further cone adaptation
Rod adaptation
Light to dark adaptation (minutes)
Retinal damage
Cones
Rods
Direct sun
Arc lamp
Bright reflected sun
Fine work
Optimum range
Reading
Moonlight
Dark
Luminance (millilamberts)

VISION FACT FILE

So important is our sense of vision that 70 percent of the body's sensory receptors are contained within the eyes. In spite of this, the visual spectrum represents only one-seventieth or 1½ percent of the entire electromagnetic spectrum.

Diameter of eyeball	1 inch (2.5 cm)
Weight of eyeball	¼ ounce (7 g)
No. of receptors in each retina	160 million
rods	150 million
cones	10 million
Wavelength of visible spectrum	380–750 nm
Greatest sensitivity in dark	500 nm
Greatest sensitivity in light	560 nm
No. of colors detectable	10 million

See also

CONTROL AND SENSATION
▶ Down the wire 46/47
▶ The range of senses 52/53
▶ The biological camera 56/57
▶ Seeing is believing 58/59
▶ Making sense 72/73
▶ A sense of self 74/75

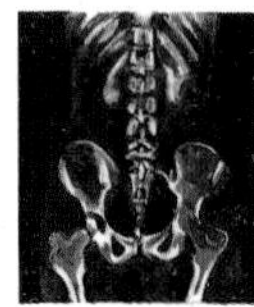

SUPPORT AND MOVEMENT
▶ Staying in shape 12/13
▶ Fine movement 24/25

The biological camera

Packed into an eyeball, a smooth sphere smaller than a golf ball, is an extraordinarily sensitive image detector.

Most of the eye is covered with a tough, whitish outer sheath – the sclera, or white of the eye. At the front the sclera is almost perfectly transparent, forming the dome-shaped cornea through which light passes. The cornea and exposed parts of the sclera are covered with a tissue-thin, see-through protective layer, the conjunctiva. It stays moist and clean because each blink of an eyelid sweeps sterile tear fluid across it.

Inside the sclera is a soft layer known as the choroid, which has many blood vessels that nourish the eye. On the inside of the choroid and toward the back of the eye is the retina, not much larger than a postage stamp and about the same thickness. The retina is the light-sensitive layer which detects the detailed, full-color view of the world.

Light entering the eye through the conjunctiva and cornea passes through a watery liquid (the aqueous humor) inside the corneal dome. Then it goes through the hole of the pupil, which is surrounded by the colored muscles of the iris. These muscles can adjust the pupil to less than 1/25 inch (0.1 cm) across in bright light, or some 1/3 inch (0.8 cm) in dull conditions. Just behind the pupil is the lens, which helps the cornea to focus light rays so they form a clear, sharp image on the retina; the cornea provides about 80 percent of the focusing power. The lens is stretched by the ciliary muscles, a ring of muscles that fine-tune the focus by changing the shape of the lens, so it becomes fat for near objects, thin for far ones. Finally, between the lens and the retina, is a clear "jelly," or vitreous humor. This forms the bulk inside the eyeball and, with the outer sclera, gives it firmness.

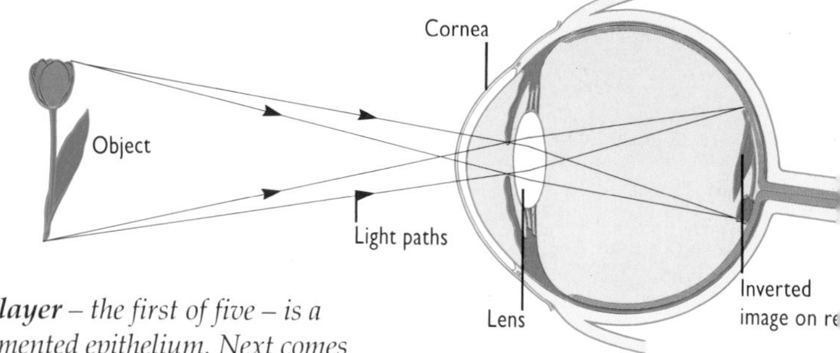

Light from objects *passes through the cornea and lens and is focused to form a sharp, upside-down image on the retina. We perceive the world the right way up because as infants we learn to link an image on a certain part of the retina with an object in a certain orientation or position in the outside world.*

The retina's outermost layer *– the first of five – is a colored membrane, the pigmented epithelium. Next comes a layer of rod and cone cells. Nerve signals from them pass inward to the third layer, the two-ended or bipolar nerve cells, and then to a fourth layer, the ganglion cells, from which fibers pass along the retina's inner surface making the fifth layer. The optic nerve, which goes to the brain, is made up of these nerve fibers, which come from across the whole retina. Horizontal, amacrine, and other nerve cells add to the network, which helps to pre-process, simplify, and code information before it reaches the optic nerve.*

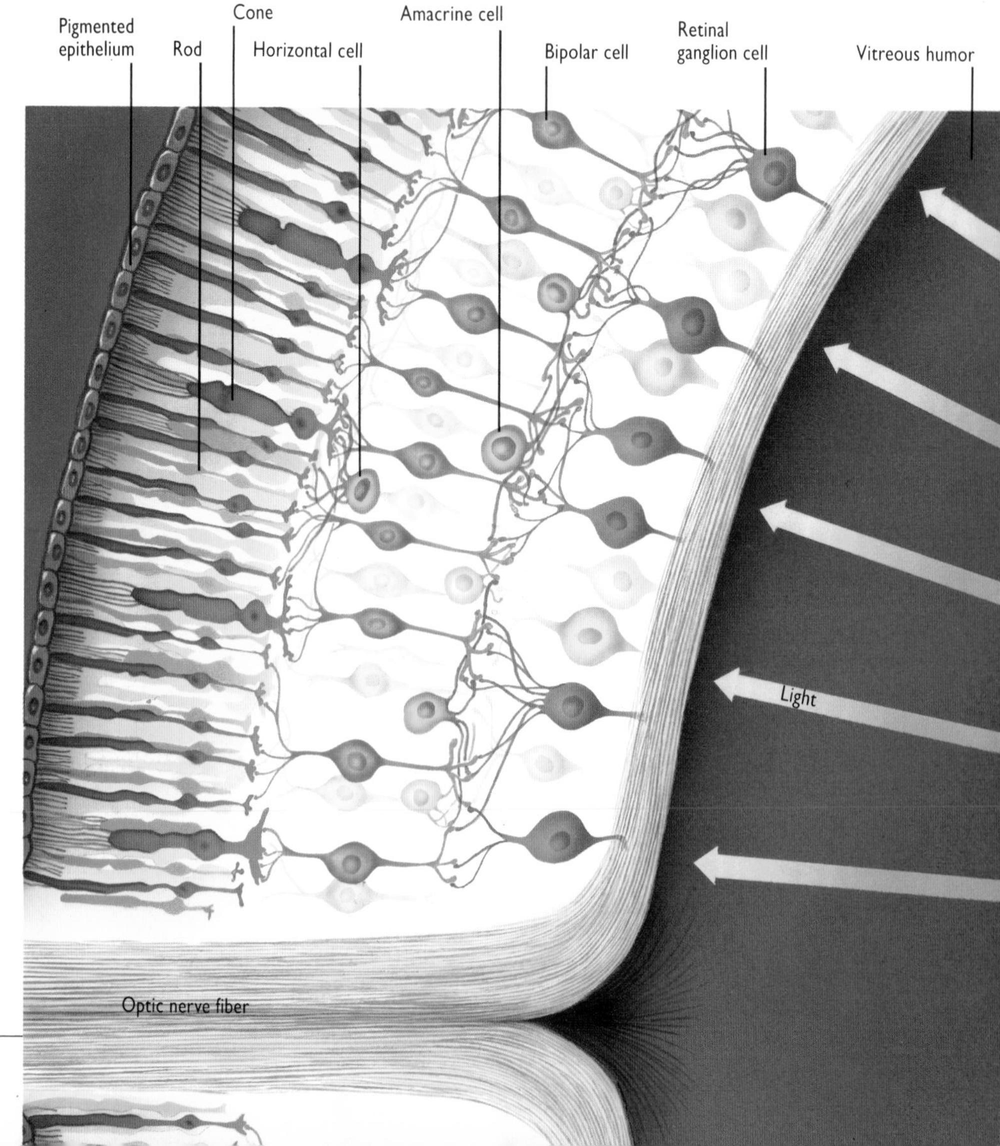

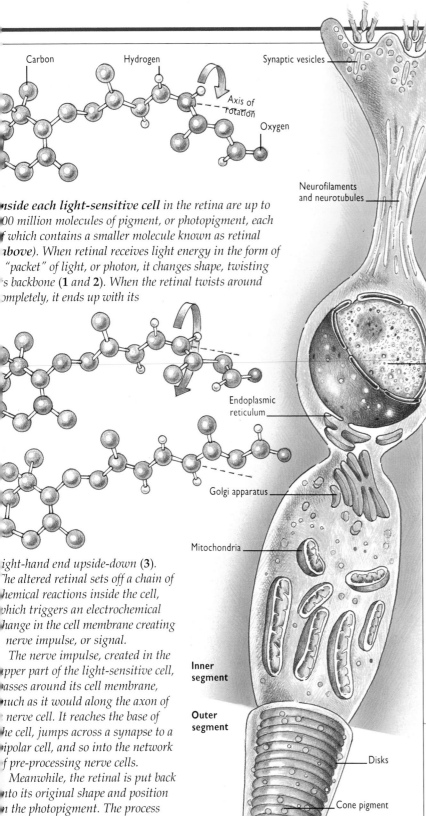

nside each light-sensitive cell in the retina are up to 00 million molecules of pigment, or photopigment, each f which contains a smaller molecule known as retinal *above*). When retinal receives light energy in the form of "packet" of light, or photon, it changes shape, twisting s backbone (**1** and **2**). When the retinal twists around ompletely, it ends up with its

ight-hand end upside-down (**3**). he altered retinal sets off a chain of hemical reactions inside the cell, hich triggers an electrochemical hange in the cell membrane creating nerve impulse, or signal.

The nerve impulse, created in the pper part of the light-sensitive cell, asses around its cell membrane, nuch as it would along the axon of nerve cell. It reaches the base of he cell, jumps across a synapse to a ipolar cell, and so into the network f pre-processing nerve cells.

Meanwhile, the retinal is put back nto its original shape and position n the photopigment. The process appens several times a second for ach of the millions of photopigment nolecules in each of the millions of od and cone cells in each eye.

***A cone cell* (*left*)** *detects color. Cone cells are sensitive to red, green, or blue light, and when light of the right color hits the outer segment, it makes the cell "fire," sending an impulse up through its inner segment's membrane. At the cell's other end, the impulse triggers the release of chemicals from synaptic vesicles which pass the impulse to cells outside the cone. Like other cells, a cone has mitochondria, a nucleus, endoplasmic reticulum, and Golgi apparatus. Most cones are packed into a small area in the center of the retina. When the image falls here, we see it in detail and full color. But cones only work in bright light. Rods, the retina's other type of detecting cells, are found mainly around the sides of the retina. They can work in low light levels and do not perceive colors.*

CAN YOU SEE IT?

Due to genetic causes, some people lack one or more of the three types of color-sensitive cone cells which react to red, green, or blue light to give our perception of color. This means they cannot distinguish colors so easily. In fact, defects of color vision affect around one person in 20, mostly males. In the red–green form, the cones that detect either red or green are missing. For such a person, the watering-can shape below might look like a mug instead. Total color blindness, where everything appears in black and white like in an old movie, is rare.

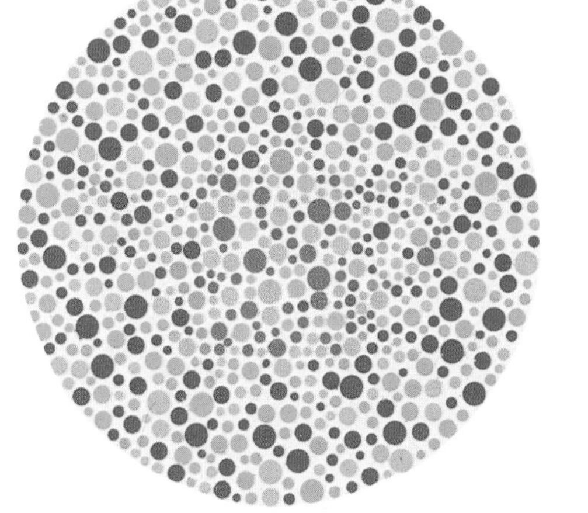

See also

CONTROL AND SENSATION
▶ Down the wire 46/47
▶ The range of senses 52/53
▶ Seeing the world 54/55
▶ Seeing is believing 58/59
▶ Making sense 72/73

SUPPORT AND MOVEMENT
▶ Fine movement 24/25

CIRCULATION, MAINTENANCE, AND DEFENSE
▶ Repelling invaders 144/145

REPRODUCTION AND GROWTH
▶ The first month 168/169

Seeing is believing

The visual perception of the outside world is an impressive demonstration of the power of the brain.

Every second, millions of nerve signals from the eyes arrive in the brain. As with the other major senses, these signals are analyzed in specific patches of the brain's outermost layer, the thin sheet of gray matter known as the cerebral cortex. They are sorted mainly at the rear of the brain in the primary and secondary visual cortices, or centers. A sharp blow to the rear of the head can jolt the visual cortex and interfere with vision, causing you to "see stars" or "black out" momentarily.

Nerve signals arriving from the eye along the optic nerve have already been processed and refined by sets of networked retinal nerve cells, including bipolar and ganglion cells, which receive raw data from the light-sensitive rod and cone cells. After passing through the optic chiasma – where the optic nerves partly cross over one another – messages from the eyes pass along diffuse pathways, or optic tracts, to areas known as lateral geniculate bodies. They then fan out to the visual areas (cortices) at the rear of the brain. The lateral geniculate bodies divert some vision nerve messages to the brain's temporal lobes. Other nerve pathways carry visual information to the spatial awareness and memory sites on the prefrontal lobes.

It seems that when you look at an object you do not decipher the nerve signals from the eyes in a single part of the visual cortex. The image is dissected into its ingredients or features – outline, contours, colors, movements – and dealt with by separate patches of the primary and secondary visual cortices. The secondary visual cortex assembles the result and, drawing on memories and experience, identifies and makes sense of it. All this happens every split second as your eyes observe a scene.

Right visual cortex
Left visual cortex
Lateral geniculate body
Right eye
Optic chiasma
Optic nerve
Left eye
Right visual field
Left visual field

Nerve signals *from the eye are not routed straight to the visual cortices. Nerve fibers from both optic nerves cross over at the optic chiasma. As a result, the brain's left visual cortex receives signals from the left side of the retina in each eye, the right visual cortex receives those from the right sides. But the eye's lens reverses the retinal image, top to bottom and left to right, from the scene in front of the eye. So the left visual cortex "sees" the right part of the visual field (blue), and the right visual cortex sees the left part of the visual field (yellow).*

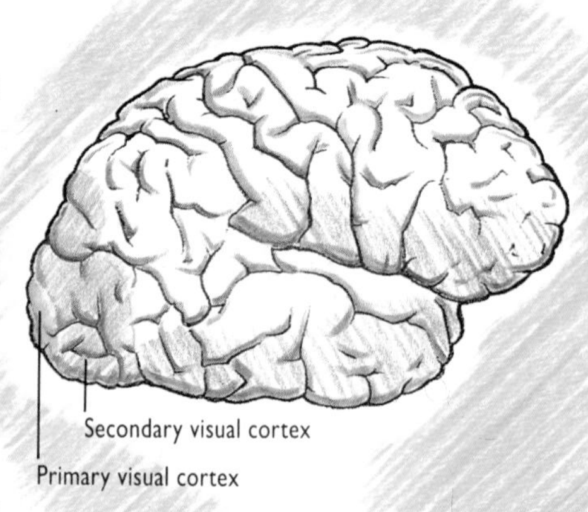

)ne way to fool the sense of ***'ision*** *is by camouflage. The* *Aalaysian horned frog* ***(above)*** *has* *eaflike skin texture and coloration* *o merge with the background.* *'o increase success, the frog's* *amouflage also has a behavioral* *lement. It seeks out the right place* *o hide – on the dead leaves it looks* *ike – and stands as still as the* *bjects around. This is necessary* *ecause the eyes of most animals,* *ike ours, pick up movement as* *listinct from color and shape, and* *he slightest mismatch in the motion* *f objects at once attracts attention.*

'he visual centers *are at the lower* *ear of the brain. Each part of the* *'ye's retina sends signals to a* *orresponding patch of the primary* *isual cortex, so the retina is* *'mapped" on the cortex. The* *econdary visual cortex assembles* *he various aspects of vision – shapes,* *olors, and movements – and links* *vith memories of images so that* *ve can make sense of what we see.*

The thin lines *in the image below seem to spiral into the center. But this is not the case; each thin line forms a circle.*

In the white junctions *between the green squares* ***(below)****, the visual system is fooled into "seeing" green spots where none exist.*

Impossible three-dimensional objects *shown in two dimensions can fool the brain's methods of discerning depth.*

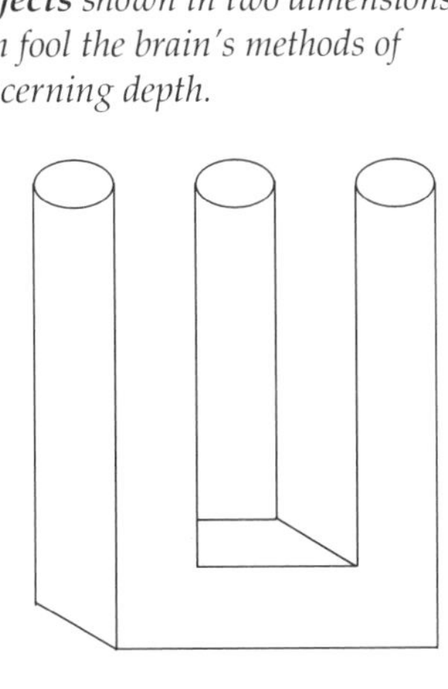

BELIEVING WHAT YOU SEE

Optical illusions and similar images are supposed to trick the eye. But the eye simply records the light patterns, does some initial processing, and sends the nerve signals to the brain. It is the brain that is tricked. In early life, as we learn to look, we learn many assumptions and short cuts to help understand what we see. For instance, if the brain is presented with a regular pattern that has a piece blocked out, it "assumes" the pattern continues into the blank portion, since in daily life this is usually the case. Many optical illusions play on these assumptions by presenting images that do not normally occur in real life. Others exploit the brain's attempts always to make sense of the messages it gets from the eyes.

An egg is an egg is an egg. *Or is it? Although the outlines of all these assorted objects may be egg-shaped, you can tell that not all have been laid by hens. You do this by using your visual memories and experiences to identify the colors and surface textures, which tell you that all but one are man-made from various substances.*

See also

CONTROL AND SENSATION

▶ Down the wire 46/47

▶ Seeing the world 54/55

▶ The biological camera 56/57

▶ Making sense 72/73

▶ A sense of self 74/75

▶ Making memories 78/79

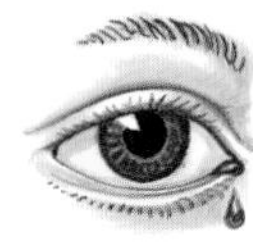

SUPPORT AND MOVEMENT

▶ Fine movement 24/25

Loud and clear

The awkward-looking ear flap belies the delicacy of the sensory organ behind it.

In animals such as horses and rabbits, the large and movable outer ear flap works like a hearing trumpet or collecting dish to gather sound waves. It can be swivelled to where sounds are loudest, thereby pinpointing their direction. After millions of years of evolution, human ears have lost this ability. However, we can still judge where sound is coming from, since sound travels relatively slowly in air, at 1,125 feet/sec (343 m/sec). So sound from the side reaches one ear about 0.001 seconds before it gets to the other – and it is louder in the first ear, too. The brain detects this time and volume difference, and figures out the direction of the sound.

Sounds come through the air in the form of pressure pulses – sound waves – which consist of countless tiny air molecules vibrating to and fro. The waves pass into the slightly S-curved outer ear canal, bounce off the eardrum, and reflect out again. But as they bounce, they set the eardrum vibrating. This membrane, with the area of your little finger's nail, is thinner than skin and more flexible than rubber. Attached to its inner surface is the first of three tiny ear bones, or ossicles.

The ossicles (and the other delicate inner parts of the ear) are protected from knocks and stray vibrations by the skull bone that surrounds them. They are connected to each other by joints that have cartilages, ligaments, and synovial (lubricating) fluid like the larger joints in the elbow or knee, but far smaller. The joints allow the ossicles to rock and move, and to pass the vibrations from the eardrum to the fluid-filled snail-shaped cochlea.

Sound energy undergoes a three-stage transformation in the hearing process. First, the vibrations of molecules in air change to vibrations in solids, namely the eardrum and the three tiny ear bones. Second, they become vibrations in the fluid inside the cochlea. Finally, they are changed into electrical nerve signals by the organ of Corti in the cochlea. The signals then pass along the cochlear nerves to the brain's hearing center – the primary auditory cortex, on the side of the large, wrinkled cerebral hemisphere. This is the "mind's ear," where the sounds finally become real in our conscious awareness.

Just as you can direct your eyes *to look at a certain player in the orchestra, you can also "listen in" to particular instruments. In vision this sensory selection is a physical process as the eyes move and focus. But with hearing it is purely mental. Whether you hear the whole orchestra or listen mainly to one instrument, your ears receive the same sound waves. The switch in auditory awareness happens solely in your brain. You may even close your eyes to "hear better" – that is, be less distracted by the dominant sense of sight.*

In the brain's primary auditory cortex, *patterns of nerve signals coming from the ear and representing sounds are decoded, processed, and compared with patterns already in the memory. This allows us to characterize and identify the sound.*

Nerve signals representing low-pitched sounds are sent mainly to the front of the auditory cortex, while the rear part deals with sounds of higher pitch. The so-called association area around the primary auditory cortex helps to compare and recall sound memories and link them to other features – the name and a mind's eye image of the person or instrument making them, for example.

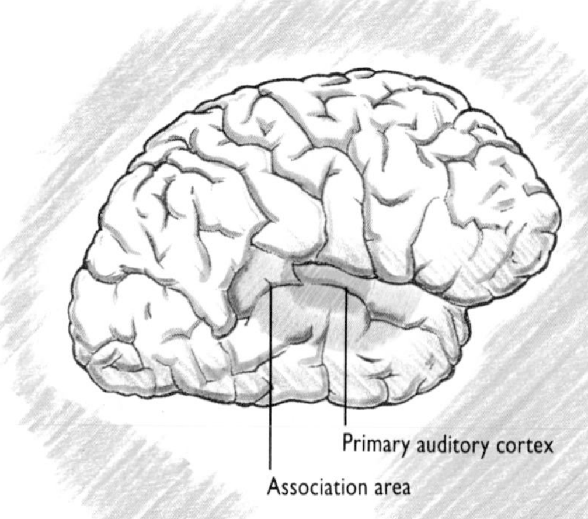

Pressure at eardrum (N/m²)

2 x 10
2 x 10⁻¹
2 x 10⁻³
2 x 10⁻⁵

Loudness (dB)

160
120
80
40
0

80 phons
40 phons
0 phons
80 dB at 1,000 Hz
40 dB at 1,000 Hz
0 dB at 1,000 Hz
Just audible
Silence
Jet engine
Chainsaw
Dripping faucet

Frequency (Hz) 20 60 250 1,000 4,000 16,000

Human hearing range
Middle C (256 Hz)
Piano frequency range
Human singing range

ound intensity – its loudness – is measured both in decibels dB) and as the force per unit area (Newtons/m²) of the sound nergy at the eardrum. A whisper is less than 10 dB, while a jet engine an be as much as 130 dB. Sound frequency, or pitch, is measured in lertz (Hz), or cycles per second. Most people can hear sounds in the ange 30–16,000 Hz. The phon is a comparative unit of loudness, based n the perceived volume of a sound at a standard frequency of 1,000 Hz. t indicates the ear's sensitivity at different frequencies. Low-pitched ounds have to be more intense – have more energy – to produce the ame sensations of loudness, compared to high-pitched sounds.

HEARING FACT FILE

Length of external ear canal		1 inch (2.5 cm)
No. of	bones in each ear	3
	winds in cochlea	2¾
	inner hair cells per ear	3,500
	outer hair cells per ear	20,000
	hairs on inner hair cell	40–60
	hairs on outer hair cell	80–100
Max. range of hearing frequencies *		20–20,000 Hz
Max. sensitivity range		300–3,000 Hz
Approx. pain threshold		130 dB
Efficiency of energy transmission into cochlea		99.9%

* The top frequencies decline with age.

Sound vibrations *follow a path along the outer ear canal to the eardrum, then via the ossicles to the cochlea. The Eustachian tube – a tunnel linking the air behind the eardrum to the air in the throat and thence to the outside air – allows atmospheric pressure changes to reach both sides of the eardrum. If air in the middle ear cavity was isolated, atmospheric pressure changes would bend the eardrum, reducing its sensitivity.*

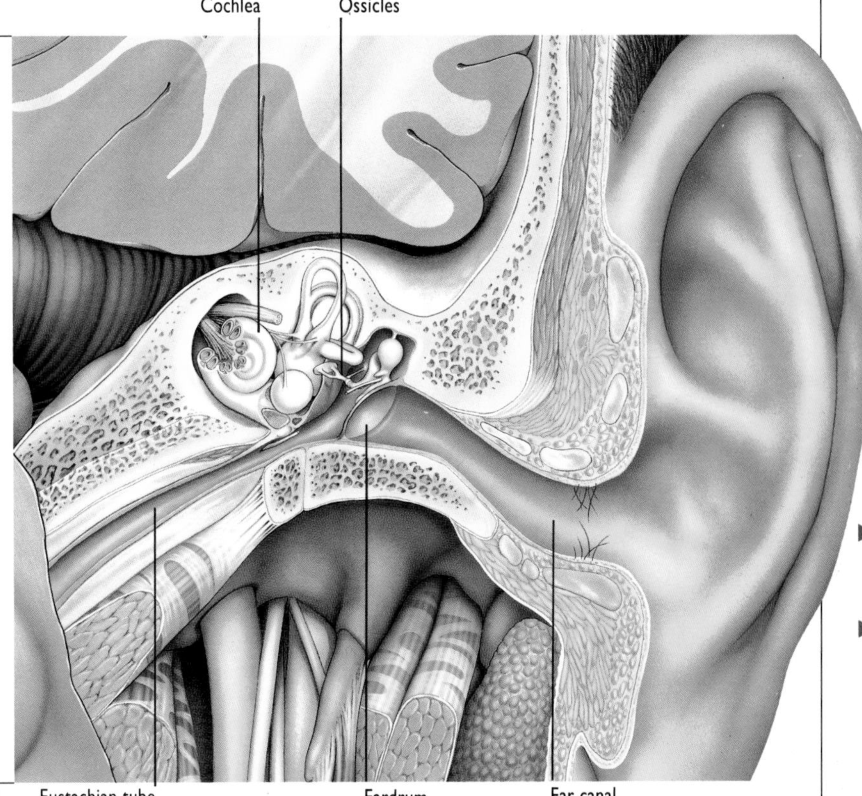

See also

CONTROL AND SENSATION

▶ Down the wire 46/47

▶ The range of senses 52/53

▶ Sound receiver 62/63

▶ In the balance 70/71

▶ Making sense 72/73

▶ A sense of self 74/75

▶ Learning skills 80/81

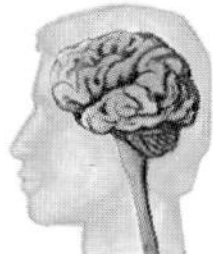

SUPPORT AND MOVEMENT

▶ Bearing the load 14/15

▶ Bone junction 18/19

Sound receiver

More sophisticated than electronic sensors, the ear's cochlea turns vibration into nerve signals.

Many delicate electronic devices, from quartz watches to miniature microphones, depend on the unusual properties of crystals – when electricity is passed through certain crystals, they vibrate, or, when vibrated, produce tiny electric currents. So does the cochlea, a spiral-shaped organ deep inside your ear, almost behind your eye. It is about ¼ inch (0.6 cm) wide and ⅖ inch (1 cm) high – so small it could sit on your little finger's nail. When vibrations arrive and make the fluid inside the cochlea vibrate, it generates electrical signals – nerve impulses. But it is far more delicate and sophisticated than any crystal.

As you thrill to an orchestra playing a thunderous finale, all of the sounds – from deepest drum notes to shrillest cymbals – are being converted from physical vibrations in fluid to electrical impulses inside the cochlea. It all takes place in a tiny strip of membrane less than 1⅕ inches (3 cm) long and 1/25 inch (0.1 cm) wide – the organ of Corti.

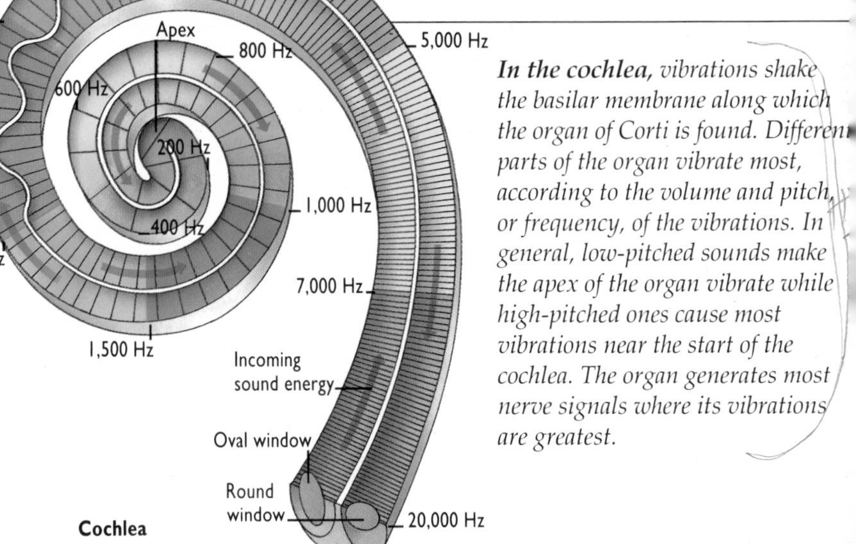

In the cochlea, vibrations shake the basilar membrane along which the organ of Corti is found. Different parts of the organ vibrate most, according to the volume and pitch, or frequency, of the vibrations. In general, low-pitched sounds make the apex of the organ vibrate while high-pitched ones cause most vibrations near the start of the cochlea. The organ generates most nerve signals where its vibrations are greatest.

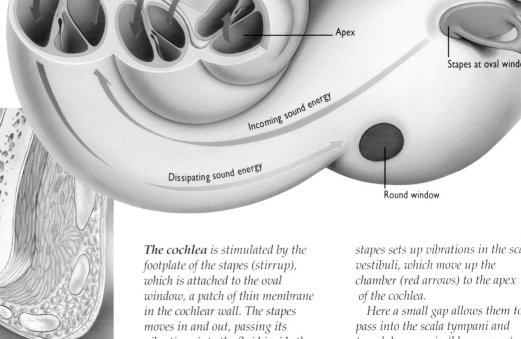

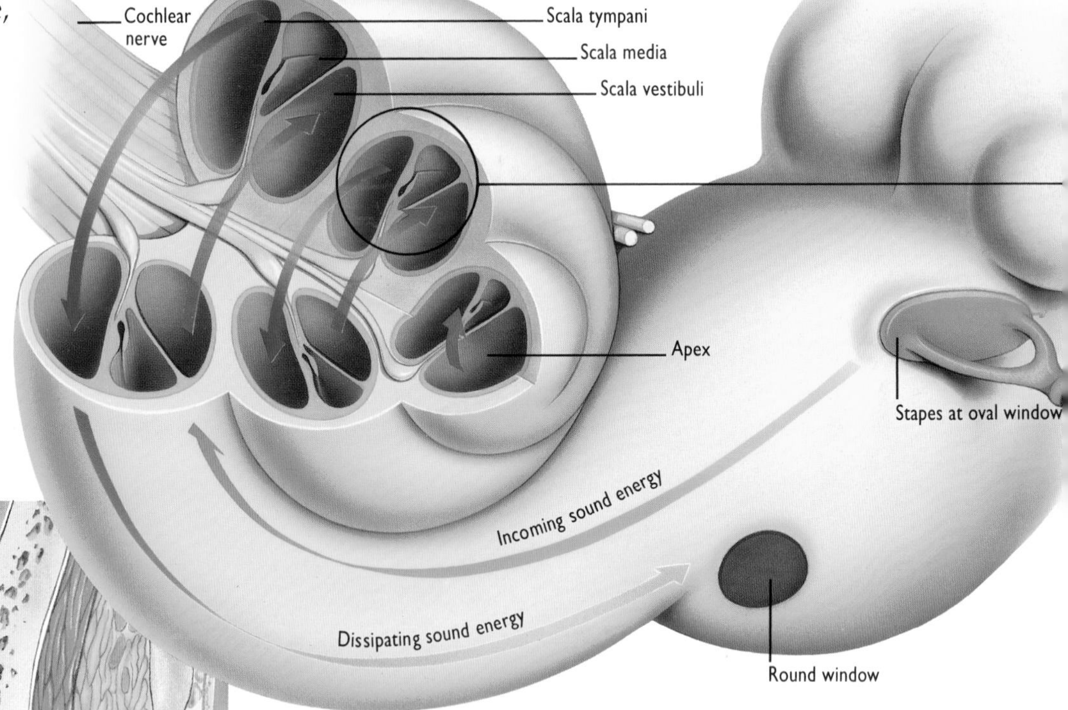

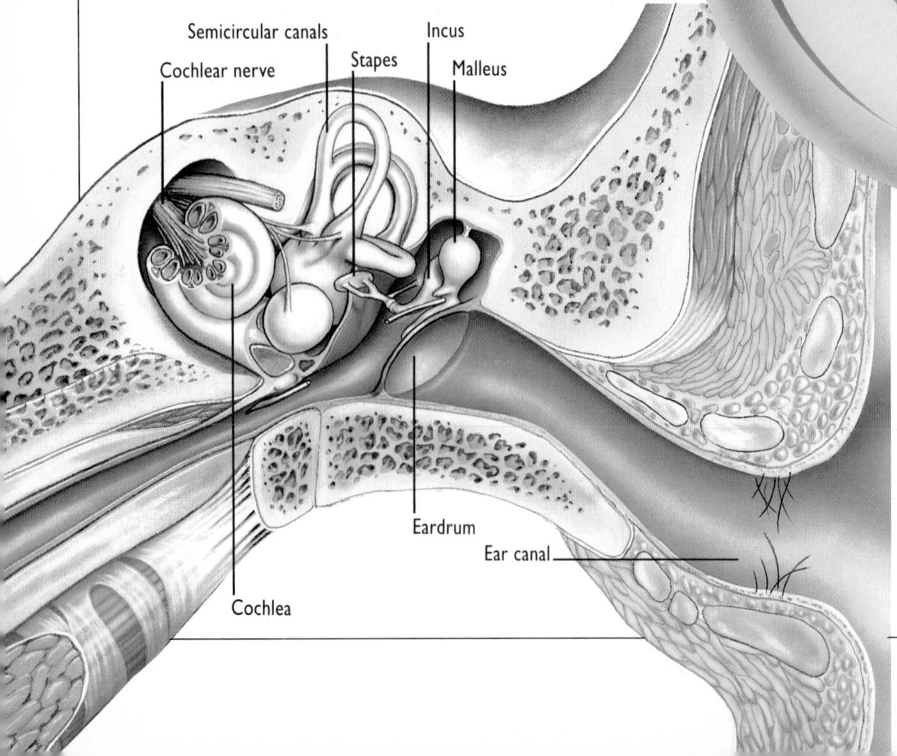

Semicircular canals
Cochlear nerve
Stapes
Incus
Malleus
Eardrum
Ear canal
Cochlea

The cochlea is stimulated by the footplate of the stapes (stirrup), which is attached to the oval window, a patch of thin membrane in the cochlear wall. The stapes moves in and out, passing its vibrations into the fluid inside the cochlea. Inside the cochlea (above) is a spiraling set of membranes arranged in a Y shape to create three fluid-filled chambers. The stapes sets up vibrations in the scala vestibuli, which move up the chamber (red arrows) to the apex of the cochlea.

Here a small gap allows them to pass into the scala tympani and travel down again (blue arrows) to the round window, or fenestra cochlea, which acts as a pressure-relief membrane as it flexes and bulges, dissipating the energy.

he organ of Corti is set onto the basilar membrane, which forms one of the arms of the Y inside the cochlea. Pressure changes in the fluid outside the scala media create vibrations in the basilar membrane and in Reissner's membrane, which forms the other arm of the Y. The vibrations pass into the fluid in the scala media and shake the hair cells.

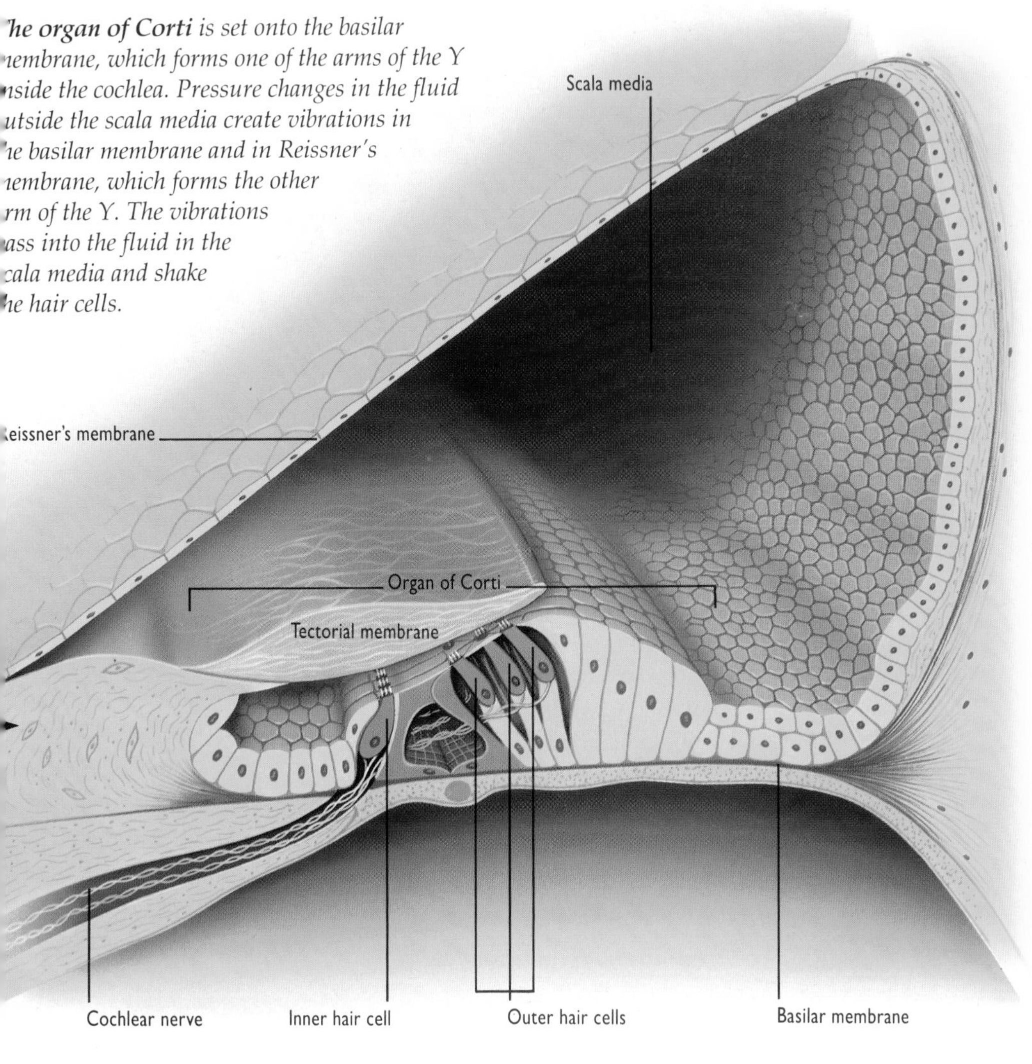

The change in energy, *from physical vibrations to electrical nerve impulses, is carried out by two sets of hair cells (**below**) along the organ of Corti. There are about 3,500 inner hair cells in a single row. Each has a rounded body and bears about 50 miniature hairs, or stereocilia, on its top. The outer hair cells, around 20,000 in total, are in three rows. They are taller and thinner, and each outer cell is topped by up to 100 stereocilia.*

The inner stereocilia are rocked by vibrations in the surrounding fluid. The outer stereocilia touch a jellylike strip, the tectorial membrane, and are bent and tilted as the basilar membrane below vibrates the hair cells. The rocking bases of the stereocilia generate nerve signals, which are channeled by nerve fibers along the organ of Corti to the main cochlear nerve, and then to the brain for processing into the sounds that we perceive.

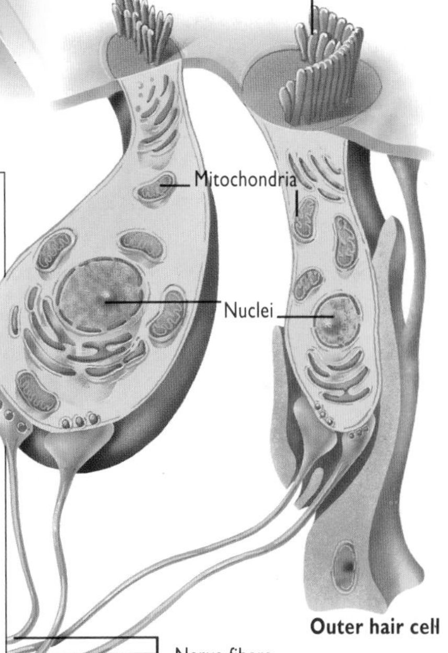

THE ELECTRONIC EAR

Researchers are developing electronic methods of restoring some hearing to people with hearing problems. A cochlear implant bypasses much of the ear's machinery and stimulates the inner ear electrically. The signals arrive from a microphone via a computer or direct from a computerized speech processor. They are sent from a transmitter attached to the skin near the ear to a receiver implanted under the skin. The signals are taken by thin wires to electrodes in the cochlea, which stimulate the organ of Corti, or the cochlear nerve directly.

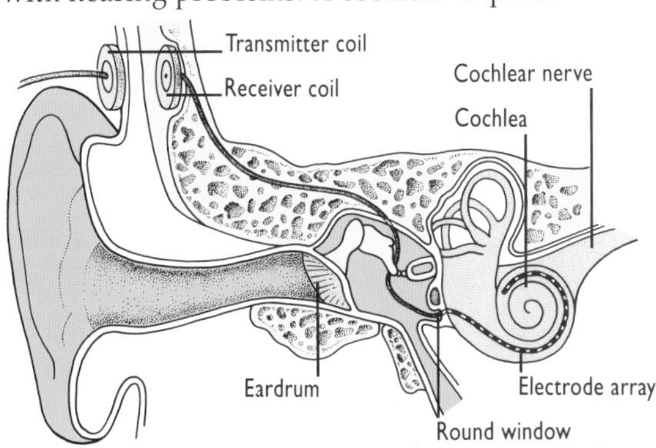

See also

CONTROL AND SENSATION
▶ The range of senses 52/53
▶ Loud and clear 60/61
▶ In the balance 70/71
▶ Making sense 72/73

SUPPORT AND MOVEMENT
▶ Bearing the load 14/15

ENERGY
▶ The cell and energy 112/113

CIRCULATION, MAINTENANCE, AND DEFENSE
▶ Maintaining the system 136/137

REPRODUCTION AND GROWTH
▶ Getting on 182/183

Something in the air

An important sense, smell, affects us at a fundamental, "primitive" level, often evoking memories and feelings.

The faintest whiff of an odor – of a person, perfume, or place – has the power to evoke long-lost memories, including images, sounds, and tastes, and with them, deep-felt emotions. This is because the nerve pathways dealing with our sense of smell – olfaction – are part of the brain's limbic system, which also deals with memories and emotions. Smell probably evolved to warn us when food is bad, and of potential dangers such as smoke from a fire. It is also used, perhaps less consciously, to detect the natural body scents called pheromones involved in nonverbal communications, such as recognition, sexual arousal, and fear.

A smell begins as chemical molecules floating in the air. When these pass through the nose on breathing in, some of them may float and swirl up to a thumbnail-sized patch of lining in the roof of each nasal cavity – the olfactory epithelium. This consists of millions of tall, slim cells jammed together in a crowd. Some of these are the olfactory sensory cells, which do the smelling. Each one has 10–20 long tiny hairs, or cilia, that stick downward into the watery mucus that covers the inner linings of the nasal cavity. Odor molecules dissolve in the mucus and stimulate receptor sites on the cilia. This generates nerve signals that pass upward, along the olfactory sensory cell body, which narrows into a wire-shaped nerve fiber, or axon. The axons from thousands of sensory cells group into bundles and convey their nerve signals to the olfactory bulb just above, where they are partly sorted before going to the brain. Thus smell, like taste, is a chemosense – it detects chemical substances. It can also only detect molecules that dissolve in the mucus of the olfactory epithelium.

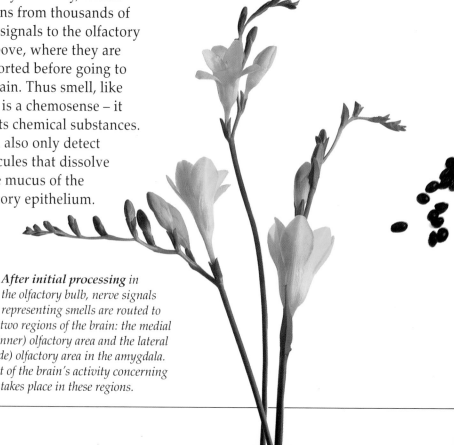

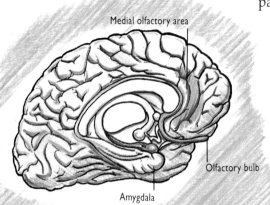

Medial olfactory area
Olfactory bulb
Amygdala

After initial processing *in the olfactory bulb, nerve signals representing smells are routed to two regions of the brain: the medial (inner) olfactory area and the lateral (side) olfactory area in the amygdala. Most of the brain's activity concerning smell takes place in these regions.*

SENSE OF SMELL FACT FILE

Smell is the sense in which habituation occurs most quickly. Habituation is the process in which a sense becomes accustomed to what it detects so that it is no longer perceived. Most odors can hardly be perceived just 30 seconds after they are first detected.

Total area of olfactory epithelium	1/3 sq. inch (2.5 cm^2) in each nasal cavity
No. of sensory cells	25 million
Life of sensory cell	30 days
Lowest concentrations detectable	1 part in 30 billion
No. of odors distinguished	10,000 or more
Decline with age	approx. 1% of sensory cells are not replaced each year

Try to imagine the smells *of the items shown here. (It may help to look and then close your eyes, to cut out the dominating sense of sight.) Some may be pleasant, such as the scent from flowers; others, such as strong household chemicals, less so. The brain can retain the "odor profile" of hundreds of smells in its memory. Like other memories, familiar odors can be recalled by a trigger such as the sight of an object or the sound of its name.*

ur noses are constantly assailed y different smells. The tiny invisible rticles that make up the odor olecules we can detect are given off y objects as diverse as coffee beans, owers, and leather and carried in e air.

Odor molecules dissolved in mucus made by Bowman's glands come into contact with the cilia, hairlike projections from the olfactory sensory cells, and trigger the sensory cells to generate nerve signals. These pass along wirelike fibers, which group into about 20 bundles and pass through holes in the bone above, into the olfactory bulb.

In the bulb, the fibers form complicated ball-shaped sets of connections (synapses) with relay cells. These connection areas, up to $\frac{1}{250}$ inch (0.1 mm) wide, are olfactory glomeruli, and there are hundreds in each olfactory bulb. Each glomerulus receives signals from more than 25,000 sensory cells and has tens of thousands of connections from the relay cells in the bulb itself. Much sorting and processing of the signals takes place in the glomeruli. The resulting nerve messages are sent along the olfactory tract to the brain.

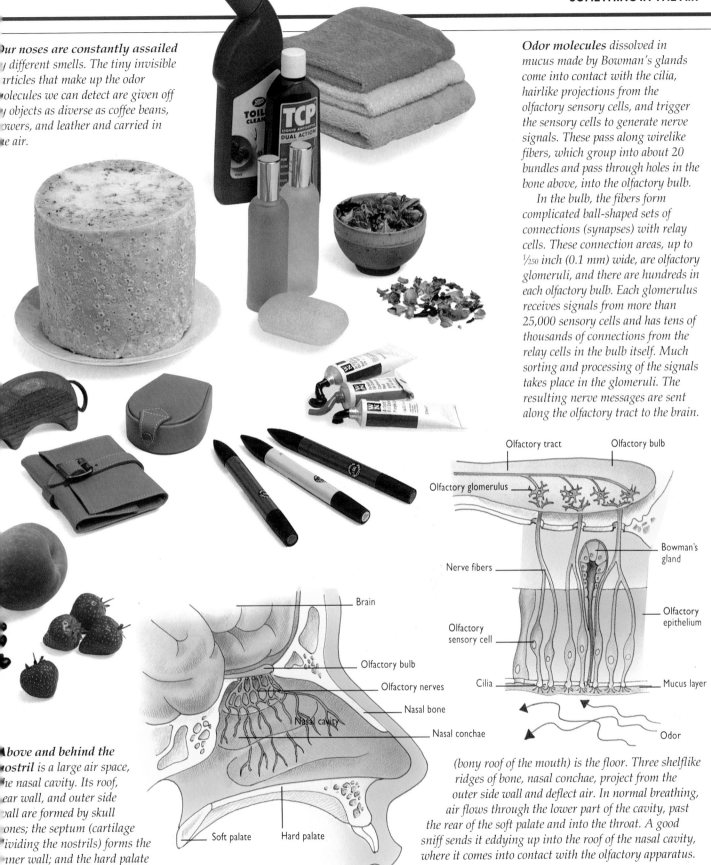

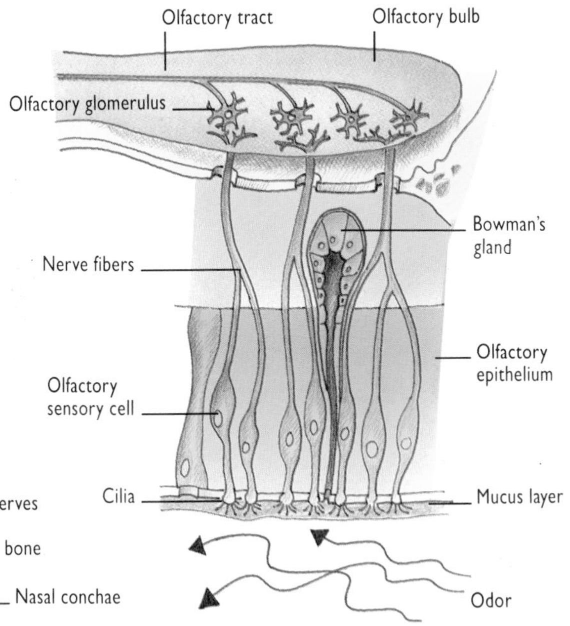

Olfactory tract
Olfactory bulb
Olfactory glomerulus
Bowman's gland
Nerve fibers
Olfactory epithelium
Olfactory sensory cell
Cilia
Mucus layer
Odor

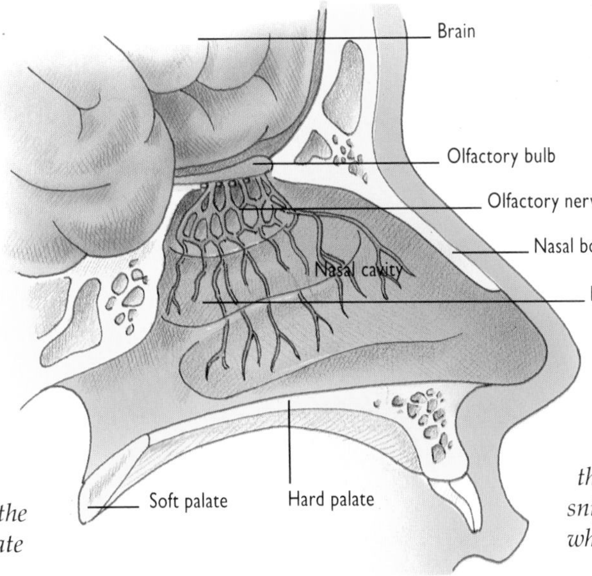

Brain
Olfactory bulb
Olfactory nerves
Nasal bone
Nasal cavity
Nasal conchae
Soft palate
Hard palate

bove and behind the ostril *is a large air space, he nasal cavity. Its roof, ear wall, and outer side all are formed by skull ones; the septum (cartilage ividing the nostrils) forms the ner wall; and the hard palate (bony roof of the mouth) is the floor. Three shelflike ridges of bone, nasal conchae, project from the outer side wall and deflect air. In normal breathing, air flows through the lower part of the cavity, past the rear of the soft palate and into the throat. A good sniff sends it eddying up into the roof of the nasal cavity, where it comes into contact with the olfactory apparatus.*

See also

CONTROL AND SENSATION

▶ The range of senses 52/53

▶ Good taste 66/67

▶ Making sense 72/73

▶ Making memories 78/79

▶ Moods and emotions 84/85

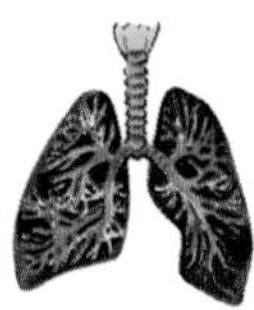

ENERGY

▶ A deep breath 108/109

Good taste

Despite the narrow range of perceived flavors, taste not only helps us enjoy food, but it also warns us about potential poisons.

The sense of taste is sited in the tongue, and it relies on thousands of microscopic cell clusters, or taste buds, embedded in its surface. There are also taste buds on the rear of the mouth's roof and at the back of the throat. If you stick out your tongue and study it in a mirror, you cannot see the taste buds, as they are far too small. But you can see dozens of small projections known as papillae. Taste buds are located on the sides and around the base of papillae.

The tasting, or gustatory, cells in the buds have hairy tips which detect chemicals in solution. When stimulated by flavor molecules, these cells generate nerve signals which they send to the taste center on the brain's cortex, and also to the hypothalamus, which is concerned with appetite and the salivating reflex.

Our discrimination of tastes is relatively unsophisticated. It is thought that there are only four basic flavor sensations, each of which is detected chiefly on a certain part of the tongue: sweet at the tip, salty on the front sides, sour on the middle sides, and bitter at the central back. There is little tasting ability on the tongue's central upper part, since there are few taste buds there.

Taste nerve signals *are carried by three nerves in each side of the head (cranial nerves) to a small part of the medulla (brain stem). The signals then travel to parts of the brain, such as the hypothalamus, the thalamus, and the gustatory part of the sensory cortex – the "taste center." In the taste center, the signals are interpreted.*

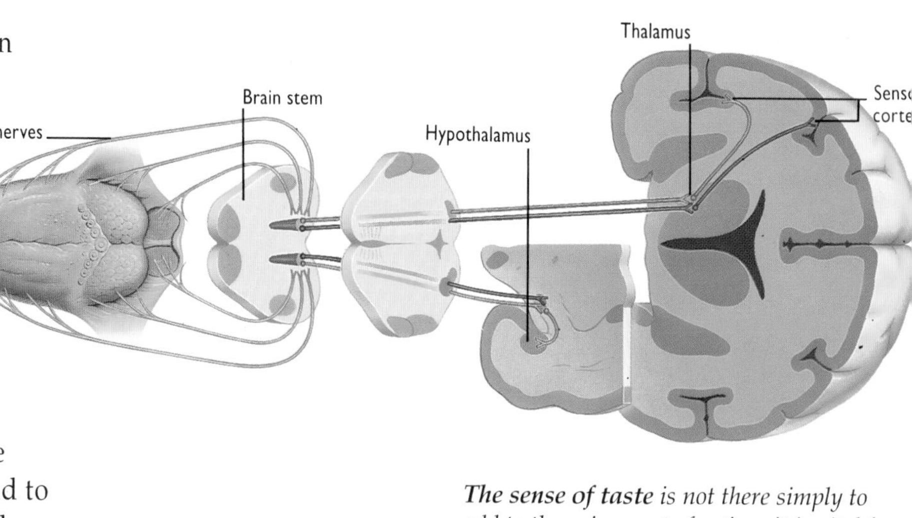

The sense of taste *is not there simply to add to the enjoyment of eating; it is vital for survival. It tells us what is good to eat and warns us against things that are not good. It evolved to help us pick out sweet, ripe fruits and other foods that contain energy-packed sugars and starches. Likewise, taste is extremely sensitive to bitter flavors – many poisonous berries, fruits, and fungi are bitter-tasting.*

Potato chips

Cheese

Barbecue

Onions and sour cream

Potato chips *often have salty and sour tastes – they have instant appeal as they hit the tongue.*

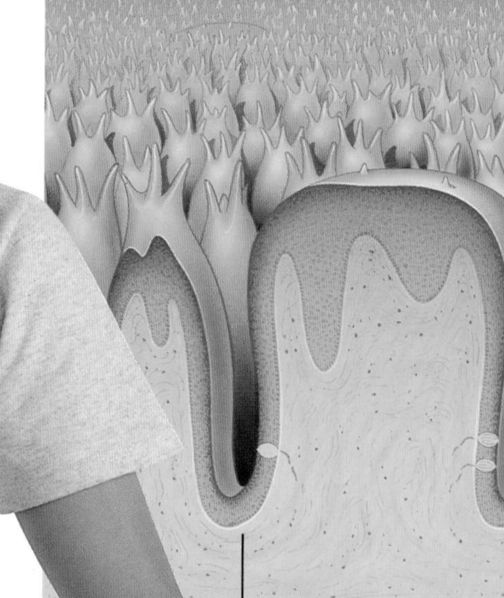

'he pleasantness, in ercentage terms, of the ›ur flavors depends on ıeir concentrations. At ›w levels, sweet tastes are npleasant, sour ones less). Weak bitter flavors are ›lerated, but soon become istasteful. Salty ones are ıore pleasing, but then ırn unpleasant. Sour avors follow the same attern, but at higher ›ncentrations. Sweet tastes nprove with strength.

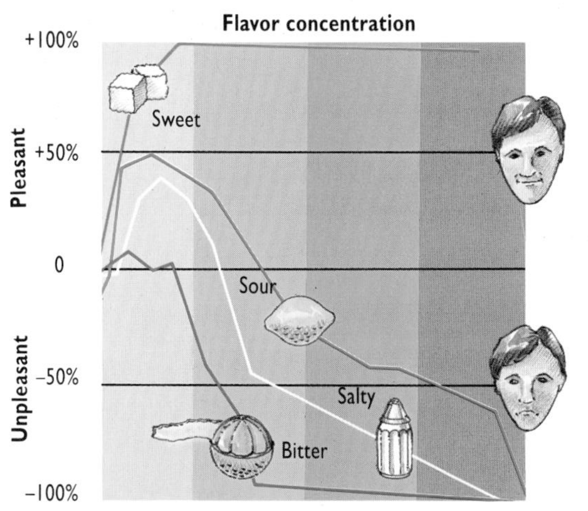

TASTE FACT FILE

The perceived strength of a strong flavor fades to one-tenth within 10 seconds.

Special sense organ	Tongue
No. of taste buds	5,000–12,000
Size of taste bud	About 1/500 inch (0.05 mm) high and wide
Gustatory cells per bud	25–40
Life of gustatory cell	7–10 days
Flavors perceived	Sweet, sour, salty, bitter
Greatest sensitivity	Bitter substances*

* The perception of bitter substances is acute. For instance, it is 1 part in 2 million for quinine.

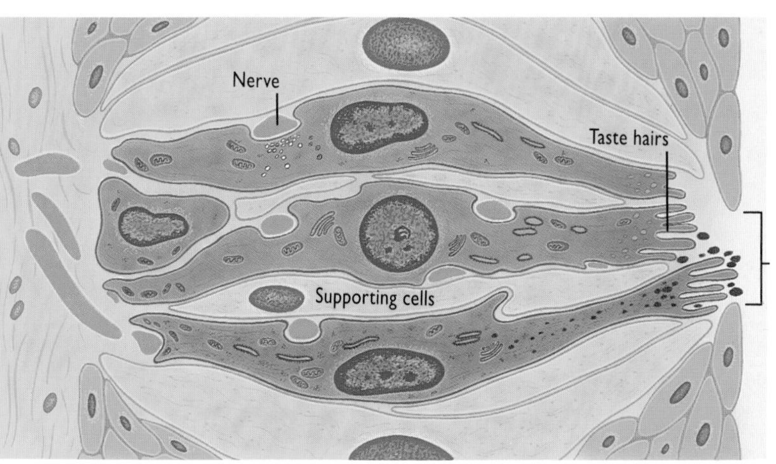

An individual taste bud *looks rather like a peeled orange buried just under the tongue's surface, with cells as the segments. Each bud contains between 25 and 40 gustatory cells, which do the tasting. They are held in place by supporting cells. Their thin projections (taste hairs) stick up into a tiny hole in the surface, the apical pore, which lets food molecules seep into the taste bud.*

A tongue-scape, *looking from the back of the tongue toward the tip, shows the three main types of papillae, or "pimples," on the tongue's upper surface. Filiform papillae are generally conical or pointed; fungiform papillae are flat-topped; vallate papillae are larger with an outer groove. Most taste buds are set in the lower parts of the vallate grooves, with mucus-secreting glands below them.*

See also

CONTROL AND SENSATION
▶ The range of senses 52/53
▶ Something in the air 64/65
▶ Making sense 72/73

ENERGY
▶ You are what you eat 96/97
▶ Chewing it over 98/99

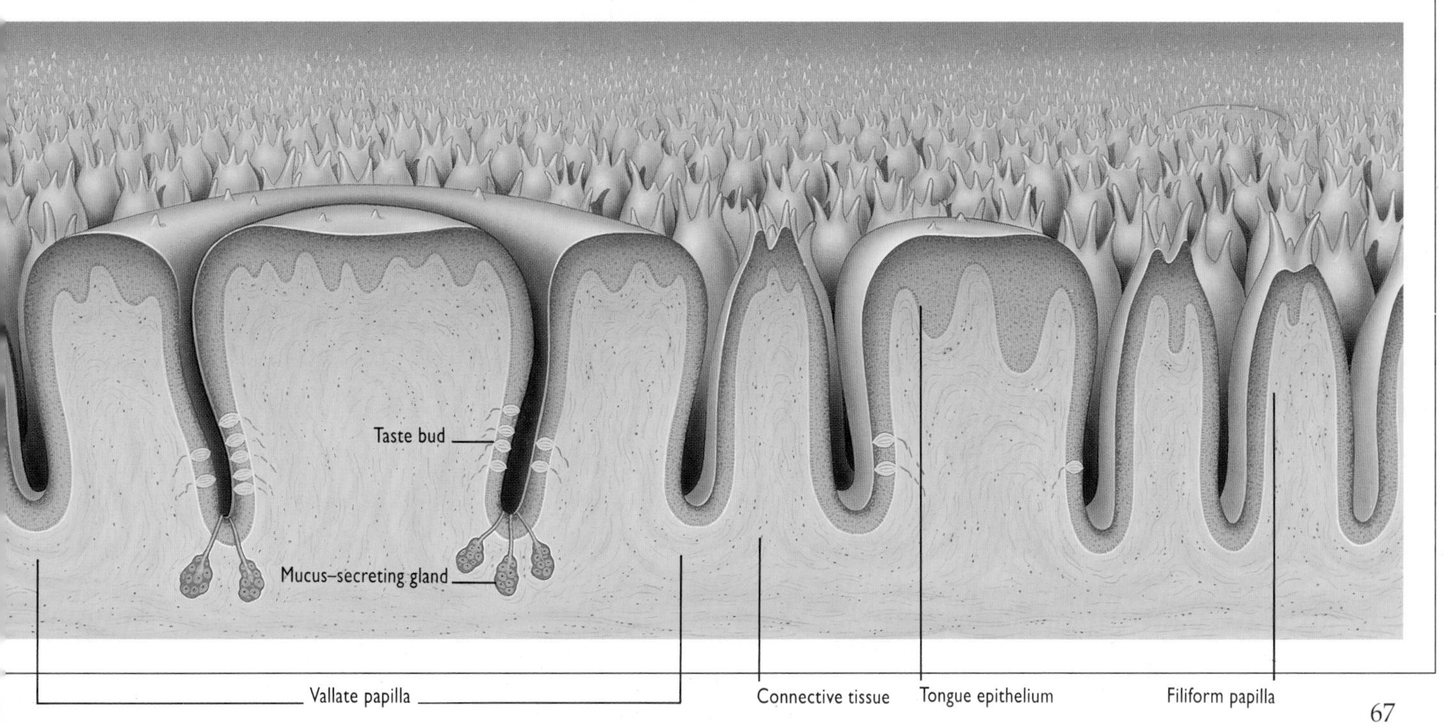

The contact senses

From the brush of a feather to the searing pain of a burn, your skin informs you about the outside world.

As you knead warm dough or grip a wriggling worm, millions of microscopic sensors buried just under the surface of your skin produce barrages of nerve signals. These go to the somatosensory cortex – the part of your brain dealing with touch. When you interpret these signals, you perceive not simply touch, but a whole range of physical information about the contacted object – hardness and resilience, surface texture, temperature, dryness, and so on. An object need not even touch the skin; if it contacts the skin hairs and rocks them, sensors wrapped around the hair root are stimulated.

Touch sensors respond in different ways. Some react to physical changes that mechanically distort their shape – they are mechanoreceptors. Some of these respond quickly, keeping up with fast vibrations, like those from a tuning fork; others are slower and react to alterations in the shape of the skin that take place over a few seconds. Some are triggered by the lightest touch; others send most signals when squashed by heavy pressure or when damaged skin releases specific chemicals. The brain figures out what is happening from the overall pattern of signals.

PHANTOM FEELINGS

People who have had a limb or other body part amputated, like one-armed British naval hero Nelson (**left**), often continue to feel sensations in the missing part, such as itching or pain. These are known as phantom feelings, although they seem real to the person concerned.

While the part's sensory endings were taken away with its skin, two components of the touch system remain. These are the sensory nerves (now cut short), which took messages to the brain, and the areas of the brain which perceived sensations from the missing part. It is thought that when the cut-short nerves are stimulated, the brain, knowing no better, interprets the resulting messages as it had learned to do when the part was present.

Skin has a thin epidermis, *which is mainly protective, and a thicker dermis below. In addition to small blood vessels, it has tiny nerves ending in the various types of touch receptors.*

Meissner's endings *are found in the uppermost part of the dermis, especially on the hands, feet, lips, and inner surfaces of the eyelids. They are shaped like eggs and are both quick- and slow-change mechanoreceptors, detecting light touch and vibrations.*

Merkel's endings *are like tiny disks stuck in the underside of the epidermis, where they feel slight changes in its shape, thereby detecting light touch. They are both quick- and slow-change mechanoreceptors.*

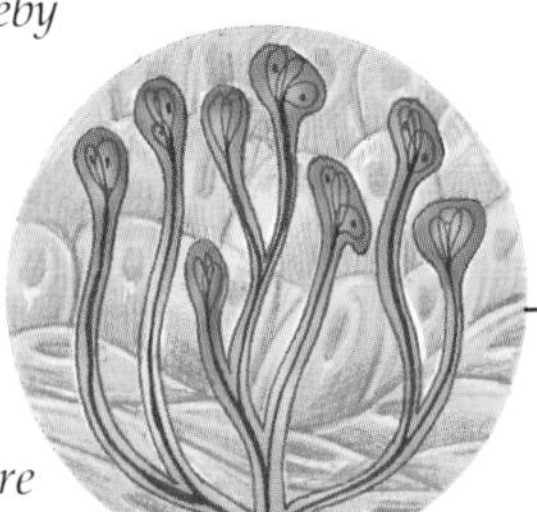

Bulbs of Krause *are multi-layered capsules with many-branched nerve endings. They are quick-change mechanoreceptors, triggered by rapid alterations in shape caused by pressure or vibrations, and may also help us to feel extreme cold.*

Meiss endin

Merkel's endings

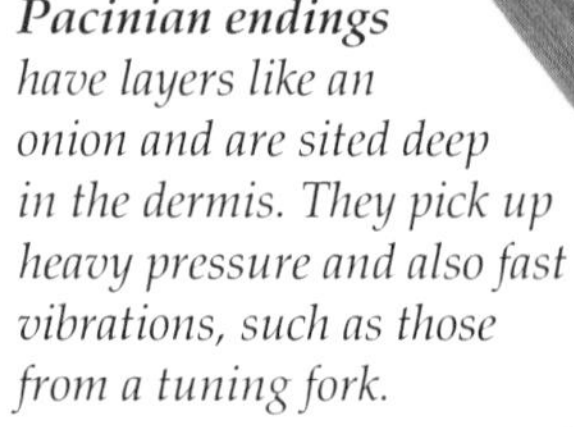

Bulb of Krause

Pacinian endings *have layers like an onion and are sited deep in the dermis. They pick up heavy pressure and also fast vibrations, such as those from a tuning fork.*

Pacinian en

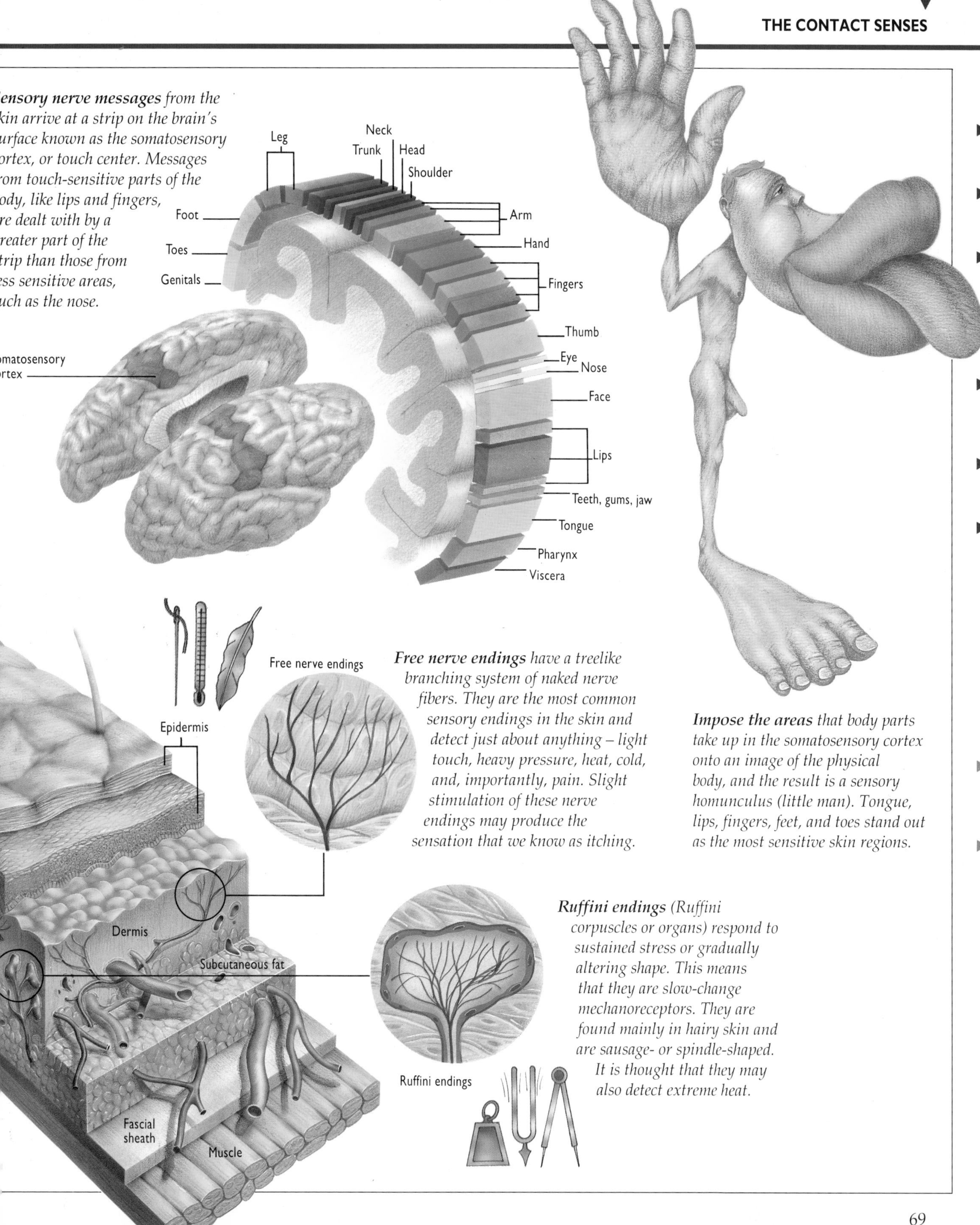

ensory nerve messages *from the kin arrive at a strip on the brain's urface known as the somatosensory ortex, or touch center. Messages rom touch-sensitive parts of the ody, like lips and fingers, re dealt with by a reater part of the trip than those from ss sensitive areas, uch as the nose.*

Free nerve endings *have a treelike branching system of naked nerve fibers. They are the most common sensory endings in the skin and detect just about anything – light touch, heavy pressure, heat, cold, and, importantly, pain. Slight stimulation of these nerve endings may produce the sensation that we know as itching.*

Impose the areas *that body parts take up in the somatosensory cortex onto an image of the physical body, and the result is a sensory homunculus (little man). Tongue, lips, fingers, feet, and toes stand out as the most sensitive skin regions.*

Ruffini endings *(Ruffini corpuscles or organs) respond to sustained stress or gradually altering shape. This means that they are slow-change mechanoreceptors. They are found mainly in hairy skin and are sausage- or spindle-shaped. It is thought that they may also detect extreme heat.*

See also

CONTROL AND SENSATION
- The nerve net 44/45
- Down the wire 46/47
- Into action 50/51
- The range of senses 52/53
- In the balance 70/71
- Making sense 72/73

CIRCULATION, MAINTENANCE, AND DEFENSE
- Maintaining the system 136/137
- Outer defenses 142/143

In the balance

Our two-legged posture is inherently unstable, so how does the balance mechanism help us stand tall?

Balance is sometimes called the sixth sense. Although it relies on sensory body parts, it is, in fact, an ongoing process that involves both sensory and motor – movement – systems. Four main sets of sensory input are involved. First, information from the skin is important, especially from the touch and pressure sensors on different parts of the feet, which tell the brain if you are leaning. Second is eyesight with which you judge verticals such as trees or buildings, to which you should be parallel, and horizontals such as the ground, to which you should be at a right angle. Third is the body's proprioceptive sense – our awareness of the position in space of parts of the body. Its stretch receptors in muscles, tendons, and joints tell the brain about the positions and angles of the arms, legs, torso, and neck.

Fourth are the sensory parts dedicated to balance, deep in each inner ear, next to the cochlea. These parts are known collectively as the vestibular apparatus and are part of the same network of fluid-filled chambers as the cochlea. They consist of the utricle, the saccule, and the semicircular canals. In certain parts of their linings are tiny hairs, whose roots are embedded in lumpy crystals or gels. The crystals or gels are attracted downward by gravity, and they are also pushed to and fro by the fluid in the chambers, which swirls as the head changes its position.

Imagine tea in a cup. Turn the cup, and the tea moves too, but lagging behind the cup. Keep turning the cup and the tea catches up. Stop the cup and the tea carries on swirling for a time. Now imagine that the hairs and crystals of the vestibular apparatus dip into the tea. As the cup speeds up and slows down (representing the movements of the head), the hairs and crystals are pushed this way and that by the motions of the tea.

The hairs grow from sensory hair cells, and when these detect their hairs moving, they fire appropriate signals to the brain. The utricle and saccule each have a concentrated patch of these hair-bearing nerve cells in their lining, known as the macula. The hairs here are embedded in a gelatinous, granular membrane, which gravity pulls downward. As the position of the head changes in relation to gravity, the hairs pull in different directions, and so the hair cells generate different patterns of signals. Individual nerves (ampullar nerves) from the utricle, saccule, and semicircular canals collect to form the vestibular nerve which goes, alongside the cochlear nerve, to the brain.

In muscles and joints *there are stretch receptors that feed data to the brain, which then works out where the limbs are in space. This lets an organist, for instance, move his fingers to the correct keys without having to look at them. This sense of knowing where body parts are is called proprioception.*

The vestibular nerve feeds its information chiefly to the cerebellum and to four structures in the medulla known as vestibular bodies. Using this data, as well as input from the other three sensory sources, the brain works out what to do, usually subconsciously. It then sends nerve signals out along motor nerves. The muscles tense or relax and constantly make small adjustments and so you keep your balance.

WHICH WAY IS DOWN?

The effects of gravity are not felt in free fall – in an orbiting space craft, for instance – so the body does not know which way is up or down. The eyes can help by using clues from the interior design of the craft's cabin, as can the proprioceptive sense, which detects body posture. With no gravity, the inner ear's vestibular apparatus can respond only to movements of the head, and the skin's pressure sensors cannot work at all. The result is an unfamiliar mass of inputs that bewilders the brain, and which may result in nausea, giddiness, and confusion, the symptoms of space sickness.

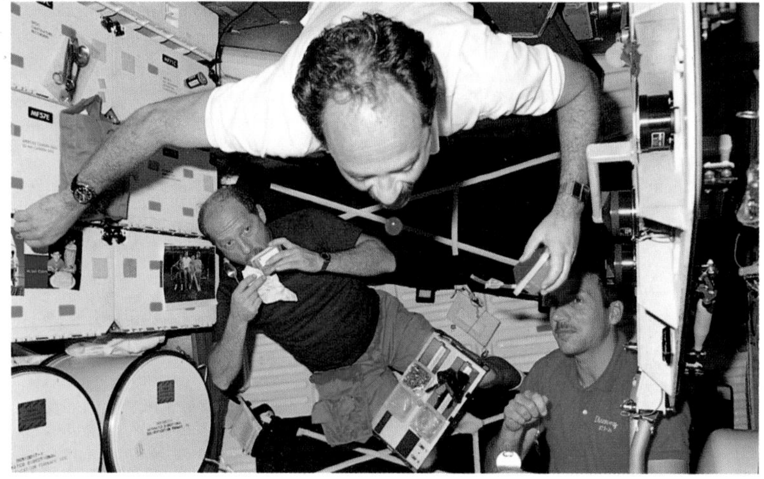

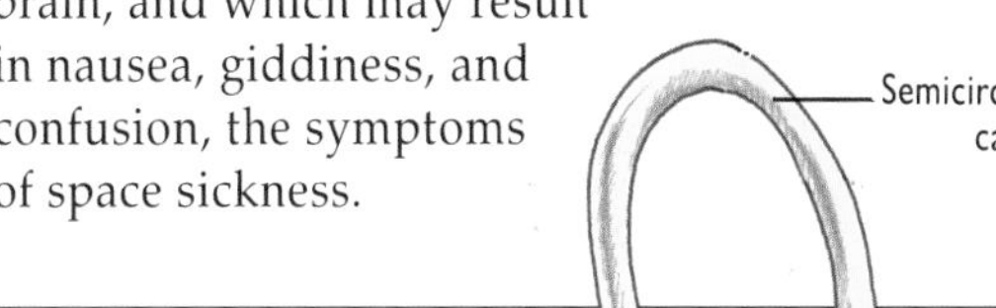

Cross section of semicircular canals

Near the organs of hearing in the inner ear are some of the major parts of the balance mechanism. These include the saccule, the utricle, and the semicircular canals. These are all fluid-filled tubes or chambers. The semicircular canals – the superior, posterior, and lateral – are each set in one of the three planes of space, like the x, y, and z axes on a three-dimensional graph. The saccule and utricle mainly detect movements of the head in relation to the Earth's gravitational field.

No matter which way the head moves, fluid in at least one semicircular canal is disturbed. Near a canal's junction with the main chamber is a bulge – the ampulla – containing a jellylike crystal lump, the crista. As the fluid (known as endolymph) is displaced, it tilts the cupulla (the crista's upper part), which moves hairs embedded in it.

These hairs are projections from hair cells, in the lower part of the crista. As the hairs pull and distort, their cells generate nerve signals that flash away along the ampullar nerve to the brain.

Detail of crista

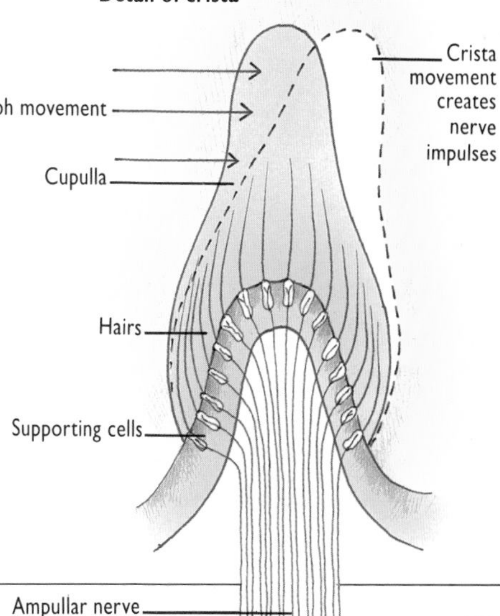

See also

CONTROL AND SENSATION
- ▶ Into action 50/51
- ▶ The range of senses 52/53
- ▶ Loud and clear 60/61
- ▶ Sound receiver 62/63
- ▶ Making sense 72/73
- ▶ Learning skills 80/81

SUPPORT AND MOVEMENT
- ▶ Bone junction 18/19
- ▶ Making a move 22/23

Making sense

Our brains literally give us a sense of ourselves while they monitor, control, and perceive.

A number of the body's processes – including those that are automatic, such as breathing and heartbeat, as well as general awareness levels, emotional matters, and some aspects of memory – are centered in the lower parts of the brain. These parts – the medulla, reticular formation, limbic system, amygdala, and hippocampus – are surrounded by the semi-sphere of the cerebrum, divided into two large, grooved cerebral hemispheres. These are linked to each other by a bridge of 100 million nerve fibers, the corpus callosum.

Each hemisphere's inner bulk is white matter – mainly the fibers of nerve cells, which carry signals to and from the surface layer, or cortex. This is a folded sheet, about $\frac{1}{8}$–$\frac{1}{5}$ inch (3–5 mm) thick, of gray matter – the cell bodies and projecting neurites of billions of super-networked nerve cells. It is the site of conscious awareness and thought processes.

Every second the senses bombard the cortex with millions of nerve signals. They are pre-processed and pre-selected in the two egg-shaped masses of nerve cells that make up the thalamus, in the center of the brain. The signals are sent on to the relevant patches or "centers" of the cortices for detailed analysis, except for smell signals, which have a direct route to the olfactory cortex. The cortex also has a motor center, which sends signals out to initiate voluntary body movements.

One anatomical oddity is that the left side of the brain deals with control and sensation in the right side of the body, and vice versa. This is because nerves cross over from one side to the other in the upper spinal cord and lower brain stem. Thus damage to the right side of the brain affects the left side of the body.

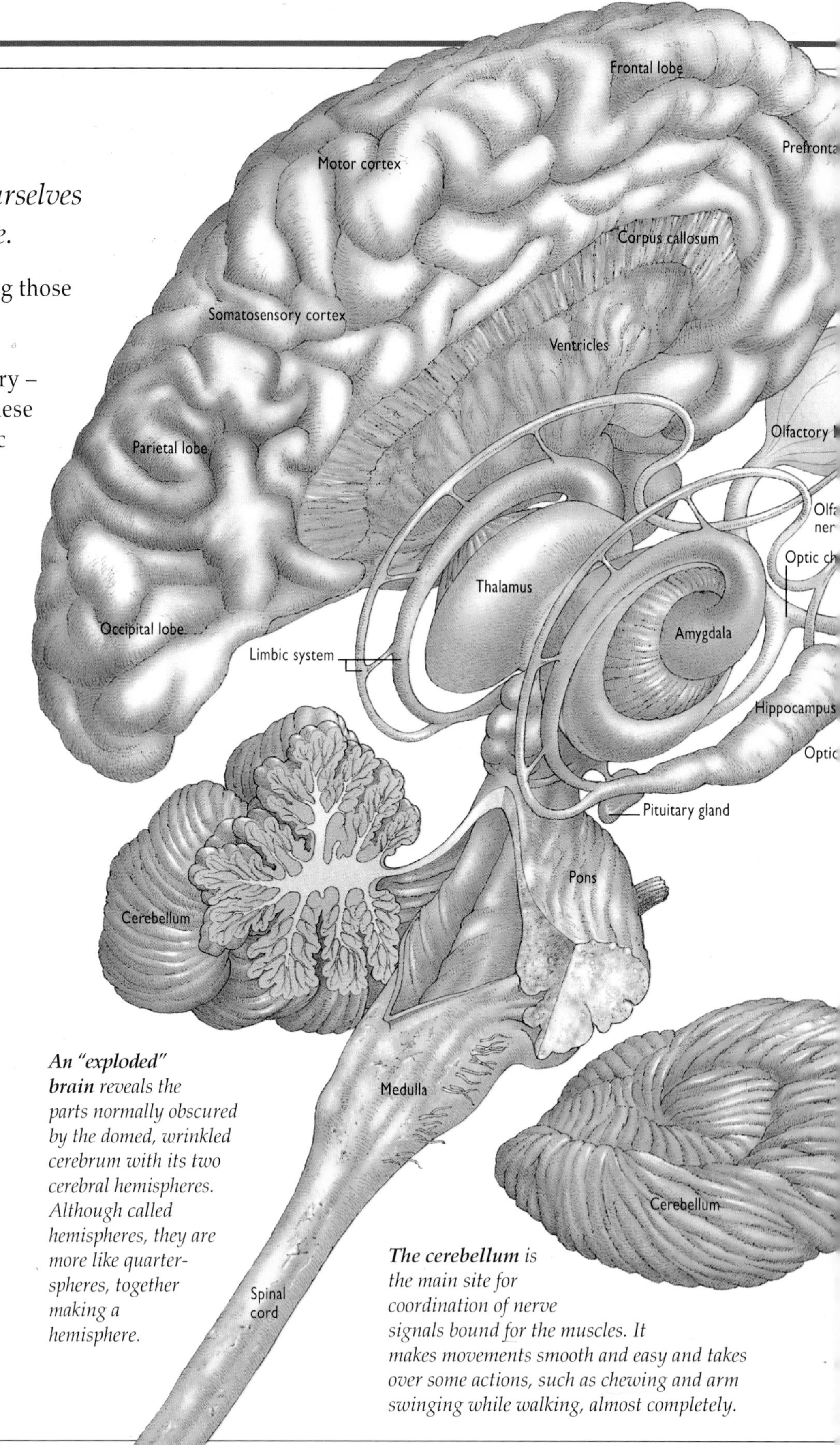

An "exploded" brain *reveals the parts normally obscured by the domed, wrinkled cerebrum with its two cerebral hemispheres. Although called hemispheres, they are more like quarter-spheres, together making a hemisphere.*

The cerebellum *is the main site for coordination of nerve signals bound for the muscles. It makes movements smooth and easy and takes over some actions, such as chewing and arm swinging while walking, almost completely.*

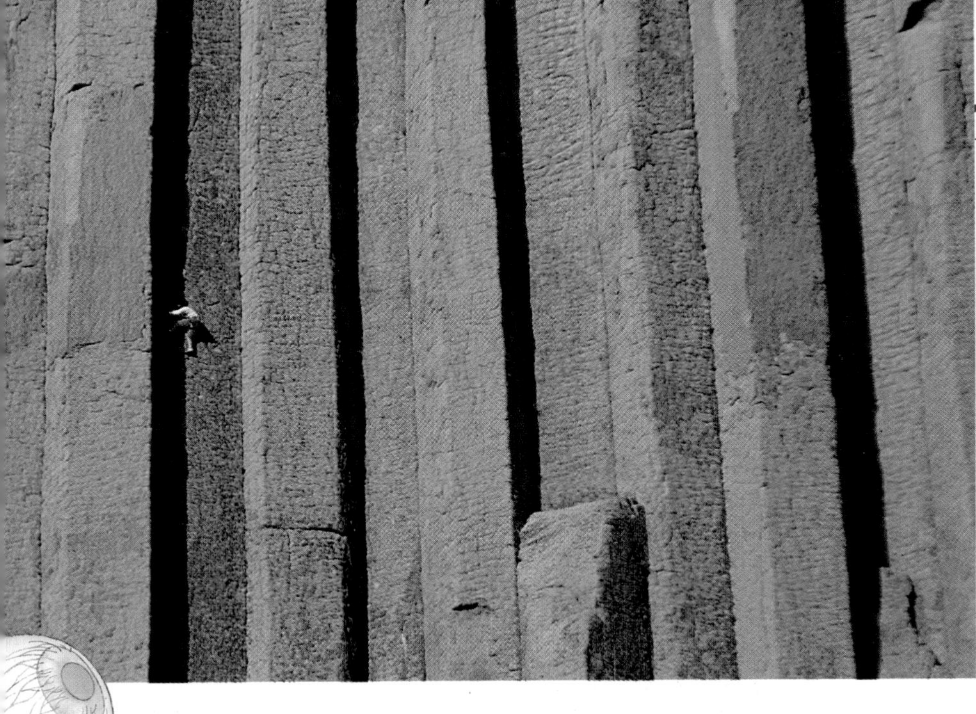

***Your eyes quickly scan** the image (**left**) to show vast columns of rock, but almost at once your attention is stimulated, and your brain directs you to study the "extra" shape near the center. Your brain's visual cortex analyzes the patterns of nerve signals from your eyes to make out the shape. It is compared with memorized images and identified as a person. Then the brain's association and reasoning areas run through the meaning of the situation, to try to understand it.*

This directing of attention to an "irregular" detail shows how the brain makes sense of a perception and focuses on what is unusual in a scene and thus requires further investigation.

yeball

***A cerebral hemisphere's wrinkles** increase its surface area, which contains the nerve cells of the cortex, making it the size of a smallish office desk.*
· this view, only one-third the area is visible. ·ie rest is down ·nong the ·ep folding.

Frontal lobe

Motor cortex

Prefrontal lobe

Somatosensory cortex

Parietal lobe

Temporal lobe

ccipital lobe

***The five lobes** on each side of the brain – prefrontal, frontal, temporal, parietal, and occipital – are named after the skull bones that cover and protect them.*

BRAIN FACT FILE

Dimensions	width	8 inches (20 cm)
	length	8 inches (20 cm)
	depth	6 inches (15 cm)
Ave. weight	male	46–49 ounces (1,300–1,400 g)
	female	42–46 ounces (1,200–1,300 g)
Effective weight*		5–7 ounces (150–200 g)
Brain to body weight ratio		
	human	1:50
	dolphin	1:100
	chimpanzee	1:120
	cow	1:1,200
No. of nerve cells	100 billion approx. (plus millions of supporting cells)	
Major parts	cerebrum cerebellum diencephalon (thalamus, hypothalamus) brain stem (mid-brain, pons, medulla) four ventricles (inner chambers containing cerebrospinal fluid)	
Blood supply	20% of the body's total supply (7–9 times requirement of most organs)	
Weight at birth		12 ounces (350 g)
Annual rate of shrinkage		
Between 20 and 60 years		1/30–1/10 ounce (1–3 g)
Over 60 years		1/10–1/7 ounce (3–4 g)
Brain cells lost between 20 and 60		10,000–100,000 per day

* Effective weight is due to buoyancy and support of surrounding cerebrospinal fluid and membranes.

See also

CONTROL AND SENSATION

▶ The nerve net 44/45

▶ Down the wire 46/47

▶ The automatic pilot 48/49

▶ Into action 50/51

▶ A sense of self 74/75

▶ Making memories 78/79

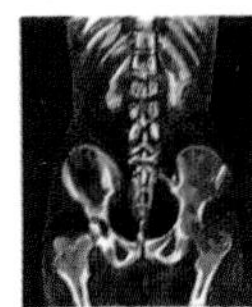

SUPPORT AND MOVEMENT

▶ Bone junction 18/19

ENERGY

▶ Fueling the body 94/95

▶ Cells at work 114/115

CIRCULATION, MAINTENANCE, AND DEFENSE

▶ Routine replacement 138/139

A sense of self

One of the great mysteries of being human is that we are aware of being alive – we have consciousness.

What sort of person are you? Perhaps you are intelligent but not annoyingly clever, strong-willed yet reasonable and able to listen, confident rather than overbearing, sensitive without being weak, and – of course – modest and honest. Maybe you are, maybe not. We have hundreds of words to describe a person's mind and mental processes, attitudes and outlooks, feelings and reactions, ethics and judgments, manners and mannerisms, motivations and morals – what is sometimes given the extremely hard-to-define name of "personality." We can even attempt to put ourselves in the position of other people and try to imagine what they think about us. This shows a high degree of self-awareness and self-consciousness in the human mind. But where exactly is this based in the brain? Where do we actually think? Where are we aware of those thoughts?

Researchers are finding that conscious mental functions are spread through several regions of the brain. The parts are "networked," rather like an ultrasophisticated modern computer installation with its linked workstations. The frontal lobes are definitely involved, and especially the front parts of the frontal lobes, the prefrontal cortex. Evidence for this has come from people who have had injuries or tumors in these parts of the brain – they may undergo personality changes. A classic case was Phineas Gage, an American building worker from the last century. In 1848 he suffered an accident when explosives went off unexpectedly and blew an iron rod, 1½ inches (4 cm) wide, lengthwise up into his cheek, through his left frontal lobe, and out through the top of his skull. Gage survived, but apparently he changed from being friendly and easy-going to being surly, moody, and foul-mouthed. Yet his basic intelligence and memories were not affected. Occasionally, surgery such as the prefrontal lobotomy – in which nerve fibers in the frontal lobe of the brain are cut – is carried out with the aim of changing the patient's personality, in cases of extreme violence, for example.

Also important in our awareness of self are two structures in the brain – the hippocampus and the amygdala. These are involved both in storing and in recalling memories, and – functionally – are part of the limbic system, the brain's main system for expressing emotions and feelings.

Most of us are aware of personality features or traits, *and of how these vary among people we know. We can sometimes "think" ourselves into another's mind and attempt to adopt the type of speech, behavior patterns, and mannerisms of that person. Actors, the professionals in this sphere, aim to suspend or suppress their own personality and adopt that of a chosen character.*

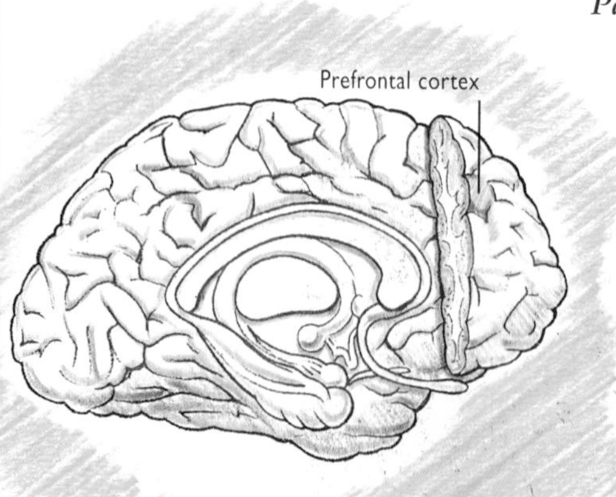

Parts of each prefrontal cortex *are thought to be important in consciousness since they deal with awareness of where we are in space, a sort of visual–spatial short-term memory. Intelligent animals – humans, apes, and dolphins, for example – all have a well-developed prefrontal cortex. Yet humans appear to be the only ones with more than a rudimentary sense of self-awareness.*

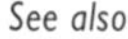

A THINKING MAN

French mathematician and philosopher René Descartes (1596–1650), known as "the father of the mind–body problem," suggested that the mind, or soul, exists separately from the physical brain and body. He claimed that the mind is a nonphysical conscious entity which can think, will, sense, understand, and imagine. The brain–body is the physical entity, which exists in normal three-dimensional space. He held that the small pineal gland, in the rear lower brain, linked the two. Scientists and philosophers still consider the question of whether physical and mental entities are fundamentally different, but Descartes's beliefs are not commonly held today.

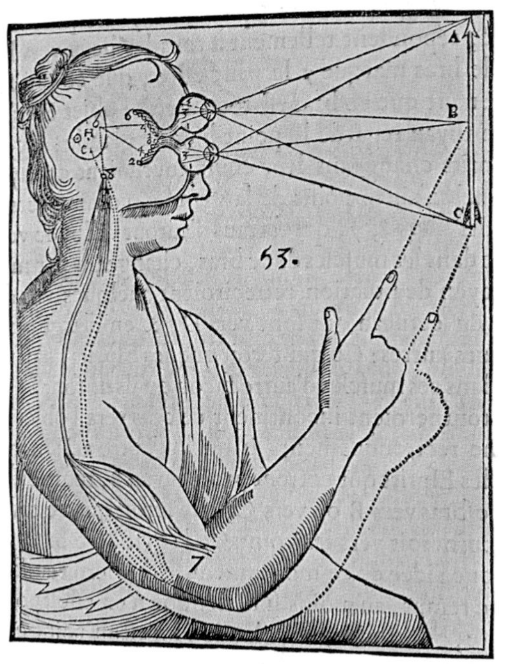

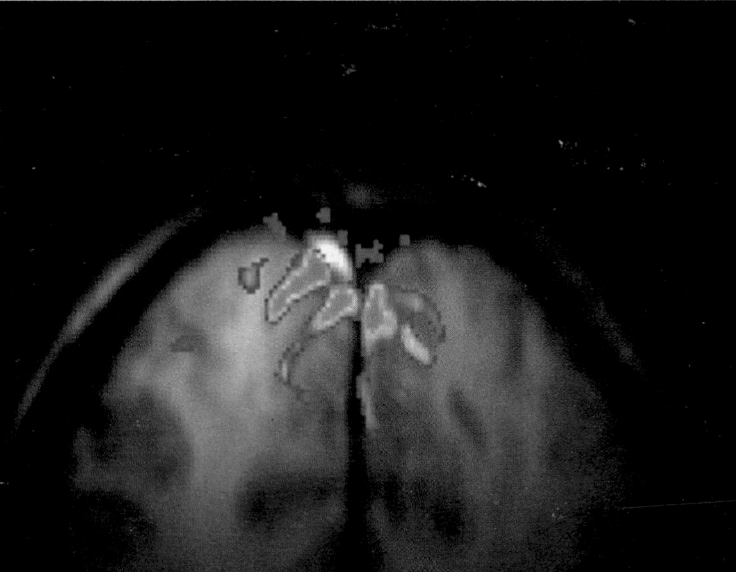

MRI can now detect thoughts. *It is so sensitive that it can sense changes in activity within the brain and nerves by locating areas where the blood flow temporarily increases. These two MRI scans show blood flow in the brain of a person who is resting, and not thinking about anything in particular (**below**), and who is then stimulated by bright lights and so is concentrating on them (**left**). The active areas are effectively those which are "thinking" hardest. Using MRI and other scanning methods, researchers may be able to pinpoint mental activity and let us see fleeting thoughts as they flash around the brain.*

he body scanning method MRI *(magnetic resonance naging) creates a strong magnetic field that makes ydrogen atoms in water line up like tiny bar magnets. he scanner then shoots radio waves at the atoms, making em wobble slightly. As they realign, they give off signals, hich the scanner detects and analyzes into an image.*

See also

CONTROL AND SENSATION

▶ Down the wire 46/47

▶ Into action 50/51

▶ Seeing is believing 58/59

▶ Making sense 72/73

▶ The active brain 76/77

▶ Making memories 78/79

▶ Learning skills 80/81

▶ Problem solving 82/83

▶ Moods and emotions 84/85

▶ Mind over matter 86/87

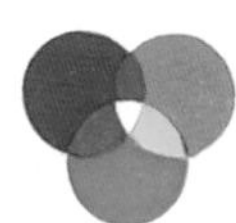

The active brain

Even when we sleep, the brain is active – it monitors the senses and body functions and, every so often, dreams.

When we go to sleep, our awareness switches off, and unless woken by a loud noise or some other sensory message, we are unaware of the world outside. In sleep voluntary muscles relax, and many body functions become less active: heartbeat, breathing, and metabolic rate all slow and core temperature falls. But others become more active – levels of growth hormone and protein synthesis involved in tissue repair both rise.

So why do we sleep? We must need it, since people deprived of it soon become confused, erratic, and violent. Since we are primarily visual animals, sleep could be a survival strategy from our distant past, a way of conserving energy and drawing least attention to ourselves during the hours of darkness. Or it might be a time when the brain clears its desk and does its filing, forgetting clutter while reinforcing that which is deemed to be important.

But there are other states of unconsciousness as well as sleep. For instance, stupor – partial loss of consciousness – has many causes, including severe lack of sleep, the effects of drugs, and brain hemorrhage. There is also fainting – loss of consciousness due to lack of blood supply to the brain – and concussion – loss of consciousness, lasting from a few seconds to many minutes, usually caused by a blow to the head that shakes the brain within. The most profound state is coma – a prolonged unconsciousness with suppression of reflexes such as coughing that normally occur during sleep. It is generally due to serious disease or damage to the brain.

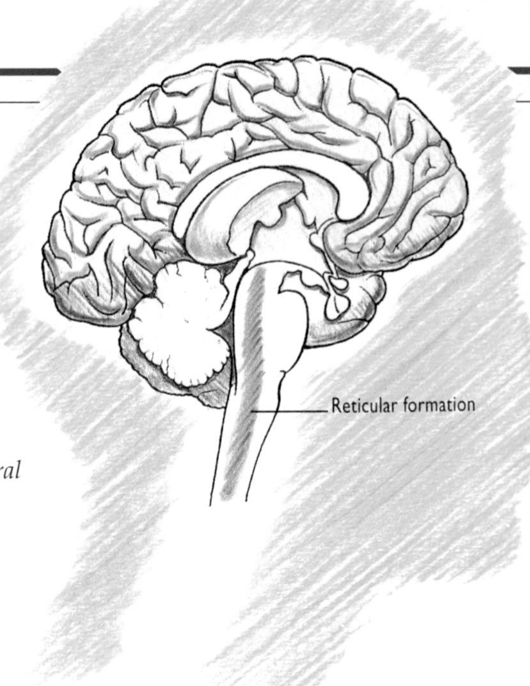

***Brain arousal level** is handled by the reticular formation, which starts at the bottom of the spinal cord and continues up through the brain stem to the thalamus. It controls the flow of signals up to the cerebral cortex – the site of awareness – as well as such processes as waking from deep sleep, or going from daydreaming to hyper-alertness.*

EEG readings

Awake

Stage 1

Stage 2

Stage 3

Stage 4

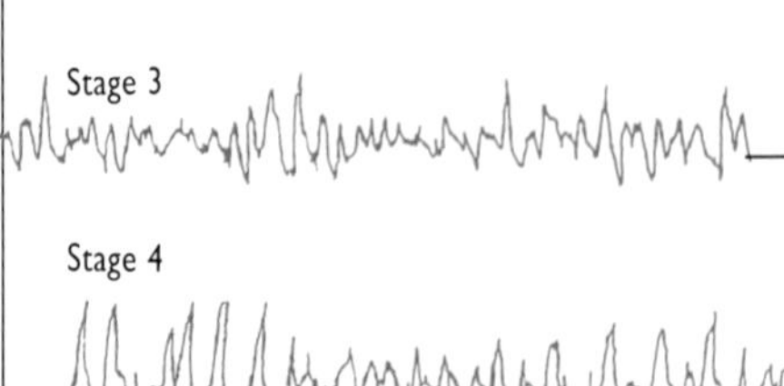

REM

REM

***Sleep can be broken down** into four main stages, shown by brain activity recorded on an EEG machine. When we fall asleep, we pass rapidly down to stage 4, when the skeletal muscles and brain are at their most inactive. After 80–90 minutes, sleep becomes lighter, reaching stage 1, at which point we tend to enter a state known as REM (rapid eye movement) sleep, which lasts for 5–15 minutes. In REM sleep, certain body parts become active, such as the eyes, hands, and feet (though other skeletal muscles stay relaxed); heartbeat and breathing speed up, and there may be sexual arousal. The name comes from the fact that the eyeballs dart to and fro under closed lids, as if watching a busy scene – this is when we dream. Sleepers usually cycle between deep and REM sleep several times in a night. Toward morning, periods of ordinary sleep get shorter and less deep, while REM sleep periods lengthen.*

In sleep laboratories, researchers track body processes such as blood pressure, breathing, heart activity, and body temperature throughout the night. They also record the electrical activity of the brain using an electroencephalogram (EEG). The EEG monitors the brain's tiny electrical nerve signals that are conducted through the skull to the scalp. The signals are picked up by electrodes stuck to the skin and are displayed as a wavy line.

Each time you sleep, you probably have dreams. Scientists know when people dream because they have woken volunteers when their EEG recordings show signs of dreaming sleep. The volunteers woken in this way report that they were dreaming. People tend not to remember a dream unless they wake up while it is happening or just after it. But you can often recall dreams from the early morning, since you wake up during or soon after them.

See also

CONTROL AND SENSATION

▶ Down the wire 46/47

▶ The automatic pilot 48/49

▶ Making sense 72/73

▶ A sense of self 74/75

ENERGY

▶ Fueling the body 94/95

LEVELS OF CONSCIOUSNESS

First-aid providers are trained to test for the level of consciousness – more correctly, responsiveness – in persons found lying still by trying to talk to or rouse them.

There are six levels of consciousness. In the first, the person responds normally and converses intelligibly. At level two, the person answers only direct questions, and says or does little else. At the next level, he or she responds to questions, but only vaguely, and may be incoherent. The fourth level person understands and obeys commands, but does not speak or make other actions. At the fifth level, a person is generally still and silent and responds only to pain, as when pinched. Finally, at level six, a person does not respond at all and their vital signs, such as pulse and breathing, should be checked.

During dreaming, *characterized by rapid eye movement (REM), the brain's electrical activity, measured by an EEG, is similar to that when a person is awake. The switch from sleep to dream activity coincides with REM onset, which is measured with an electrooculogram (EOG). This detects eye movements.*

REM EOG reading

REM EEG reading

REM

REM

Making memories

There are, it seems, many types of memory, from the memory of names and faces to that of abstract ideas.

We use memory effortlessly most of the time when, say, a person's name comes to mind, a skill is exercised, or a scene from the past is recalled. But remembering is a hugely complex process that is only slowly being understood. Scientists have now devised systems to classify the different types of memory and their functions. They are also making progress in finding out which parts of the brain are involved in the different types of memory, and the physical basis of how memories are stored.

It seems that a thought or memory exists as a set of nerve signals that flash around circuits or pathways between the nerve cells (neurons) via their connections, or synapses. We may learn and memorize something by developing a new set of connections between the nerve cells, and setting aside or dedicating this circuit to a given memory. When each unique circuit is reactivated, we recall that memory. Working (short-term) and long-term memories appear to be based in the cerebral cortex. The hippocampus, toward the center of the brain, mediates between these two types of memory. If it is damaged, a person can no longer make new memories and so cannot recall recent events, or even retain information long enough to answer a question. Yet memories of events that took place before the damage are relatively unaffected.

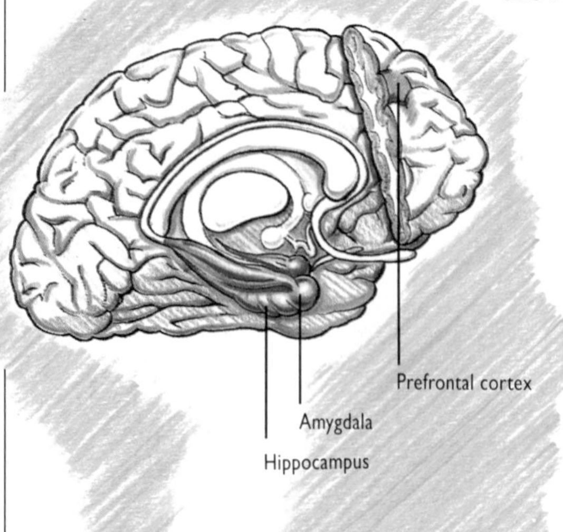

__The hippocampus__, on the lower, inner side of the brain's temporal cortex and close to the amygdala, seems to take short-term memories from the thoughts that are in the cortex. It processes them over a period of days or weeks and then sends them back to other sites in the cortex where they are stored over the long term.

__Long-term memories__ can last a lifetime. Events that happened when the body was alert and in a high state of awareness, such as a wartime scene or a wedding, birth, or death, tend to be memorized and recalled most strongly. It also helps to recall and run over the information occasionally. This seems to "refresh" the nerve-cell connections and circuits, and delay them from fading or being assigned to other functions.

***Advertising and promotion** rely on evoking and associating memories. Companies hope to link the product with pleasant memories – of relaxation, good times, warm sunshine, and vacations, for example – if not from our own experience, then from the advertisements. Then we will recognize a particular brand and remember to ask for it in preference to others.*

One type of memory, semantic memory, *deals with knowledge of words and language – and the way we think about and understand the world and its events in words or similar "units," such as symbols. The meanings of the words and images on a poster must be in our semantic memory before what they represent can be understood. This poster advertises a rock 'n' roll show in a stadium. To know what it means, you have to recall what the words "rock 'n' roll" and the image of "stadium" mean. Then if the idea appeals, you might decide to go.*

Episodic memory *concerns the happenings in an episode or time span, such as a day or a well-delineated event. So when you see the poster, you recall a previous rock 'n' roll concert and associated memories, such as how you got there and what the show was like. Episodic and semantic memory are sometimes included in a larger group – declarative memory – which covers recall of both concepts and events.*

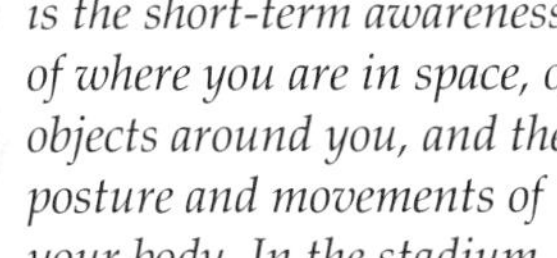

Visual–spatial memory *is the short-term awareness of where you are in space, of objects around you, and the posture and movements of your body. In the stadium, after a quick scan of the scene, you use this memory to find your way and avoid tripping over steps or seats.*

Non-declarative memory *involves chiefly physical or active skills which you have learned through a period of practice. As you jive and bop to the music, you use this type of memory, which is also known as procedural memory. The instructions for dancing may be planned in the brain's frontal lobes, sorted by the premotor and motor cortex, and fine-tuned by the cerebellum at the rear of the brain, to produce smooth, coordinated movements.*

Phonological loop memory *lets you temporarily retain sequences of words, sounds, or symbols. Typically the sequences are six to eight items long, and you retain them for several seconds, up to a minute or two, by running through or looping the information in your mind, saying it again and again. This is how you remember your ticket's row and seat number.*

Working memory, *the brain's "scratch pad," includes visual–spatial memory and phonological loops, plus the information you need to function second by second. It analyzes sensory inputs, such as sights and sounds; processes thoughts, words, and emotions; and plans and monitors movements. It would help you figure out if you had enough cash for a snack.*

See also

CONTROL AND SENSATION
- The nerve net 44/45
- Down the wire 46/47
- The range of senses 52/53
- Making sense 72/73
- Learning skills 80/81
- Problem solving 82/83

CIRCULATION, MAINTENANCE, AND DEFENSE
- Maintaining the system 136/137

REPRODUCTION AND GROWTH
- The growing child 178/179

Learning skills

Acquiring a new skill, from walking to talking, is hard, but once learned it becomes second nature.

If you practice something often enough, you will probably remember how to do it again. This ability to learn and repeat relies on memory. Learning continues through life, and we acquire new information, knowledge, and mental and physical skills on an ongoing basis, from the early stages of walking and talking, to mastering such skills as reading and writing, doing math and making friends, to learning about life, how to fit into society, achieve success, and so on.

Watch any child making the transition from crawling to toddling, and it is obvious just how difficult it is. There is intense concentration on each step, and on holding the head and body correctly and moving the arms and legs to produce smooth forward motion. Most people cannot remember learning to walk, yet they too went through all the same stages. The ability to walk is now so firmly etched into the brain that they hardly ever think about it. But it is an important skill acquired in much the same way that an adult might begin to learn something new involving physical and mental activity, such as playing the harp, windsurfing, or speaking another language.

But humans begin learning even before birth – infants remember and are soothed by recordings of the sounds they heard inside the womb. Later, they realize that crying brings warmth, comfort, and nutrition. In fact, throughout life we learn by making a thought, movement, or action, and assessing the results. With time, we acquire memories of the action involved and the effects it produces. These memories and thought patterns probably exist as particular circuits and pathways for nerve signals through the extraordinarily complicated maze of the brain's billions of nerve cells and their connections.

Much insight into the human learning process has come from animal studies. For instance, "smart" creatures such as chimpanzees have been taught human sign language to ask for food or playthings and to comment on information from their keepers. They cannot copy our spoken words because their larynx (voice box) and the parts of the brain which control it have not evolved for this purpose. However, they can use purpose-designed keyboards to express ideas and wants, even showing an awareness of simple grammar.

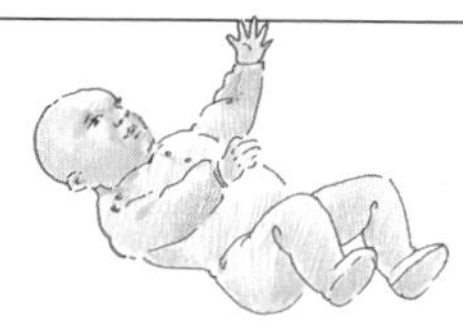

Each developmental milestone *reached in a child's life is the culmination of an incredible amount of learning and skill acquisition. For instance, the ability to communicate with verbal language – one of the jewels in the crown of the range of human accomplishments – represents an extraordinary coming together of physical and intellectual skills.*

As newborn babies, humans are equipped with the physical apparatus of speech, including the voice box and regions of the brain allocated to speech. They also have the basic built-in skills and behavior patterns that start them off on the road to language acquisition.

Before a baby makes any sound other than crying and the natural sounds of living such as burping, it has experienced one-way verbal

dialogue with its parents, who talk to it, often comment on its every move, and coo hen the child smiles. But then, at rhaps three or four months, a baby akes speechlike sounds – it babbles. Researchers believe that babbling innate behavior and that the ild is not imitating the sounds those around it. Telling evidence r this is the fact that even deaf bies babble and that babies of l cultures make very similar pes of noises.

Babbling reaches a peak at around 10 months, when babies start to string together sounds like "ba-ba-da-go." Now comes the hard part, the harnessing of this innate ability for communication into real language.

At about one year, babies start to pronounce specific words deliberately. These first words are learned by copying the sounds made by parents and others, and associating them with certain objects drawn to the child's attention, classically "mama," "dada," "doggy," and so on.

By the age of about two, a child will have learned perhaps 50 or so words and will know roughly what they mean. But concepts are somewhat blurred.

For instance, "moon" may be used both for the moon in the night sky and as a label to describe round things.

It is between about two and three that children start to put words together to make sentences. The complex rules of grammar are gradually picked up: the difference between action now and action in the past, for instance, is well known by the time a child is five or so.

***At elementary school** a child learns the meanings of 10 to 20 new words daily, as well as their spellings and their grammatical types. In fact, during the early years, the brain is memorizing, recalling, and learning at an incredible rate. This is helped by "plasticity" which allows the brain physically to change and adapt its circuits and nerve signal pathways as they are being established around its vast nerve-cell network.*

In a child of 12 or 13, the brain is physically almost fully grown. It is moving to new methods of learning, relying more on abstract thought, experience, observation, and deduction. The fast-changing "plastic" phase is ending. This is one reason why learning a second language in the teens seems so difficult. First time around, the brain's wiring was more adaptable.

***The brain's cerebral cortex** is involved in the learning, understanding, and production of language. Consider the word "book." Memorized concepts about this word are in different sites: the typical size and shape of a book in one place, common book colors in another, the idea of a book as a store of information, knowledge of how book pages are read and turned in order in still others, and so on.*

When it comes to speech production, meaningful sentences are assembled in the posterior speech center (Wernicke's area). This forwards the information to the anterior speech center (Broca's area). Here, the sentence is broken down into speech-sound elements (phonemes), and signals go out to the muscles of the chest, larynx, mouth, and lips, in order to produce the spoken words.

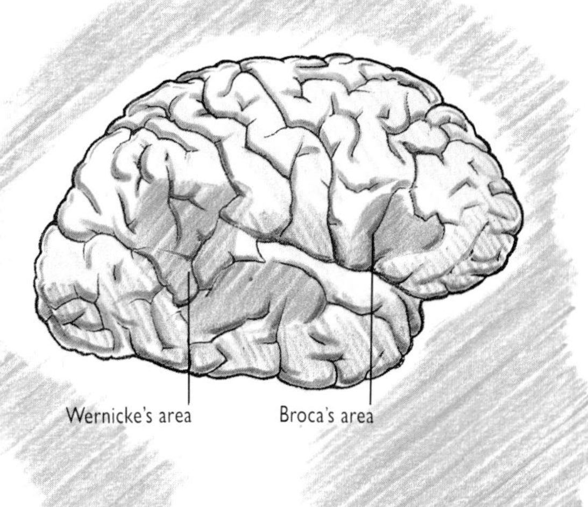

See also

CONTROL AND SENSATION
- ▶ Down the wire 46/47
- ▶ Into action 50/51
- ▶ Seeing is believing 58/59
- ▶ Making sense 72/73
- ▶ A sense of self 74/75
- ▶ Making memories 78/79
- ▶ Problem solving 82/83

REPRODUCTION AND GROWTH
- ▶ The newborn baby 176/177
- ▶ The growing child 178/179

Problem solving

You can deal with problems in a purely mental way, but your creative brain may be in two minds about it.

Have you tried the puzzle where you have to put different shaped pieces into the appropriate shaped holes in a board or container? A two-year-old might grab a piece and jab it at each hole in turn until, with luck, it fits. Toward the age of three, the child may be able to pick up each piece and match it to the correct hole, then turn the piece into the correct orientation so that it fits. By four, the child can probably pick up the piece and hold it in the correct position so that it slots in first time. An older child would look at the pieces, look at the holes, mentally fit each piece to its hole, and complete the puzzle – without making a move. Most people undergo this process of increasing intellectualization, changing from the very physical world of the young, to the more mental world of the adult. In the brain, the basics of the change happen in the hardware, or wiring, and the software, or programs, of the cerebral cortex.

While some educational and developmental psychologists believe that intellectual development is a continuing and overlapping process, others claim that children pass through well-defined stages at certain ages. For instance, the first two years, or sensorimotor stage, are when intelligence and interaction are primarily physical, using the senses and motor skills (muscle movements). Next comes the pre-operational stage, up to about seven, when children learn to use words and begin to manipulate objects mentally in a more experimental way, using trial and error, and building on experience.

Third is the concrete operational phase, up to about 12, when we become increasingly skilled at seeing similarities and differences, classifying, and using logic and deduction to predict outcomes. Fourth is the formal operational stage, which is essentially adult in nature. We can assess and analyze situations in the abstract; propose actions and follow them through by thinking; make choices and see their results – all in the mind. This applies not only to the physical world of objects, but also to social and emotional areas such as relationships.

In spite of such differences of opinion about stages of development, however, experts do agree that in structure and appearance the two sides of the brain are mirror images. In terms of function, many processes, including seeing and initiating movements, also occur roughly bilaterally – equally on either side of the brain. The corpus callosum, the "bridge" of 100 million nerve fibers between the left and right cerebral hemispheres, lets each side know what the other is doing.

But despite appearances, there are differences in function between the two sides of the brain. Evidence comes partly from "split-brain patients" who have undergone surgery to cut the corpus callosum, usually to treat severe epilepsy. It seems that the left cerebral hemisphere dominates in spoken and written language, and in mental skills involving analysis, reasoning, logic, and deduction, from playing chess to computer programming or planning complex scientific or engineering projects. In most people, speech tends to come under the control of the posterior and anterior speech centers of the left hemisphere by the age of seven or eight.

The right cerebral hemisphere is complementary to the left. It takes the lead in grasping the meaning of shapes and symbols, creating and appreciating art and music, using the imagination, and having ideas. Neither hemisphere's domination is complete, but is a matter of degree and emphasis.

Map reading *is a prime example of the brain's problem-solving power. The skilled map-reader looks at a pattern of colored lines and shapes on a piece of paper, and from it constructs a mental image of the landscape it represents. This is related to the real landscape in view, to see if one fits the other.*

Once the map-reader has "locked into" the way map and countryside relate, he or she can navigate even with the map upside-down. The whole scene is rotated mentally to make it fit, and the next stage of the route is chosen by "thinking" along it and imagining the terrain, the views, and the difficulties and rewards it will bring.

A puzzle *that requires both use of logic and appreciation of spatial relationships in its solution is an especially good test of overall problem-solving ability. For instance, the nine matches below – laid out to form three triangles – can be reconfigured to make five triangles by moving just three matches. See if you can figure out how to do it. The puzzle is not that hard, but it does need some thought.*

The solution to the puzzle is shown elsewhere in this book; see if you can recall where, using your memory. Failing that, look for the reconfigured matches – and thus the answer – using your visual recognition skills.

See also

CONTROL AND SENSATION
- Down the wire 46/47
- Into action 50/51
- Making sense 72/73
- A sense of self 74/75
- Making memories 78/79
- Learning skills 80/81

CIRCULATION, MAINTENANCE, AND DEFENSE
- Maintaining the system 136/137

THE PROBLEM OF CREATIVITY

Many aspects of human thought and behavior seem, from the point of view of survival, to serve no real purpose. Take the arts, for instance. Why, throughout the existence of our species, should humans not only appreciate and be emotionally moved by images, sculptures, attractively designed objects, music, and stories, but also be driven to create them? The answer to this question is one of the conundrums of existence.

Are creativity, imagination, and an appreciation of what they produce incidental spin-offs from having the big brains that provide the problem-solving power that has put humans in such a dominant position on our planet? Or are they in themselves useful in survival? Whatever the case, creativity and imagination come as part of the territory that we all inhabit by being human. Interestingly, machines made by humans can now create what many might call works of art, like this computer-generated image (**right**).

Moods and emotions

Love them or loathe them, you cannot ignore emotions. But where do they come from and what do they do?

The feelings and emotions that can well up inside us often seem to develop beyond our conscious control. But they are central to our survival in the face of threat, and to the forging of bonds and relationships with family, loved ones, and friends. Besides mental and behavioral effects such as crying, uncontrollable laughter, and even fainting and collapse, they also have physiological effects on the body, such as increased heart and breathing rates, dilated pupils, sweaty palms and face, flushed skin, dry mouth, and "butterflies" in the stomach (due to increased stomach and intestinal squirmings).

Emotions are based at a deep, "primitive" level in the brain's mental–intellectual hierarchy. Nerve fibers run from the thinking parts of the cortex (especially the prefrontal lobes), where awareness and consciousness are based, down to structures in the brain which are collectively termed the limbic system. In the center of the limbic system is the hypothalamus, which controls basic drives and motivations such as hunger, thirst, and sexual arousal, as well as automatic processes such as heart rate, breathing, pupil dilation, and sweating. The hypothalamus and the adjacent hormone-making pituitary gland are the link between the subjective mental feelings of emotion based in the cortex and the simultaneous bodywide physical and physiological changes.

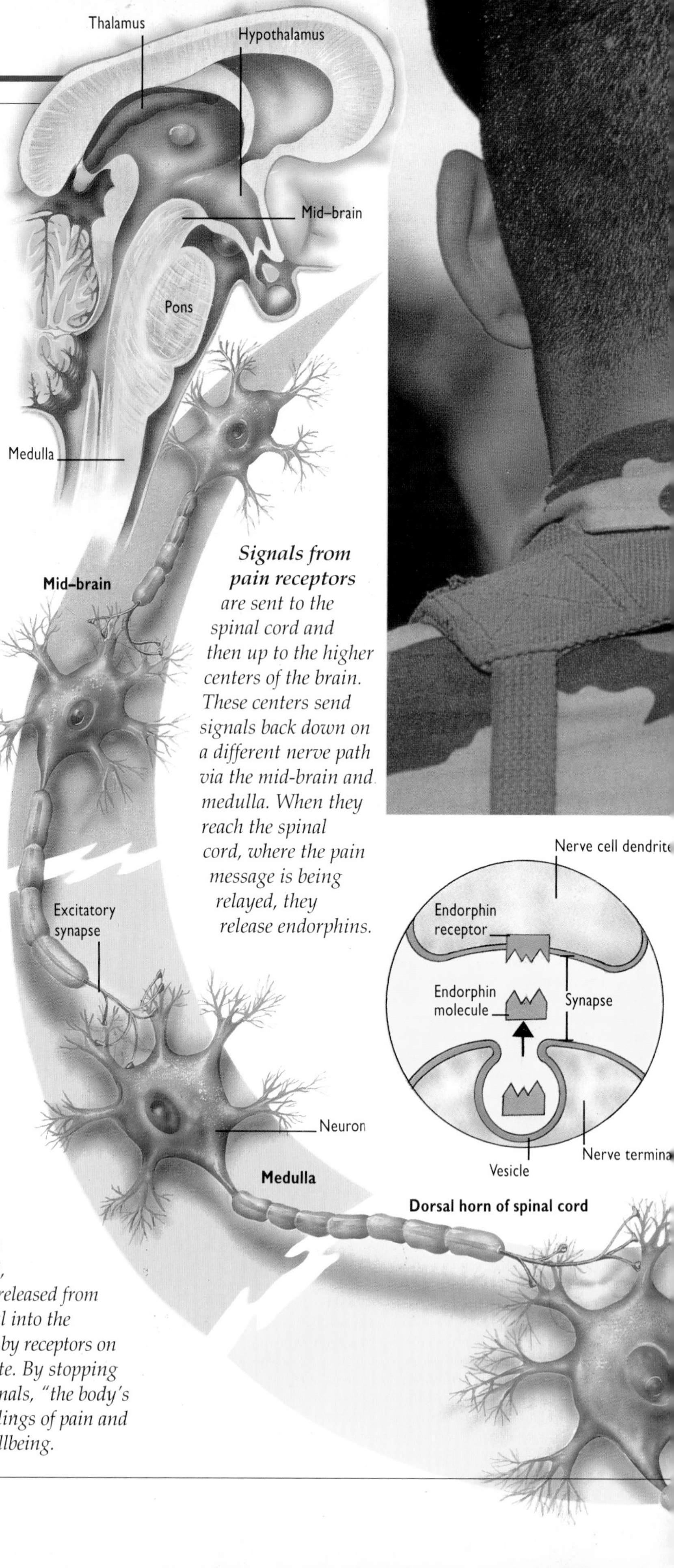

Thalamus
Hypothalamus
Mid–brain
Pons
Medulla

Mid–brain

Signals from pain receptors *are sent to the spinal cord and then up to the higher centers of the brain. These centers send signals back down on a different nerve path via the mid-brain and medulla. When they reach the spinal cord, where the pain message is being relayed, they release endorphins.*

Excitatory synapse
Neuron

Medulla

Nerve cell dendrite
Endorphin receptor
Endorphin molecule
Synapse
Vesicle
Nerve terminal

Dorsal horn of spinal cord

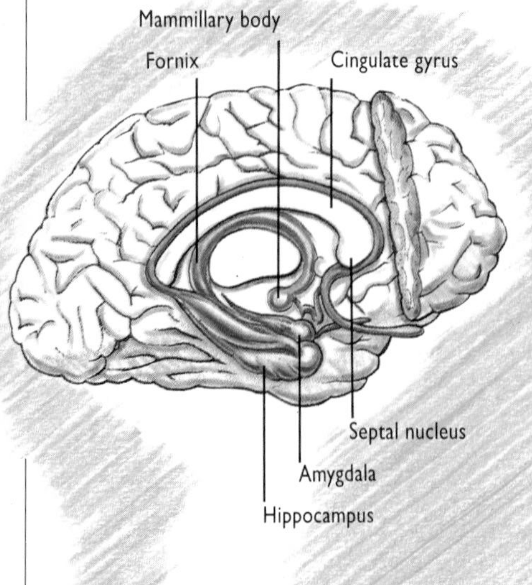

Limbic system

The structures of the limbic system *deal with pain, pleasure, and emotional balance.*

Several body chemicals *can change moods and emotions. For instance, endorphins – like opium-derived drugs – can change our perceptions, moods, and emotions. After being released from vesicles in a nerve terminal into the synapse, they are picked up by receptors on an adjacent nerve cell's dendrite. By stopping that cell from transmitting signals, "the body's natural painkillers" reduce feelings of pain and induce a temporary sense of wellbeing.*

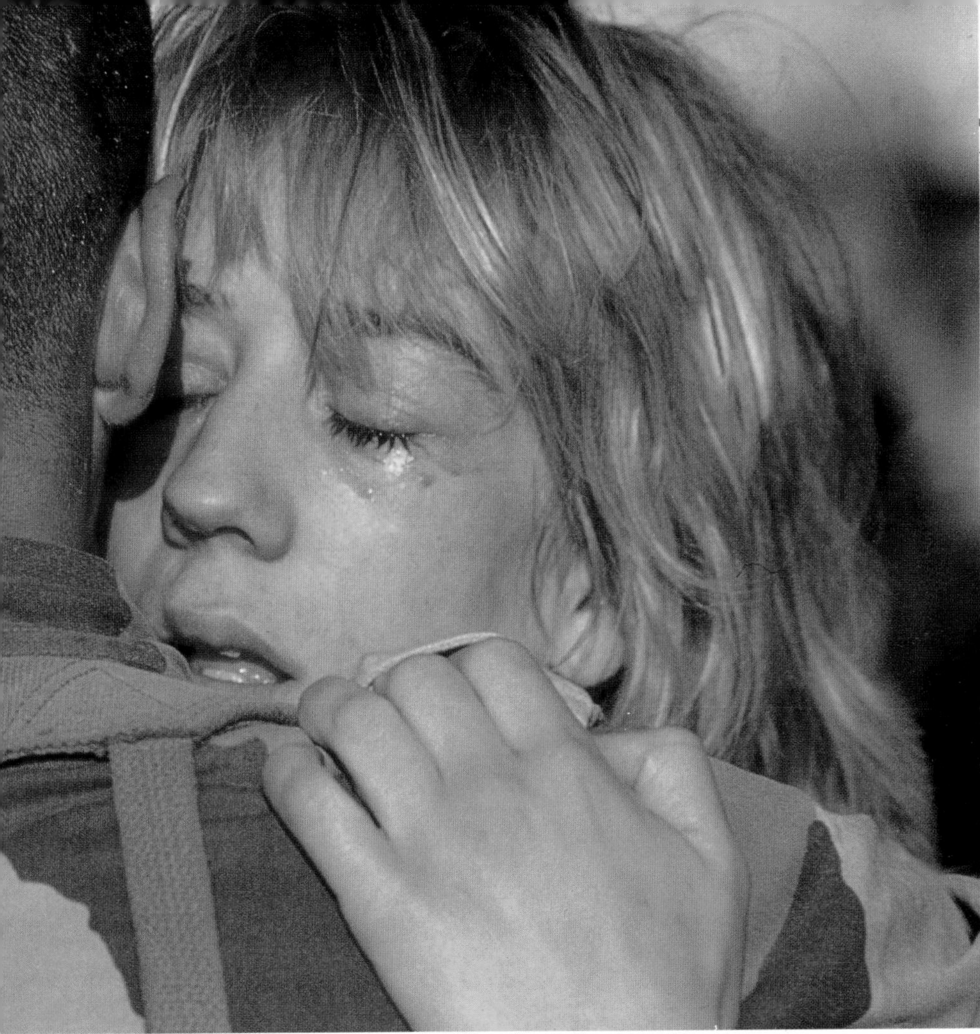

Emotion *would appear to throw a wrench into the works of the analogy of the human body as a machine. After all, machines do not cry with happiness or sadness, neither do they laugh, fall in love, or at times of great stress, have feelings that "get the better of them." And yet ultimately it seems that feelings must be the product of the complex mechanisms in the brain and nervous system, involving the interplay of nerve signals between neurons.*

Drugs *that are used to treat conditions with a large emotional component include (clockwise from top) chlorpromazine, fluoxetine, temazepam, diazepam, and lithium.*

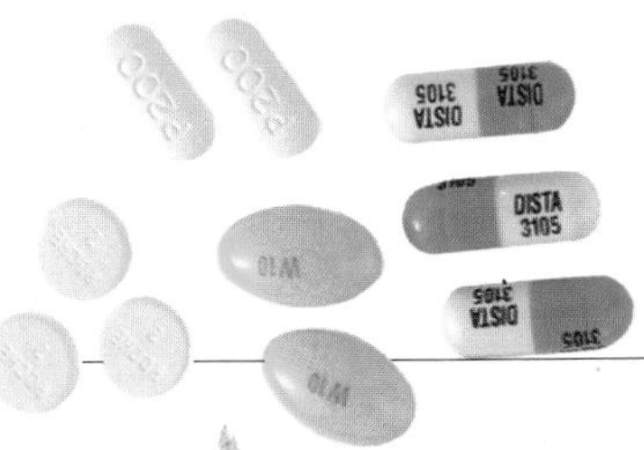

See also

CONTROL AND SENSATION

▶ Control systems 36/37

▶ The chemicals of control 38/39

▶ Hormones of change 40/41

▶ Down the wire 46/47

▶ Making sense 72/73

▶ A sense of self 74/75

▶ Making memories 78/79

▶ Mind over matter 86/87

way to the brain

Axon containing endorphin receptors

Pain receptors

Cell body

Spinal cord

MOOD-ALTERING DRUGS

Medical science has come up with drugs for most unwanted mental states, from depression to psychosis. Most work by altering the amount of a given chemical at the synapses between nerve cells at various sites in the brain. For instance, the antipsychotic chlorpromazine reduces the action of the stimulating neurotransmitter dopamine. The antidepressant fluoxetine (Prozac) increases the activity of the excitatory neurotransmitter serotonin. Diazepam and temazepam – tranquillizers that are used to treat anxiety and insomnia respectively – both promote the action of gamma-aminobutyric acid (GABA), which inhibits brain-cell activity in the emotion-handling areas of the brain stem. Lithium reduces the mood swings of manic depression, otherwise known as bipolar disorder.

Mind over matter

As the brain's complexity is unraveled, even more mysterious links are found between mind and body.

Do you feel tired, stressed, and generally put upon? If so, you are probably more at risk from illness than if you were cheerful, positive, and generally on top of the world. Researchers are finding ever more ways in which life's events and the resulting mental states influence the body's physiology, including susceptibility to infections, digestive ulcers, heart conditions, and even cancers.

But does the mind really have such marked effects on the body? Doubters need look no farther than the well-known "placebo effect." In some medical trials, patients are divided into two groups. One group is given a new drug, the other an inactive dummy tablet (placebo). Neither the patients nor the doctors administering the drug know which group is which. Those in the placebo group almost always show some real improvement, apparently because they believe they are receiving a form of treatment.

Physical illnesses that seem to be caused by, or at least linked to, mental factors are termed psychosomatic (mind–body) states. In the past, many of these conditions were dismissed as being "all in the mind." But in reality there seems to be complicated interplay between mind and body. It is known that some people who have experienced stressful events, or who tend to depression or anxiety, have lowered levels of certain antibodies – the natural germ-killers – in their blood. Their germ-fighting lymphocyte white cells are also less active and efficient. Prolonged stress may produce various effects, via high levels of the fight-or-flight hormone epinephrine and increased nervous system activity. These include extra secretion of the digestive acids in the stomach, which can lead to ulcers, and increased heart activity, which can be at the root of high blood pressure. But it gets more complicated. People respond to similar stresses in many different ways. Personality, attitude, and outlook are all important. These, in turn, link to family background, education, job demands, exercise and fitness, diet, and other facets of lifestyle.

The other side of the mind–body coin is the beneficial aspect. Patients who have a positive approach to life – fighting back against disease and believing they will recover – actually do better than those who give up.

Yoga is one of the meditational activities *that emphasizes the harmony of body, mind, and spirit. The postures, breathing, and mental approach to these activities can help the body to achieve profound relaxation, causing the heartbeat and breathing rates to slow, muscles to relax, and blood pressure and metabolic rate to reduce.*

In some forms of transcendental meditation, the participant focuses on a simple sound or thought to become relaxed. Various forms of massage, and exercises such as t'ai chi that concentrate on posture, have similar effects. All these activities can help to counteract the stresses and anxieties of daily life and lead to improved health.

Biofeedback involves the use of monitoring devices such as pulse or muscle tension gauges. The participant concentrates inwardly on achieving calm and peace, which will register indirectly on the biofeedback device.

FOOD AND CONTROL: THE MIND AND EATING DISORDERS

Psychological conflicts show themselves in many ways. One manifestation in the west is in eating disorders such as anorexia nervosa and bulimia nervosa. Anorexics avoid food or refuse to eat, lose weight – sometimes becoming critically undernourished and weak – and may develop a distorted self-image of being overweight, even when skeletally thin. Bulimics eat huge quantities of food in binges and control weight gain by inducing vomiting.

Anorexia affects mainly adolescent girls – up to 1 in 100 in some areas. It appears most common in relatively well-off families where achievement expectations are high. In boys, the incidence is 1 in 2,000. Bulimia is rarer and also affects chiefly young women. There are several possible reasons for these disorders. For instance, an anorexic girl might feel she does not live up to fashion ideals and is too fat. Or she may use eating as a way of controlling others, especially parents; mealtimes become a contest of wills. She may subconsciously desire to stay a child, rather than go through adolescence into womanhood, so she starves to retain a more childlike shape. These disorders can also be a cry for help, in the same way as a suicide attempt.

__Medical and life-insurance companies__ have surveyed the circumstances of their policy-holders and their ongoing state of health over many decades. Some results of this monitoring are the so-called life events charts (__below__). These show which life events are likely to produce negative effects on physical and mental health. The scale is comparative, with the various circumstances ranked relative to the event generally agreed to be the most stressful: death of a spouse or partner. Even supposedly happy events, such as outstanding achievement at work, or things ordinarily reckoned to be relaxing, such as taking a vacation, still add points, so even in a happy, successful year a person can still score a significant number. A person who totals more than 300 points in a year has an increased chance of developing some serious illness.

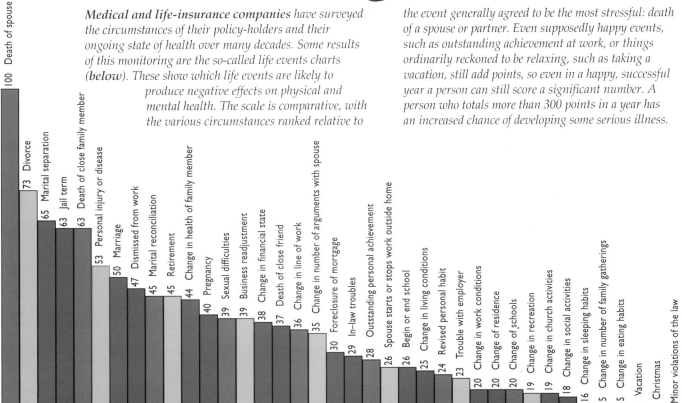

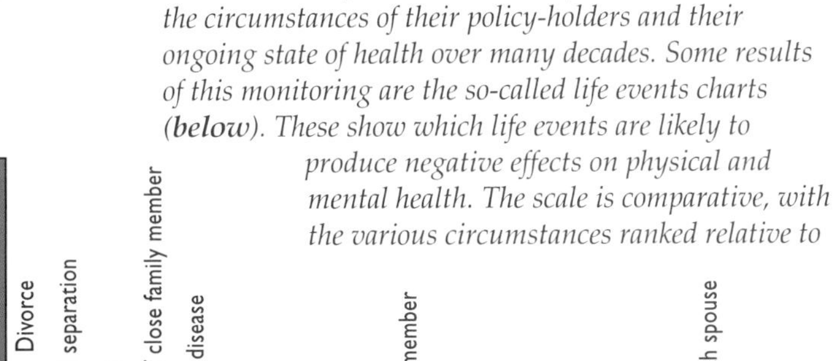
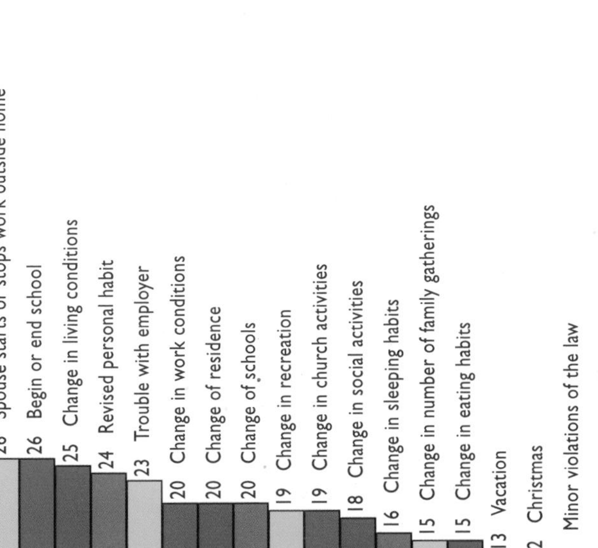

See *also*

CONTROL AND SENSATION
▶ Making sense 72/73
▶ A sense of self 74/75
▶ Moods and emotions 84/85

SUPPORT AND MOVEMENT
▶ The heart's special muscle 28/29

ENERGY
▶ You are what you eat 96/97

CIRCULATION, MAINTENANCE, AND DEFENSE
▶ Sentinels of immunity 146/147
▶ Helpers and killers 148/149

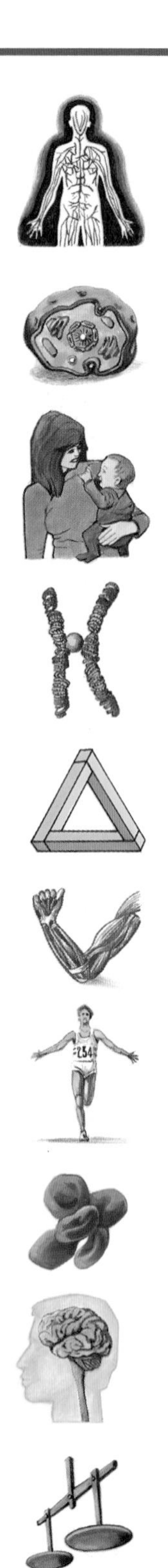

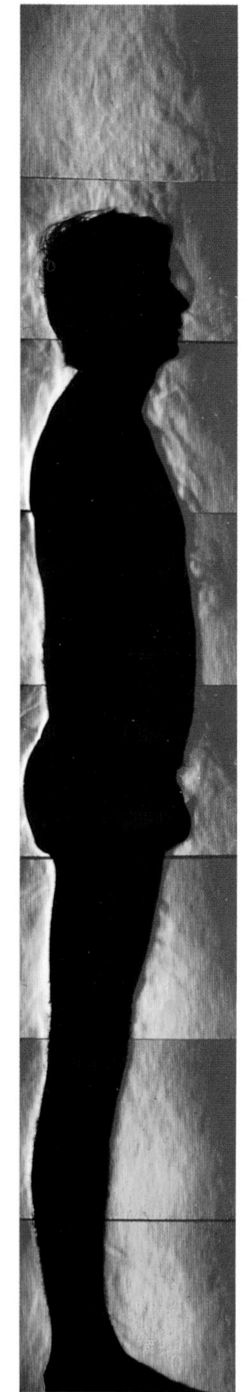

Energy

On the scale with which we are familiar – that of the everyday world – humans are great users of energy, driving around in cars, illuminating the night with electric lights, flying in aircraft, and manufacturing products as diverse as bread to eat and rockets to send satellites into space. But we also use energy on a less familiar scale – a molecular one – to power the chemical processes that occur within the body's cells.

Our rate of internal energy use depends on the level of activity of the community of cells that makes up the body. To match the demand for energy, we take in fuel in the form of food. This is broken down into smaller chemical components and absorbed, some to be stored and used later, and some to provide the building blocks for assembly into living tissue. Our lungs take in oxygen, which is necessary for the series of complex chemical reactions that take place in the heart of each cell to release energy from the fuel derived from food.

***Left (clockwise from top)**: inside the intestine; fiber; lung volume; heat energy.* ***This page (top)**: liver, duodenum, and pancreas; (**right**) energetic sperm.*

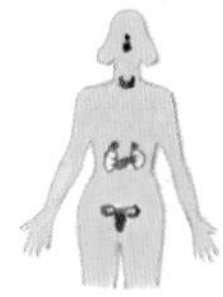

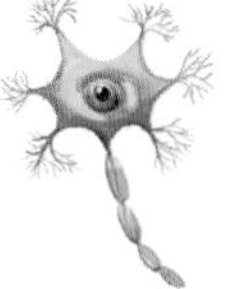

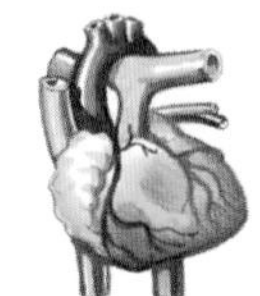

The energy cycle

Humans are voracious consumers of energy, using it in two quite different ways.

Every living thing is a user of energy, and as energy is used, it is transformed from one form to another. Thus plants which use light energy for photosynthesis turn electromagnetic radiation from the Sun into stored energy. When humans eat plants and digest them, molecules containing the stored energy are broken down and the energy is released and turned into, for example, movement via chemical reactions in the muscles. Thus energy which started as light became chemical energy and eventually kinetic, or movement, energy, which in time becomes heat energy.

Human bodies are complex metabolic machines that need energy to work – whether that work is the pull of a muscle, secretion by a gland, or the transmission of a nerve impulse in the brain or spinal cord. Ultimately, that energy supply comes from food – fuel for the body machine. Parts of it are broken down in body cells in ways that release the chemical energy that drives all the other activities of the body. This energy-releasing metabolism, which normally requires oxygen for its completion, is termed internal respiration.

We also use external sources of energy to cook our food, heat our homes, power our computers, fuel transportation systems, drive mechanical machinery, and manufacture materials and finished goods. These external sources include fossil fuels such as coal, oil, and natural gas; nuclear power; and other sources such as solar, wind, geothermal, wave, and tidal power, all of which maintain our energy-hungry existence. In industrialized societies, most of these sources are used to produce electricity.

Flying high in a jet aircraft, *passengers are served food by the cabin crew. The passengers' internal energy needs are being taken care of, and at the same time, they are flying in a plane which consumes external energy at a high rate. Not only is aviation fuel needed to keep a plane running, but a great deal is also used in its manufacture. The aluminum for its body is extracted from ore using vast amounts of electricity, for instance. Even the plastic insulating its wiring and the fabric covering the seats are made using energy, usually electricity generated from fossil fuel.*

Different energy-generating techniques have varying indirect impacts. Burning fossil fuels, for instance, produces vast volumes of carbon dioxide, which has contributed to the slow but steady increase in atmospheric levels of carbon dioxide in the 20th century. Carbon dioxide is one of the "greenhouse gases" that may induce global warming if its concentration in the atmosphere increases.

When *Homo sapiens* emerged more than a million years ago, primitive human energy usage was almost entirely metabolic, driven by the food that was consumed. With the discovery of fire, however, the first rung on the ladder of external technological energy use was attained. Since then, the amount of external energy use has increased exponentially – the average energy consumption in the United States today is the equivalent of burning 8 tons of oil per person per year.

Almost every energy source *that humans consume is, in the final analysis, derived from solar energy, which comes in turn from nuclear fusion reactions deep in the heart of the Sun, the star at the center of our solar system.*

All our foodstuffs depend on that energy because, apart from some bizarre deep-sea food webs centered on bacteria that

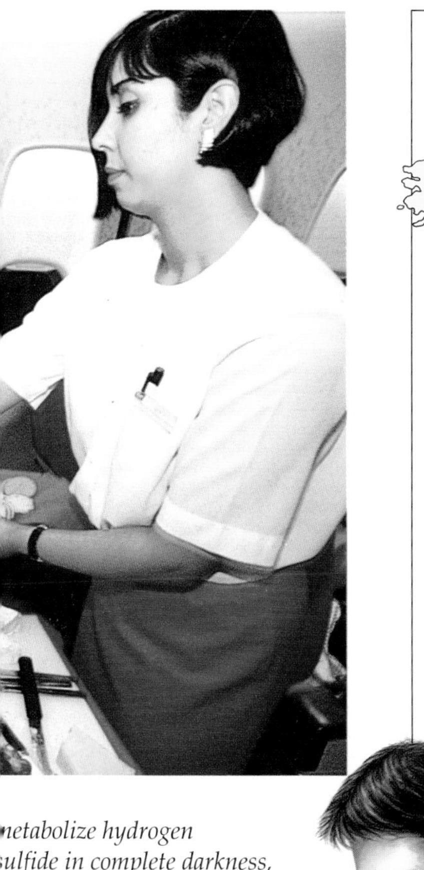

LIVING ALL OVER THE WORLD

Humans have made permanent homes in all regions of the Earth except for the polar areas.

Wherever humans go, we use energy. We use some to stoke the internal "fires" of metabolism, and some to make changes to the world around us. In fact, if we have access to energy we can live in virtually any environment. The key extra ingredients are our intelligence, the ability to pass on information from generation to generation, and the aptitude to make and use tools. In the colder regions of the Earth, we can learn how to catch animals such as seals, thus providing food for internal use, skins that can be made into garments to keep out the cold, and fat for fuel. Now other fuels can also be used to heat buildings. In hot regions, buildings can be designed to keep cool in the fiercest sun, using air-conditioning units run on electricity.

etabolize hydrogen sulfide in complete darkness, ll other land- or sea-based cosystems are constructed around he ability of green plants and other hotosynthesizing organisms to rap light energy.

Plants use light energy *from the Sun to make organic substances such as sugars, fats, and amino acids from inorganic ones like water, carbon dioxide, and mineral salts. Human foods include light-trapping plants themselves or products made from them, such as bread from grain; animals that eat plants (herbivores such as sheep or chickens); animals that eat other animals (carnivores such as salmon); or decomposer organisms such as fungi that can break down the bodies of dead plants and animals. Humans are thus omnivores (we eat almost anything), and all our foods ultimately owe their existence to solar power. Even fossil fuels, which supply us with external energy, can be thought of as stores of ancient solar power. Coal, oil, and natural gas are derived from the transformed remains of terrestrial or marine photosynthesizers that lived millions of years ago. The chemical energy locked up in their organic molecules came originally from light energy trapped by ancient living things. The energy from fossil fuels, and from nuclear and other sources, is used in countless ways in modern technological life.*

See also

ENERGY
- The speed of life 92/93
- Fueling the body 94/95
- You are what you eat 96/97
- The energy storehouse 106/107
- The cell and energy 112/113
- Cells at work 114/115

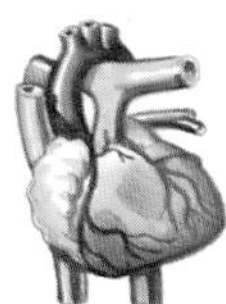

SUPPORT AND MOVEMENT
- Muscles at work 26/27
- The heart's special muscle 28/29

The speed of life

A variety of factors determines the rate at which you use energy. So how fast does your body run?

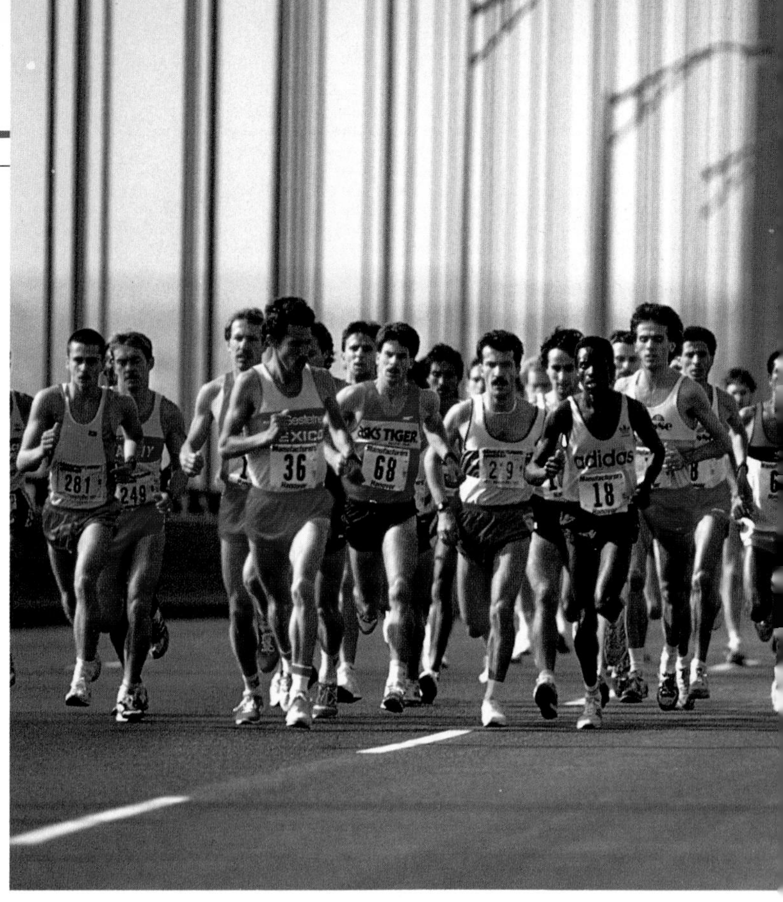

Put simply, the human body is a chemical machine in which countless billions of chemical reactions take place every split second. The sum total of all these reactions is known as metabolism, which is usually divided up into the reactions that build up complex molecules from simpler ones (known as anabolism) and those that break down large molecules into smaller ones (catabolism). In general, building processes use up energy when they take place, while many types of breakdown reactions release energy. In fact, one of the key roles of the food we eat is to provide us with nutrients that can be broken down in chemical reactions to supply us with energy; this energy then powers the body's metabolic machinery.

The amount of energy that the body uses over a given time is the metabolic rate. The rate of energy use depends on many factors, including body size, age, and a person's typical level of activity. To find a metabolic rate that can be compared between different people, the rate of energy used when they are at rest is measured, thus eliminating such variable factors. The rate obtained under these conditions is known as the basal metabolic rate (BMR) and is usually expressed as the number of kilocalories (kcals or Calories) – units of energy – used per square meter of the person's body surface per hour. A BMR for a typical 20-year-old man would be about 26 kilocalories per square meter per hour (2.5 kcals per square foot per hour). The BMR is often calculated by measuring a person's rate of oxygen use, since almost all the energy produced in a body is obtained from chemical reactions that use up oxygen.

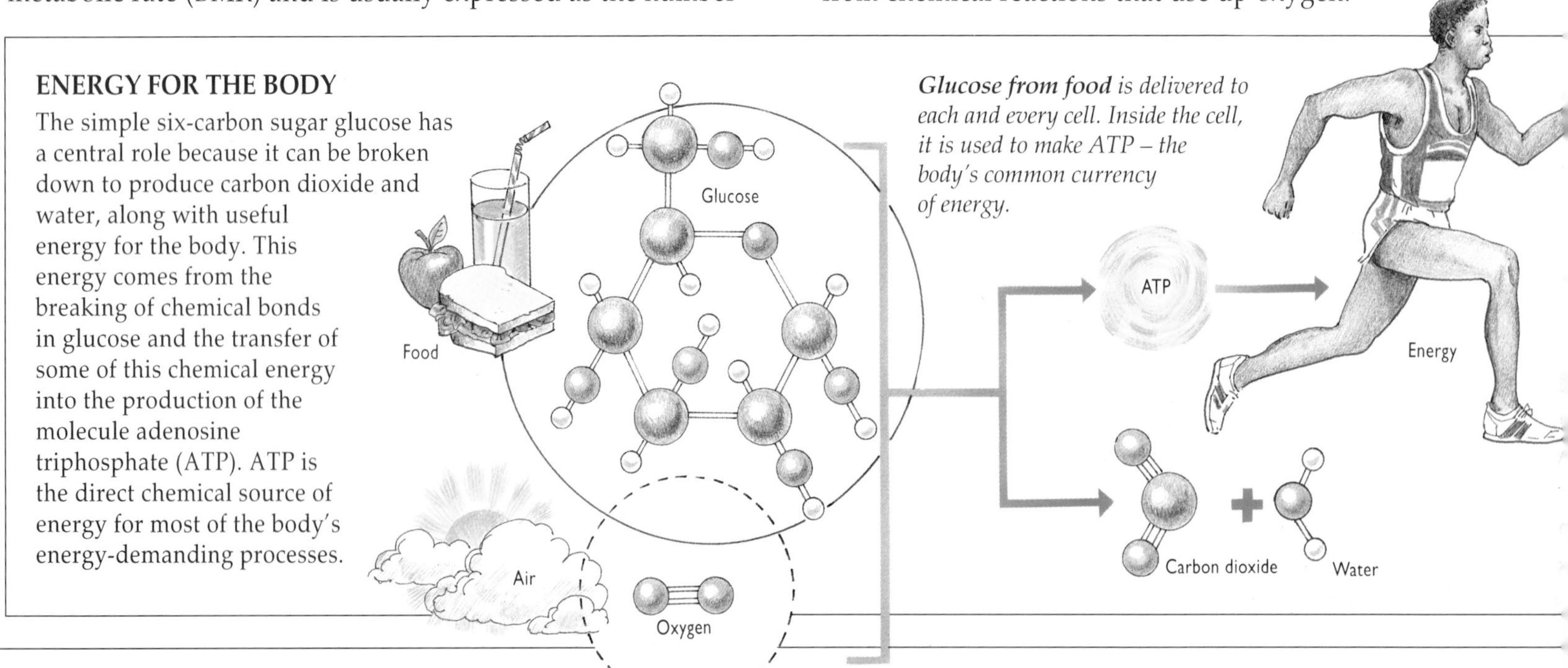

ENERGY FOR THE BODY

The simple six-carbon sugar glucose has a central role because it can be broken down to produce carbon dioxide and water, along with useful energy for the body. This energy comes from the breaking of chemical bonds in glucose and the transfer of some of this chemical energy into the production of the molecule adenosine triphosphate (ATP). ATP is the direct chemical source of energy for most of the body's energy-demanding processes.

***Glucose from food** is delivered to each and every cell. Inside the cell, it is used to make ATP – the body's common currency of energy.*

or short periods the body can power its activity ·ith energy-providing ATP stored in its cells. This ·eans that a 110-yard (100-m) sprint can be ·nished on a single deep breath. But longer periods ·intense activity require high-efficiency aerobic ·xygen-using) phases of cell energy production. ·o during a marathon, runners will be breathing hard to get ·s much oxygen to their muscle cells as possible so that glucose ·an react with oxygen to make more ATP.

In addition to age, body size, and general levels ·f activity, there are a number of other factors that ·an have an effect on BMR and the metabolic rate ·enerally. For instance, for each degree Fahrenheit ·hat the body temperature rises, there is a 5.5 ·ercent rise in metabolic rate, so people with ·evers burn up energy at a higher rate than normal. ·he metabolic rate is also increased and largely ·ontrolled by the production of thyroxine in the ·hyroid gland. And increased food intake, especially ·f proteins, can raise the metabolic rate as the body ·vorks hard to digest it. But if increased food intake ·xceeds the body's metabolic needs, a person will ·ut on weight; conversely, not eating enough can ·esult in weight loss. Ideally, metabolic rate and ·nergy intake from food should balance.

Kilocalories/30 minutes

The amount of energy a person uses depends on a number of factors, including activity level. For instance, sprinting requires much more energy than, say, sleeping.

For an average-sized person who takes little or no exercise the rate of energy use is about 2,000 kcals (Calories) a day for a woman and 2,500 kcals a day for a man (men typically use more energy for any given activity than women). This is the amount of energy the body needs to keep the heart beating, to digest the food it eats, to perform many other functions – from thinking to breathing – and to account for minimal physical activity such as sitting, some standing, and a little walking in the course of 24 hours.

700
600
500
400
300
200
100
0

748
Sprinting 516
321
Mountain climbing 241
221
173
Playing tennis 166
Golfing 130
112
Walking 84
44
Eating 33
38
Sleeping 28

Women — Men

See also

ENERGY
▶ The energy cycle 90/91
▶ Fueling the body 94/95
▶ You are what you eat 96/97
▶ Enzymes: chemical cutters 102/103
▶ The gas exchange 110/111
▶ The cell and energy 112/113
▶ Cells at work 114/115

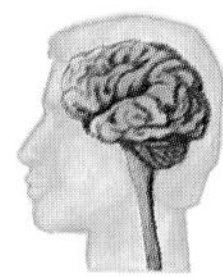

SUPPORT AND MOVEMENT
▶ Muscles at work 26/27

CONTROL AND SENSATION
▶ Key chemicals 42/43
▶ Making sense 72/73

Fueling the body

All parts of the body have a constant need for fuel to power their metabolic processes.

A space shuttle sits on its launch pad at Cape Kennedy. It is full of potential chemical energy, some in the form of liquid hydrogen and liquid oxygen in its main engine fuel tanks. When ignition begins, the potential chemical energy of this fuel (hydrogen) and oxidant (oxygen) is converted, in a controlled explosion, into superheated water vapor. The energy released from this reaction helps lift the craft into orbit.

Energy production in the human body performs the same basic chemical trick, but in a less violent, more everyday, fashion. The potential chemical energy locked up inside nutrients derived from food – especially the sugar glucose, the body's most-used primary fuel source – is liberated to power the functions of the body. This process, known as cellular respiration, requires the presence of oxygen and happens inside cells. Examples of work done using released energy are the muscle contractions that move the body, metabolic activities such as the construction of proteins from amino acids, the transmission of nerve impulses, and the generation of body heat.

In fact, heat production is perhaps the most obvious sign of energy-releasing activity – and hence the use of fuel – in the body. We are hot-blooded mammals able, like birds and a few specialized fish such as tuna, to maintain our bodies at a higher temperature than that of the surroundings and to hold that raised temperature within narrow limits. It is the stability of the internal body temperature that allows precise nervous and hormonal control of activity and metabolism.

In addition, the high, constant body temperature (around 98.6°F/37°C) means that body physiology – essentially the chemical reactions in cells – can be carried out at a fast rate, since a higher temperature increases the rate at which chemical reactions take place. This has evolutionary benefits. Generally speaking, the speed of chemical reactions in the cells allows mammals, including humans, to respond rapidly to the outside world, whether the need is to pursue prey quickly or to run from a threat. Maintaining a constant temperature is a fuel- (and thus food-) demanding process. But while a cold-blooded creature may need less fuel to keep itself going, if it gets too cold it dies.

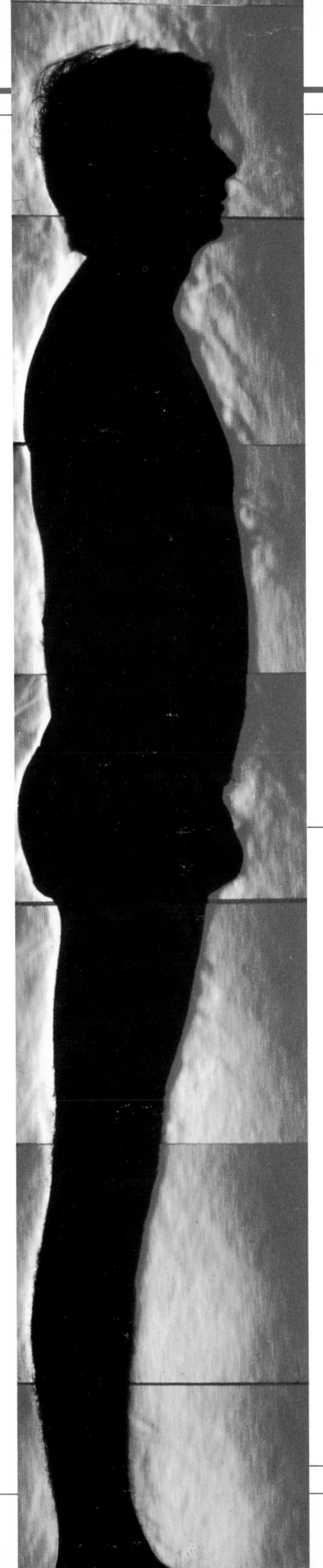

As warm-blooded mammals, *humans generate heat through metabolic activities in order to hold body temperature at a constant 98.6°F (37°C).*

The heat is generated in myriad chemical reactions that take place constantly in all the body's cells. The fuel for this heat production is ultimately the food we eat.

Images made using the so-called Schlieren method graphically illustrate the body's production of heat. The technique takes advantage of the fact that air of different temperatures has different refractive indices (it bends light to different extents). The specialized optics of the cameras used in the Schlieren method can capture the heated air around a person as it rises in smokelike plumes, warmed by body heat. In this image, red is coldest and light yellow warmest.

THE SMALL LIVE FAST

Metabolic rate in warm-blooded mammals is closely linked to the size of the animal. The larger the animal, the lower the rate; conversely, the smaller the animal, the higher its metabolic rate. This is mainly because warm-blooded animals have to maintain a constant, relatively high temperature, and the ratio of surface area to volume is much bigger in a small object than in a large one of the same general shape.

The amount of heat an animal generates as a by-product of energy-using actions such as muscle movement is related to its bulk, or volume. Its rate of heat loss is proportional to its skin size, or surface area. An

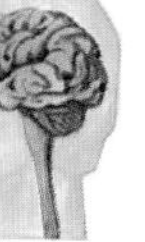

***The brain** accounts for about one-fifth of the total energy use of the body at rest. This is a remarkable statistic for a medium-sized organ that has no muscles and thus does not perform an energy-hungry process such as moving about. Much of the energy-requiring activity in the ain, which has a continual need for glucose (a sugar) as fuel, linked with powering the electrochemical nerve impulses in its ny billions of nerve cells.*

So important is the brain that the first place a lack of fuel is ticed is in the effects it has on the brain: concentration goes and edness and listlessness set in. The brain's need for fuel is more less constant. It uses just as much when you are concentrating rd as it does when you are "thinking of nothing" or even asleep.

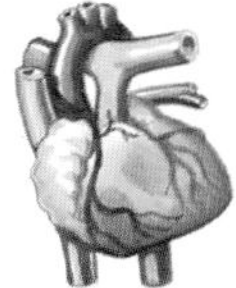

***The heart's cardiac muscles** have a self-evident energy demand: they pump continuously from "cradle to grave." This energy powers the ceaseless rhythmic contractions of the muscular walls of the two atria and two ventricles that push blood under pressure into the lungs and the rest of the body. Even without the extra work done when the body undertakes arduous physical activity, the action of the pumping heart at rest uses about one-tenth of the body's total energy requirement. The heart's need for fuel and oxygen, like the brain's, is soon noticed by its absence. If an artery carrying supplies of glucose and oxygen to the heart is blocked, the intense pain of a heart attack is experienced as the heart muscle supplied by that artery effectively dies for want of nutrients.*

***The body's main muscles,** such as those that move the arms and legs, do much physical work. The energy to power them comes from respiration in the cells of the muscles. Muscle activity, like brain activity, counts for about one-fifth of the body's total energy use. However, rate rises dramatically when the muscles work hard – when body is at extremes of physical effort, muscles generate several nes as much energy as all the other body systems put together. e muscles do not immediately require oxygen when they start work, since they carry a small store of fuel. When this is hausted, though, oxygen is needed for energy-releasing reactions.*

***Oxygen consumption** is a good measure of the total energy use – and thus fuel use – in a tissue, since oxygen is used in all energy-producing reactions in cells. In the human body, muscles and the liver use most energy, followed by the brain.*

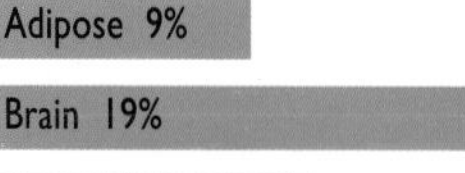

Percentage of total oxygen consumption

imal one-ninth the bulk of other has a skin surface area ly one-third that of the other. needs to generate roughly ree times more heat pound for und than the larger animal to aintain the same temperature. harvest mouse, for instance, oduces energy at a rate some times higher than an elephant. it used energy at the same te as the elephant, it would be able to stay warm because its in surface area, through which loses heat, is proportionately uch greater.

***When metabolic rate** – the sum total of all the chemical activity of an organism's cells – is estimated by the typical rate of oxygen consumed per unit of body weight per hour, the remarkable relationship between metabolic rate and body size shows up clearly.*

What this means is that the smaller the creature, the faster on average all its cells are using energy as they perform their chemical functions. The champion mammal in terms of how fast its cells operate is the shrew, which has a rate 35 times faster than a human's.

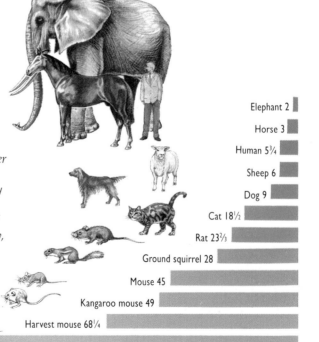

Elephant 2
Horse 3
Human 5¾
Sheep 6
Dog 9
Cat 18½
Rat 23⅔
Ground squirrel 28
Mouse 45
Kangaroo mouse 49
Harvest mouse 68¼

Shrew 201¾

Oxygen consumption (in³/lb per hour)

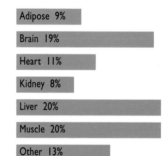

See also

ENERGY

▶ The energy cycle 90/91

▶ The speed of life 92/93

▶ You are what you eat 96/97

▶ The energy storehouse 106/107

▶ The cell and energy 112/113

▶ Cells at work 114/115

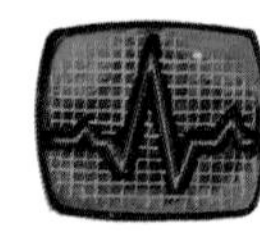

SUPPORT AND MOVEMENT

▶ Muscles at work 26/27

▶ The heart's special muscle 28/29

CONTROL AND SENSATION

▶ Steady as you go 34/35

▶ Making sense 72/73

CIRCULATION, MAINTENANCE, AND DEFENSE

▶ The living pump 120/121

You are what you eat

Food gives us both energy and the materials to build and maintain our bodies.

The food we eat *should contain all the fuel and raw materials the body needs. In this meal, the fish has protein and fat, and the rice has carbohydrate for energy as do the vegetables and fruit, which also contain fiber and vitamins*

There are two basic ways for organisms to obtain the nutrition they need. They can, like plants and some bacteria, use light energy or the energy of chemical reactions to build the nutrients they require from scratch. Alternatively, like animals (including humans), fungi, and some other bacteria, they can grab their nutrients ready formed from other organisms which they eat and digest.

A plant constructs each component of its nutrition from the simplest beginnings – photosynthesis in a leaf builds a component like glucose (a simple sugar). Using energy from sunlight, carbon dioxide from the air, and water from the soil, a plant can construct a molecule of glucose atom by atom. Humans, and other animals, cannot do this. Our glucose nutrients arrive pre-synthesized in food. For instance, the glucose in honey is made by a plant and put into its nectar; bees collect the nectar; we collect the honey from the bees' hive.

Glucose is one of the most basic nutrients, but our food nutrients come mostly in the form of giant, complex organic molecules – macromolecules – themselves made out of large numbers of organic sub-unit molecules, like glucose, linked together in complex ways. Thus carbohydrates such as starch are polysaccharides – mac of many (poly) sugar molecules (saccharides) joined together Proteins are made of similarly large numbers of sub-units – amino acids. Fats are combinations of glycerol and fatty acid

When macromolecular foods pass through the stomach and intestine, they are disassembled into their sub-units by digestive processes and absorbed in that form. Within the bod the sub-units are used as building blocks to construct human macromolecules in cells under the instruction of human gene

Complex carbohydrates *like those found in grains are the best basis for the main supply of energy (Calories) in the diet. Vegetables and fruits, which provide minerals, vitamins, and fiber, should also be eaten in large quantities, whereas smaller amounts of animal protein and fats are needed. Indeed, simple sugars, oils, and fats should form only a small part of the diet.*

This can be summed up in the food pyramid, the width of the pyramid representing the proportion of the type of food that should be eaten in a healthy diet.

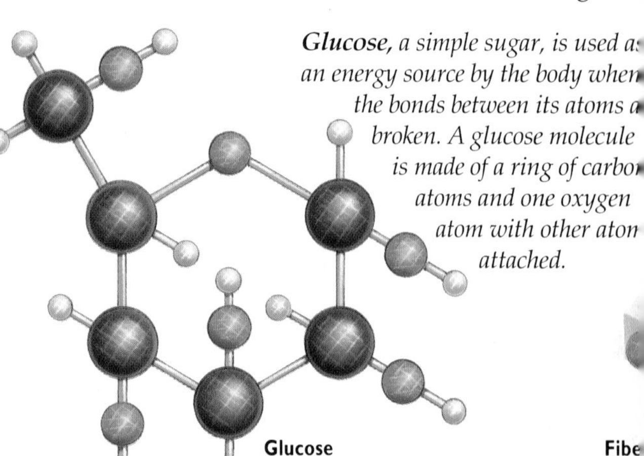

Glucose, *a simple sugar, is used a an energy source by the body when the bonds between its atoms a broken. A glucose molecule is made of a ring of carbo atoms and one oxygen atom with other aton attached.*

oteins in food are built of long folded ains of sub-units known as amino acids. ting a varied range of protein-rich foods, ch as fish, meat, nuts, eggs, legumes, and iry products, guarantees that all types of the different amino acids are released by protein gestion in the stomach and small intestine and en absorbed for use by the body. The average adult eds something like 2 ounces (60 g) of protein per day. Amino acids are the building blocks used in the anufacture of all body tissues, since they are the mponents strung together under the instruction of NA in cells. There are, in fact, thousands of different oteins made by cells.

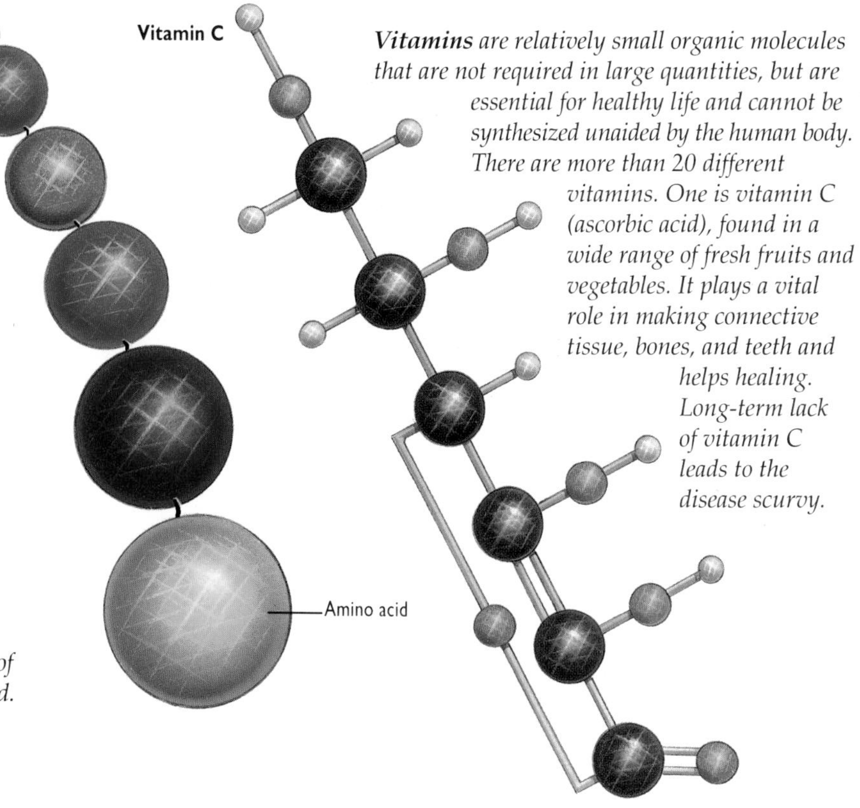

Vitamins are relatively small organic molecules that are not required in large quantities, but are essential for healthy life and cannot be synthesized unaided by the human body. There are more than 20 different vitamins. One is vitamin C (ascorbic acid), found in a wide range of fresh fruits and vegetables. It plays a vital role in making connective tissue, bones, and teeth and helps healing. Long-term lack of vitamin C leads to the disease scurvy.

ugar molecules, like glucose, link together to form rge carbohydrate molecules, or polysaccharides. In the et the most important ones are the starches, which are gested into their simple sugar sub-units and absorbed, d the cellulose fiber (roughage) – found in parts of getables, fruits, grains, and legumes – which is made of polysaccharides that cannot be digested. This indigestible material provides the bulk in the intestinal contents and feces, enabling peristalsis – the movement of matter through the intestines – to operate efficiently. The indigestible matter gives the contracting muscles that make up a peristaltic movement something to push against. A lack of fiber in the diet can lead to constipation and in the long run make a person more prone to bowel disease.

Hydrogen

Oxygen

Carbon

Minerals are found in foods in the form of ions (charged particles) and are directly absorbed from the intestines. There are two classes, the macrominerals and the microminerals, or trace elements. Macrominerals are those that have to be taken in at a rate of more than 100 mg a day. They include ions of sodium, calcium, magnesium, potassium, chlorine, sulfur, and phosphorus. Microminerals include copper, chromium, iron, obalt, fluorine, dine, manganese, d zinc.

Minerals

A CLOSE-UP ON GOOD FATS AND BAD FATS

***In saturated fatty acids** all the carbon atoms are linked by single bonds. Unsaturated fatty acids have some double bonds between the carbon atoms.*

Unsaturated fat

Saturated fat

The body needs fats, just as it needs carbohydrates, proteins, minerals, and vitamins. The fatty acid portion of the fats and oils from animal and plant foods in our diet is used after digestion and absorption to make our own fats. The fat cholesterol, for instance, is a key part of cell membranes.

Research indicates that diets containing a high proportion of saturated fatty acids, found typically in many animal fats, may be relatively harmful in relation to cardiovascular disease. But the unsaturated fatty acids, found typically in plant oils, are relatively protective, reducing the chance of illness.

See also

ENERGY
▶ The energy cycle 90/91
▶ Fueling the body 94/95
▶ Chewing it over 98/99
▶ Enzymes: chemical cutters 102/103
▶ Absorbing stuff 104/105
▶ The cell and energy 112/113

CONTROL AND SENSATION
▶ Good taste 66/67
▶ Mind over matter 86/87

CIRCULATION, MAINTENANCE, AND DEFENSE
▶ Routine replacement 138/139

REPRODUCTION AND GROWTH
▶ Language of life 158/159
▶ Building bodies 160/161

Chewing it over

Open your mouth and look in a mirror to see the first part of the body's food processing system.

Most of the digestive tract is hidden from us and works automatically and involuntarily (without being consciously controlled). The input sections of the tract, however, can be seen and are controlled in a largely voluntary way. The initial processing of foodstuffs starts with the lips, which are mobile for grasping food. The lips close to form a seal, and the teeth chop and grind the food into small fragments. Meanwhile, the tongue manipulates the chewed food, compacting it into a rounded lump, or bolus. This is then pushed to the back of the throat for swallowing – it passes down the esophagus (gullet) on the way to the stomach.

This breaking up of food is known as mechanical digestion. It changes food into fine particles which are further mechanically digested in the stomach and then chemically digested in the stomach and small intestine. The smaller the chewed particles, the faster chemical digestion occurs.

Saliva, the watery secretion of the salivary glands, is made in vast quantities when food is in the mouth. It aids chewing, lubricates the movement of food boluses, and contains the digestive enzyme amylase that begins the chemical digestion of starch even before food is swallowed.

Three pairs of salivary glands, *with ducts on each si of the face that empty into the mouth, produce saliva. This not only moistens the mouth, but also helps chewir and swallowing by wetting food so it does not stick.*

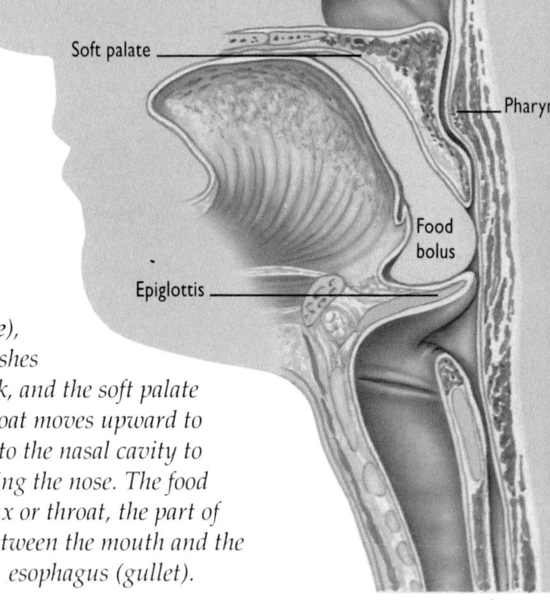

In phase two of swallowing (above), *the rising tongue pushes the bolus farther back, and the soft palate at the back of the throat moves upward to seal off the entrance to the nasal cavity to stop food from entering the nose. The food is now in the pharynx or throat, the part of the digestive tract between the mouth and the esophagus (gullet).*

In phase one of swallowing (above), *food (yellow) that has been chewed into small fragments and processed into a rounded soft ball, called a bolus, is propelled by complex movements of the tongue to the back of the mouth cavity. During the biting, chewing, and grinding stage, the food is combined with saliva from the salivary glands until the bolus is a uniform fine mixture of food particles suitable for swallowing.*

A cross-sectional view of the mouth (left) *shows how the tongue can act as a muscular piston that pushes a food bolus. The bolus, positioned between the tongue and the roof of the mouth, or the hard palate, starts to be squeezed toward the throat when the tip of the tongue rises to the front of the palate. The section also reveals how the upper and lower sets of molar teeth at the rear of the mouth can meet one another for grinding food.*

***phase three of swallowing (below)**, the combined*
tion of the tongue moving backward and the muscles in
pharynx squeezing the bolus move it farther on. The
chea, or windpipe, rises up and the epiglottis swings
wn to close off its upper end like a spring-loaded
pdoor so that no food or liquid can enter. Swallowing
a reflex action, and the coordination of the complex
gue and throat muscle contractions is automatic.

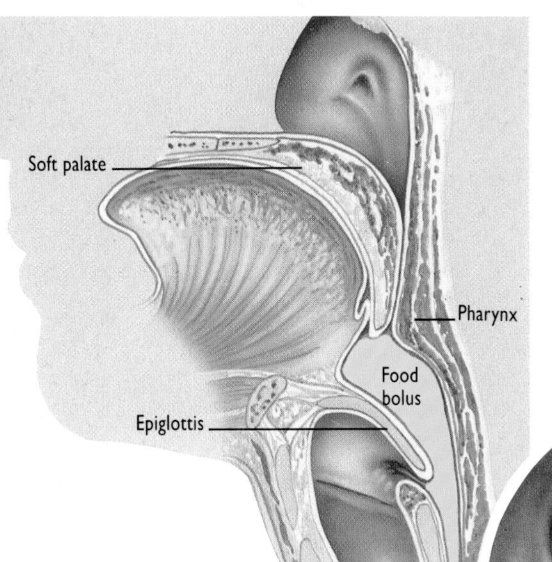

***In phase four of swallowing (right)**, more backward pressure by the tongue and constriction of the muscles of the pharynx push the food bolus past the epiglottis and down into the esophagus, or gullet. Muscle contractions (peristalsis) in the esophagus carry the food down into the stomach for the next stage of food processing.*

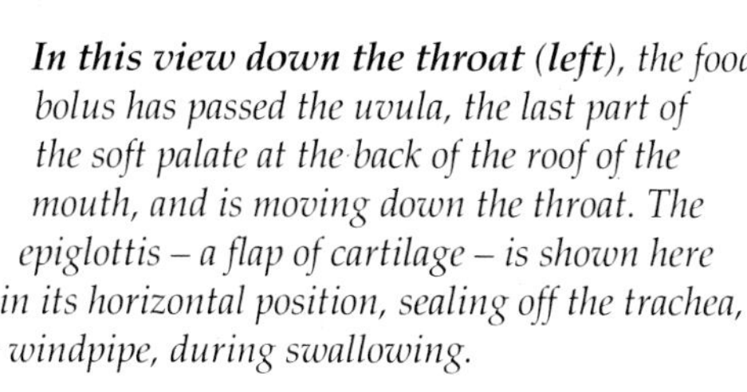

***In this view down the throat (left)**, the food bolus has passed the uvula, the last part of the soft palate at the back of the roof of the mouth, and is moving down the throat. The epiglottis – a flap of cartilage – is shown here in its horizontal position, sealing off the trachea, or windpipe, during swallowing.*

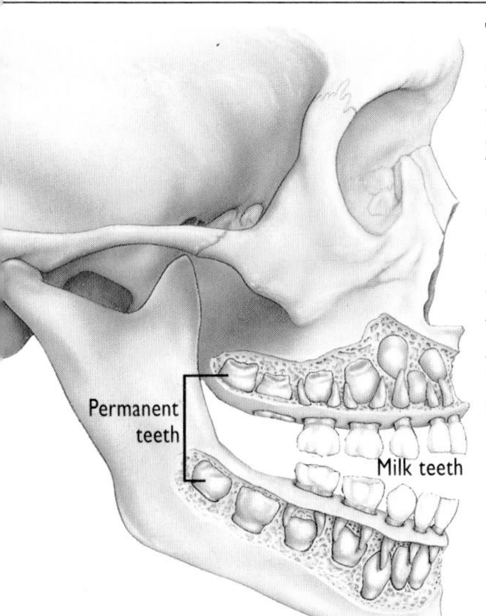

TEETH

Permanent teeth (blue), already formed in cavities within the jaw bones, eventually replace a child's 20 first teeth – the milk, or baby, teeth.

Extra-hard white enamel forms the tough outer layer of a tooth; within that is a layer of softer dentine, which itself encloses the pulp cavity containing blood vessels and nerves. Canines and incisors tear and cut bite-sized chunks of food; molars and premolars, or bicuspids, then grind them to a pulp. Adults have 4 canines, 8 incisors, and 16 bicuspids and molars; many have 4 extra molars: wisdom teeth.

***Molars are flat**, with grooved tops. Canines are pointed. Incisors are flat topped and look like shears.*

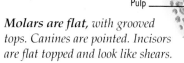

See also

ENERGY
▶ You are what you eat 96/97
▶ The food processor 100/101
▶ Enzymes: chemical cutters 102/103
▶ Absorbing stuff 104/105

SUPPORT AND MOVEMENT
▶ The living framework 16/17
▶ Smooth operators 30/31

CONTROL AND SENSATION
▶ Good taste 66/67

CIRCULATION, MAINTENANCE, AND DEFENSE
▶ Repelling invaders 144/145

REPRODUCTION AND GROWTH
▶ The growing child 178/179

The food processor

More than just a store for food, the stomach is an active organ of digestion.

Swallowed food is broken up into ever smaller bits by the churning action of the stomach, which also blends the food with a mixture of chemicals, known as gastric juice, secreted by the stomach's thick wall. The main component is strong hydrochloric acid, which kills most of the bacteria and other dangerous microorganisms that might be swallowed and helps dissolve food. The stomach secretes pepsinogen as well, and this is converted into the protein-splitting enzyme pepsin by the acid. It also gives out small quantities of a fat-digesting lipase and a protein called intrinsic factor which enables the body to absorb and use vitamin B_{12}.

After several hours of churning, acid treatment, and chemical digestion, the partly digested food, now called chyme, passes in controlled spurts into the upper part of the small intestine – the duodenum. Here, secretions from the pancreas and gall bladder continue the digestion process. Since enzymes below the stomach work only in non-acid conditions, pancreatic secretions contain alkalis which neutralize the acid chyme. Pancreatic juice also contains the enzymes amylase, chymotrypsin, trypsin, and lipase, as well as nucleases which break down or digest various components of food.

Since fats do not dissolve in water, it would be difficult for the water-soluble digestive enzymes of the intestines to deal with them without help. But bile enters the duodenum via the bile duct and acts as a biological detergent that emulsifies fats into minute droplets. These become suspended in the watery contents of the intestines so that the enzymes can work on them to split them up.

Cross section of esophagus

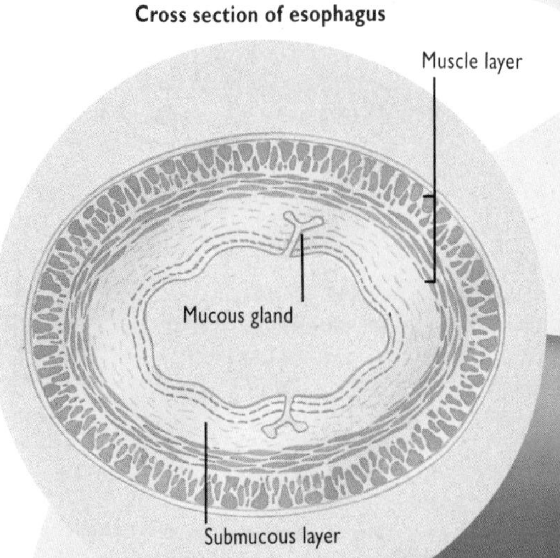

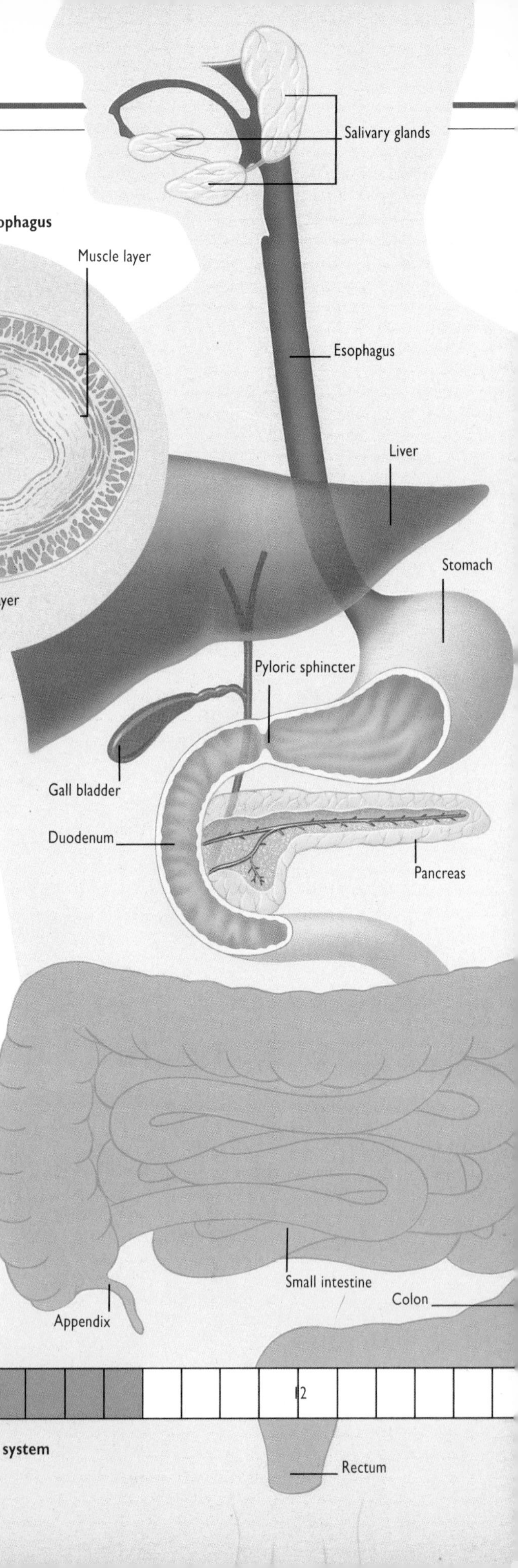

It is possible to swallow food *even while hanging upside-down. This is because powerful muscle layers in the esophagus, which joins the throat to the stomach, generate strong contractions, or peristaltic waves, that push the food along. Mucus from glands in the esophagus wall acts as a lubricant.*

On its way down, food spends only minutes or less in the mouth and esophagus (dark blue). It can spend a total of eight hours in the stomach and duodenum (red): four in the stomach before passing through the pyloric sphincter into the duodenum, where it spends a further four hours before going farther into the small intestine.

Hours spent in digestive system

n the stomach wall (below), surface ells make mucus which protects the ining from acid. Cells in pits in he wall secrete hydrochloric cid, other digestive hemicals, and the hormone astrin which helps ontrol gastric juice roduction.

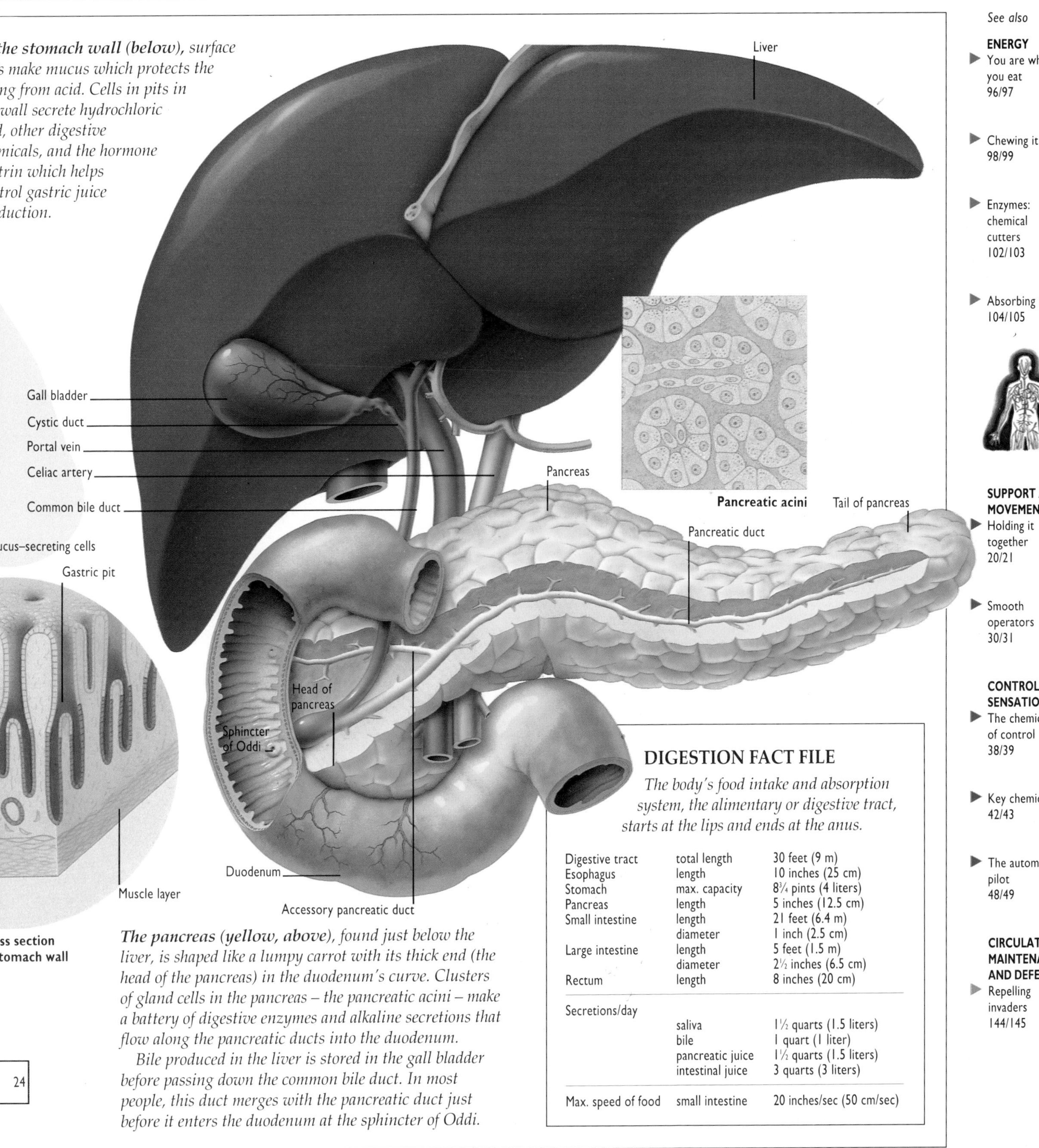

ross section f stomach wall

The pancreas (yellow, above), *found just below the liver, is shaped like a lumpy carrot with its thick end (the head of the pancreas) in the duodenum's curve. Clusters of gland cells in the pancreas – the pancreatic acini – make a battery of digestive enzymes and alkaline secretions that flow along the pancreatic ducts into the duodenum.*

Bile produced in the liver is stored in the gall bladder before passing down the common bile duct. In most people, this duct merges with the pancreatic duct just before it enters the duodenum at the sphincter of Oddi.

24

DIGESTION FACT FILE

The body's food intake and absorption system, the alimentary or digestive tract, starts at the lips and ends at the anus.

Digestive tract	total length	30 feet (9 m)
Esophagus	length	10 inches (25 cm)
Stomach	max. capacity	8¾ pints (4 liters)
Pancreas	length	5 inches (12.5 cm)
Small intestine	length	21 feet (6.4 m)
	diameter	1 inch (2.5 cm)
Large intestine	length	5 feet (1.5 m)
	diameter	2½ inches (6.5 cm)
Rectum	length	8 inches (20 cm)
Secretions/day		
	saliva	1½ quarts (1.5 liters)
	bile	1 quart (1 liter)
	pancreatic juice	1½ quarts (1.5 liters)
	intestinal juice	3 quarts (3 liters)
Max. speed of food	small intestine	20 inches/sec (50 cm/sec)

See also

ENERGY
- You are what you eat 96/97
- Chewing it over 98/99
- Enzymes: chemical cutters 102/103
- Absorbing stuff 104/105

SUPPORT AND MOVEMENT
- Holding it together 20/21
- Smooth operators 30/31

CONTROL AND SENSATION
- The chemicals of control 38/39
- Key chemicals 42/43
- The automatic pilot 48/49

CIRCULATION, MAINTENANCE, AND DEFENSE
- Repelling invaders 144/145

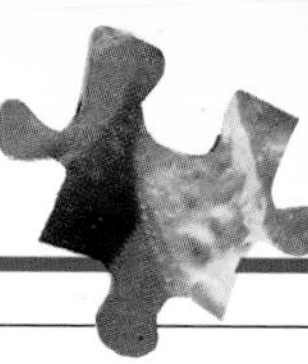

Enzymes: chemical cutters

Many body processes use highly specialized chemical tools to cut up or assemble molecules to order.

The human body depends on innumerable precisely controlled chemical reactions, many of which need catalysts to speed them up and facilitate them. Evolution has produced thousands of biological catalysts in the form of enzymes, each of which is normally capable of stimulating only one specific chemical reaction. Some enzymes join small molecules together to form larger ones, as when amino acids join to form the proteins that make up the body. Others break down large molecules into smaller ones, as in the digestion of food into small, absorbable sub-units, or in the severing of the bonds holding a sugar together to release energy.

Most enzymes are large, globe-shaped protein molecules with a complex surface shape. This surface usually has a precisely patterned groove or cavity – the enzyme's "active center." It is shaped so that the substance on which the enzyme acts to stimulate a reaction – the substrate – fits precisely into it. Fitting like a chemical key in the lock of the active center, the substrate is activated, bringing about the desired reaction. It then leaves the enzyme, freeing up the active center to receive a new substrate – and the process begins again. Enzymes are dependent on conditions being constant; the enzymes in the body generally work best at about 98.6°F (37°C) – the normal body temperature maintained by homeostasis.

Digestive enzymes *break down large food molecules into small nutrient molecules that can be absorbed by the wall of the small intestine. There are three major types of large food molecules – the polysaccharides (starches), lipids (fats and oils), and proteins.*

Polysaccharides, the starch in rice for instance, are made up of long chains of individual (monosaccharide) sugars, such as glucose, fructose, and galactose. Starch-splitting enzymes digest them by breaking the links between the sugars. The process starts in the mouth with the enzyme salivary amylase. But most is done in the small intestine, where polysaccharide fragments are attacked by pancreatic amylase. This converts them into two-sugar disaccharides such as sucrose, lactose, and maltose. Specific disaccharidases (disaccharide-splitting enzymes), such as sucrase, lactase, and maltas[e] on the surface of the small intestin[e] split disaccharides into mono-saccharide sugars. These are then absorbed by the gut wall.

Proteins *are made of chains of amino acids joined together by links known as peptide bonds. The digestive enzymes that work on proteins such as those in fish do so by breaking the bonds. Such digestion takes place in the stomach and small intestine. The main enzyme in the stomach is pepsin, which can operate in the high acid levels found there. Pepsin breaks down proteins into smaller chains of amino acids known as proteoses and peptones and also some individual amino acids. In th[e] small intestine the potent enzyme trypsin, along with enzymes calle[d] aminopeptidases, completes the breakdown of proteins, peptones, and proteoses into single amino aci[ds] which can be absorbed by the body.*

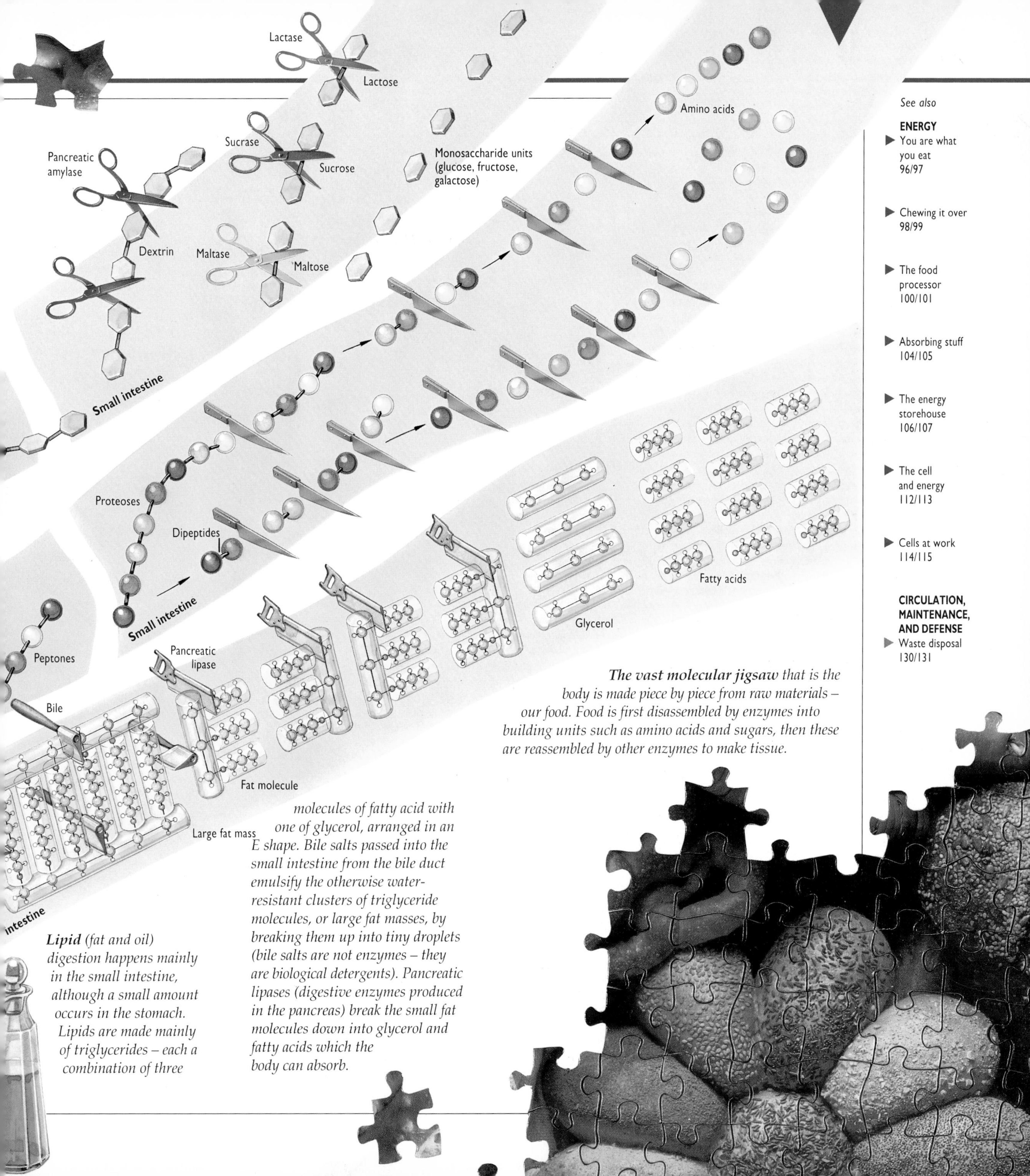

See also

ENERGY
▶ You are what you eat 96/97
▶ Chewing it over 98/99
▶ The food processor 100/101
▶ Absorbing stuff 104/105
▶ The energy storehouse 106/107
▶ The cell and energy 112/113
▶ Cells at work 114/115

CIRCULATION, MAINTENANCE, AND DEFENSE
▶ Waste disposal 130/131

The vast molecular jigsaw *that is the body is made piece by piece from raw materials – our food. Food is first disassembled by enzymes into building units such as amino acids and sugars, then these are reassembled by other enzymes to make tissue.*

Lipid *(fat and oil) digestion happens mainly in the small intestine, although a small amount occurs in the stomach. Lipids are made mainly of triglycerides – each a combination of three molecules of fatty acid with one of glycerol, arranged in an E shape. Bile salts passed into the small intestine from the bile duct emulsify the otherwise water-resistant clusters of triglyceride molecules, or large fat masses, by breaking them up into tiny droplets (bile salts are not enzymes – they are biological detergents). Pancreatic lipases (digestive enzymes produced in the pancreas) break the small fat molecules down into glycerol and fatty acids which the body can absorb.*

Absorbing stuff

The primary function of the alimentary tract is to absorb nutrient molecules from digested foods.

Once food has been chewed and churned and broken down into tiny bits of organic nutrients by enzymes, it can be absorbed into the body. This is done extraordinarily efficiently – up to 95 percent of fats and 90 percent of amino acids (protein building blocks) produced by digestion are absorbed in the small intestine, which lies between the stomach and large intestine.

The small intestine has three parts: the duodenum, where most digestion occurs, and the jejunum and the ileum, where most absorption takes place. During the one to six hours (the average time is two hours) that it takes for digested food to pass through the small intestine, it is pushed along by the peristaltic action of muscles in the intestinal walls.

Uptake of nutrients into the blood vessels and lymph vessels of the wall is aided by an amplification of the surface area of the intestinal lining. This expands the simple inner surface of the small intestine by about 600 times – if the small intestine had a smooth inner surface, it would need to be not 20 feet (6 m), but 2¼ miles (3.6 km) long.

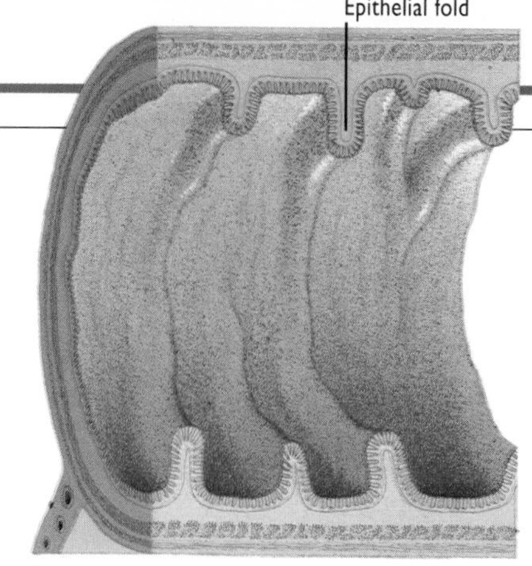

Section of small intestine

The inner wall of the small intestine (left) *has folds, or plicae circulares, which extend around the inner lining like circular shelves. These shelves are the first step in increasing the area available for the uptake of nutrients. Muscles in the intestinal wall contract rhythmically to drive the contents along. The contractions help to swirl matter around so that it comes into contact with the absorbing cells.*

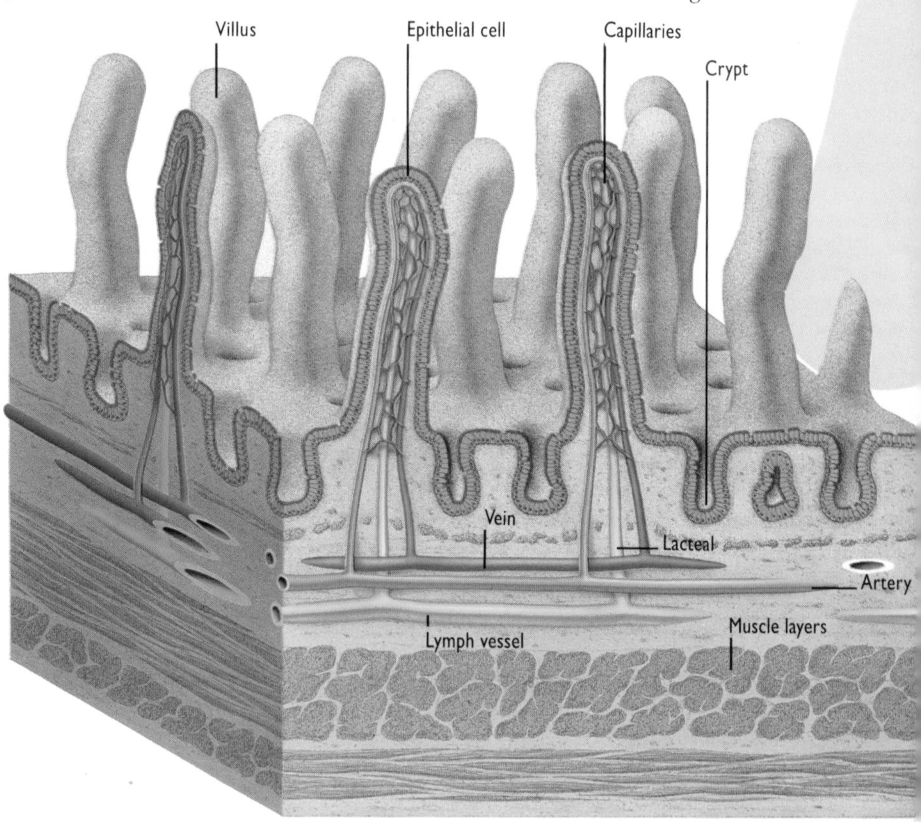

The folded surface *of the intestinal walls is expanded by millions of finger-shaped projections, or villi (**right**), and by intuckings, called crypts, between the villi. The villi wave around, helping to stir the intestinal contents. Each villus is covered by absorptive epithelial cells (**below**) and contains a tiny arteriole and venule, linked by capillaries, and a lymph vessel known as a lacteal.*

Detail of epithelial fold

Villi

Nutrients such as sugars and amino acids *pass across the epithelial cells into the blood in the capillaries to be transferred to the rest of the body. Products of fat digestion merge with the cell membrane, eventually drifting into the lacteals, which in due course empty into the blood system in a vein near the heart.*

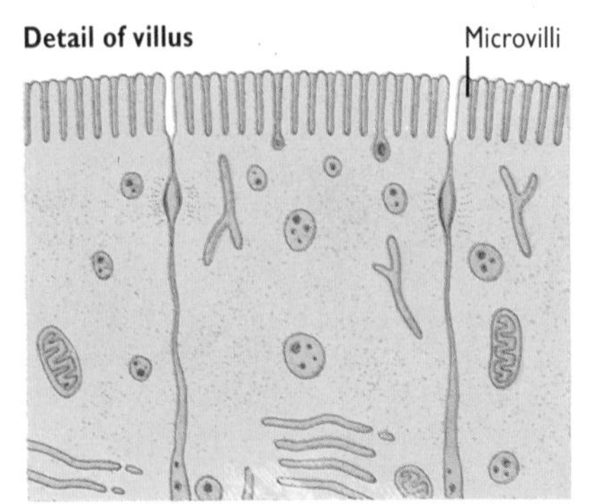

A fuzzy coat *– the so-called brush border – covers the outer surface of each villus. The border is made up of thousands of microvilli (**left**), which are thin, membrane-bound extensions of the cell. Nutrient absorption at an ultimate, molecular, level occurs across the outer membranes of the microvilli.*

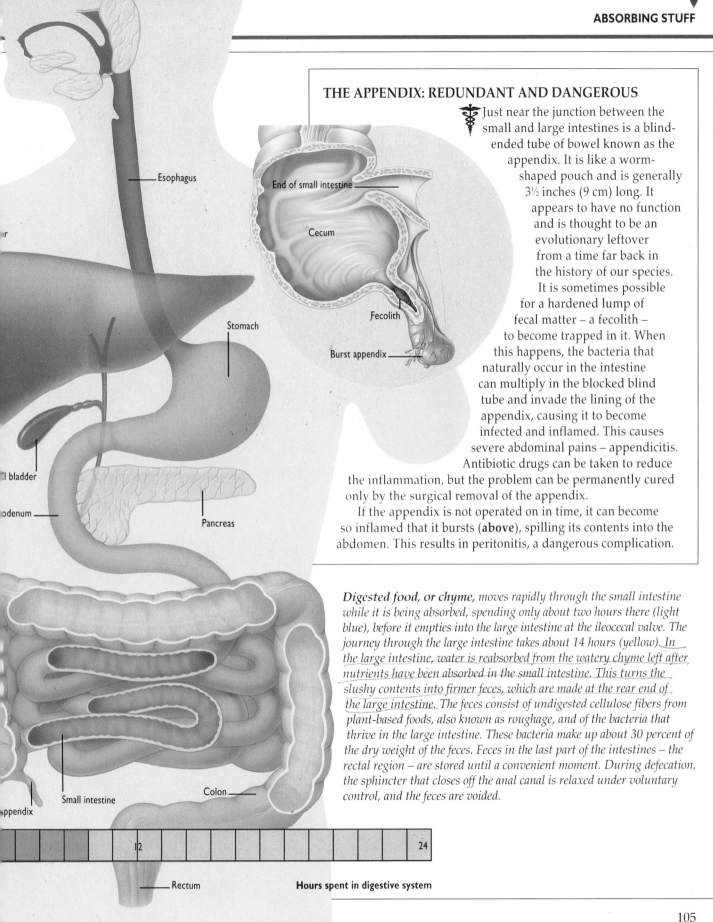

THE APPENDIX: REDUNDANT AND DANGEROUS

Just near the junction between the small and large intestines is a blind-ended tube of bowel known as the appendix. It is like a worm-shaped pouch and is generally 3½ inches (9 cm) long. It appears to have no function and is thought to be an evolutionary leftover from a time far back in the history of our species. It is sometimes possible for a hardened lump of fecal matter – a fecolith – to become trapped in it. When this happens, the bacteria that naturally occur in the intestine can multiply in the blocked blind tube and invade the lining of the appendix, causing it to become infected and inflamed. This causes severe abdominal pains – appendicitis. Antibiotic drugs can be taken to reduce the inflammation, but the problem can be permanently cured only by the surgical removal of the appendix.

If the appendix is not operated on in time, it can become so inflamed that it bursts (**above**), spilling its contents into the abdomen. This results in peritonitis, a dangerous complication.

Digested food, or chyme, *moves rapidly through the small intestine while it is being absorbed, spending only about two hours there (light blue), before it empties into the large intestine at the ileocecal valve. The journey through the large intestine takes about 14 hours (yellow). In the large intestine, water is reabsorbed from the watery chyme left after nutrients have been absorbed in the small intestine. This turns the slushy contents into firmer feces, which are made at the rear end of the large intestine. The feces consist of undigested cellulose fibers from plant-based foods, also known as roughage, and of the bacteria that thrive in the large intestine. These bacteria make up about 30 percent of the dry weight of the feces. Feces in the last part of the intestines – the rectal region – are stored until a convenient moment. During defecation, the sphincter that closes off the anal canal is relaxed under voluntary control, and the feces are voided.*

See also

ENERGY
▶ Fueling the body 94/95
▶ You are what you eat 96/97
▶ Chewing it over 98/99
▶ The food processor 100/101
▶ Enzymes: chemical cutters 102/103
▶ The energy storehouse 106/107
▶ The cell and energy 112/113

CONTROL AND SENSATION
▶ Key chemicals 42/43

CIRCULATION, MAINTENANCE, AND DEFENSE
▶ The body's drain 128/129
▶ The water balance 132/133

SUPPORT AND MOVEMENT
▶ Smooth operators 30/31

The energy storehouse

It is difficult to find any aspect of bodily functions not influenced by the many activities of the liver.

Lying at the top of the abdominal cavity, nestling under the diaphragm and over the stomach, the liver – the body's largest gland – plays a truly central role in the body's nutrient transportation and metabolic processes. For instance, it takes the key energy-supplying molecule glucose (a sugar) from blood arriving from the intestine and converts it into the starchlike carbohydrate glycogen for storage. When glucose levels in the blood fall, it then converts the glycogen back into glucose to transport around the body. The liver can also store fats and amino acids and convert them into glucose. It forms the waste product urea from waste proteins and amino acids and manufactures the key molecules – lipoproteins, cholesterol, and phospholipids – that make cell membranes.

In addition to these storage and conversion functions, the liver is also responsible for keeping the body at the correct temperature by warming blood as it passes through its inner spaces. It even defends the body by changing harmful chemicals like poisons, drugs, pesticides, and environmental pollutants into harmless products which can then be removed from the body in bile or urine.

In a container port, *materials are stored ready for access and transport to other regions. Similarly, the liver stores nutrients from the digestive system ready for distribution and use in the body.*

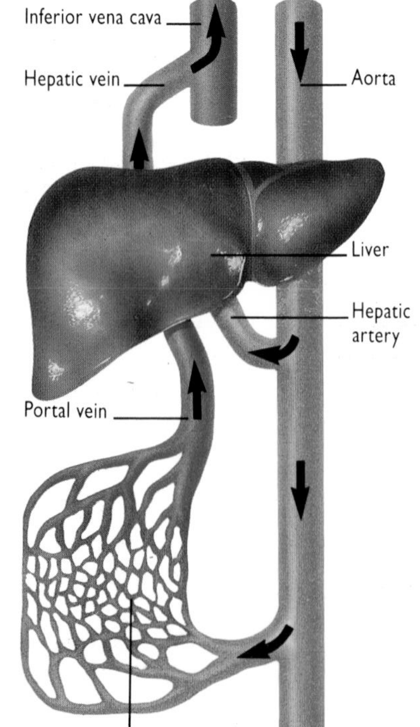

The liver is supplied *(via the hepatic artery) with blood which brings the oxygen and nutrients that enable it to do its work.*

It also receives blood for processing that comes, via the portal vein, direct from capillaries that drain those parts of the digestive tract that absorb nutrients from the intestinal contents. All blood leaves the liver via the hepatic vein.

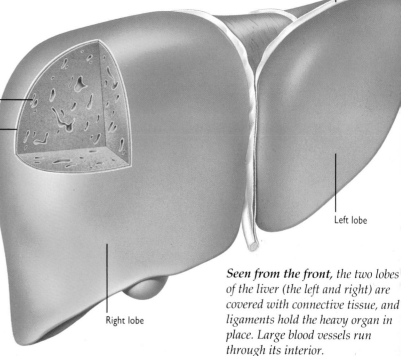

Supporting ligame
Section through blood vessels
Peritoneum
Left lobe
Right lobe

Seen from the front, *the two lobes of the liver (the left and right) are covered with connective tissue, and ligaments hold the heavy organ in place. Large blood vessels run through its interior.*

Left lobe
Portal vein
Inferior vena cava
Right lobe
Supporting ligament
Hepatic artery
Gall bladder

Cross section of lobule

Branch of bile duct
Hepatic artery
Bile canaliculi
Liver cells
Sinusoids
Branch of portal vein
Central vein (leads to hepatic vein)
Branch of hepatic artery

__The gall bladder__ temporarily stores bile, which contains water, sodium bicarbonate, bile salts, bile pigments, and cholesterol. It concentrates the bile, extracting 90 percent of its water, making it mucusy. About 30 minutes after a meal, when the duodenum first contains partly digested food, the gall bladder contracts to push bile into the duodenum, where it turns fat droplets into an emulsion so that they can be digested.

obule
Branch of bile duct
Branch of hepatic artery
Sinusoid
Branch of portal vein
epatic vein
Central vein

__ach lobe of the liver__ is made up of housands of small lobes, or lobules. lobule is made of liver cells, or epatocytes, which are supplied with lood by tiny branches of the hepatic rtery and portal vein. Blood flows in hannels known as sinusoids.

__A liver lobule's hepatocytes__ are piled up in irregularly shaped rows, radiating outward from a central vein. They are surrounded by blood which is carried in channels, or sinusoids, from the outside of the lobule toward the central vein. The blood comes from a branch of the portal vein and, after traversing the lobule, it drains into a branch of the hepatic vein.

As the blood passes through the lobule, hepatocytes get to work on it. One of their many roles in dealing with nutrients arriving from the intestines is to convert the sugars galactose and fructose into glucose, the body's essential fuel.

They also make bile – essential for the digestion of fats – which is collected by tiny bile ducts, or canaliculi, and taken to the gall bladder via the bile duct.

LIVER FACT FILE

The liver acts as a store for glucose, in the form of glycogen, the vitamins A, D, E, K, and B_{12}, and minerals such as iron as well as fats and amino acids.

Dimensions	width	8–9 inches (20–22 cm)
	height	6–7 inches (15–18 cm)
	thickness	4–5 inches (10–13 cm)
Weight of liver		3 pounds (1.5 kg)
Percentage of adult body weight		2.5
Percentage of infant body weight		4
No. of lobes		2
Diameter of lobules		1/25 inch (0.1 cm)
Blood supplied to liver		1½ quarts/min (1.5 liters/min)
Blood supply from hepatic artery		20%
Blood supply from portal vein		80%
Rate of bile secretion		1quart/day (1 liter/day)

See also

ENERGY
▶ Fueling the body 94/95
▶ You are what you eat 96/97
▶ The food processor 100/101
▶ Absorbing stuff 104/105
▶ The cell and energy 112/113

SUPPORT AND MOVEMENT
▶ Holding it together 20/21
▶ Making a move 22/23

CIRCULATION, MAINTENANCE, AND DEFENSE
▶ Blood: supplying the body 126/127
▶ Waste disposal 130/131
▶ Maintaining the system 136/137
▶ Routine replacement 138/139

A deep breath

Life depends on a continuous supply of oxygen, which we get from the air around us when we breathe.

The air we need is pumped in and out of two lungs, which sit protected inside the rib cage of the chest. In the lungs, oxygen is removed from the air and passed into the blood; the waste gas carbon dioxide leaves the blood to be eliminated in exhaled air. Breathing can be consciously controlled, but more usually this vital function is taken care of and coordinated automatically by the nervous system.

During breathing, air enters through the nose or mouth. It then passes down the trachea, or windpipe, which is supported by tough cartilage hoops that keep it open at all times. On the way, it is warmed and humidified.

From the windpipe, air goes down the left or right bronchus (airway) into one lung or the other. The airways divide into finer and finer branches, or bronchioles; the narrowest bronchioles end in alveoli – tiny blind-ended air sacs where most of the gas exchange between air and blood takes place.

The whole inner lining of the breathing, or pulmonary, system is kept moist by secretions of mucus from the epithelial (lining) cells of the air spaces. And specialized cells in the airway wall have many cilia (short, hairlike projections) which beat and move the mucus along so it flows upward from the bronchioles. The system works like a watery conveyor belt, removing inhaled particles such as dust from the lungs to the top of the trachea from where the mucus can be coughed up or swallowed.

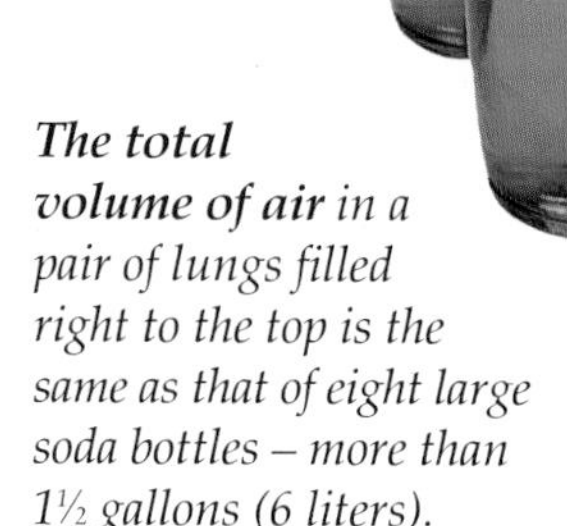

The total volume of air *in a pair of lungs filled right to the top is the same as that of eight large soda bottles – more than 1½ gallons (6 liters).*

The outside of the right lung, *seen from the left, has indentations for central chest organs, including the trachea and esophagus. Pulmonary veins take blood to the heart, and the pulmonary artery brings blood from the heart. The fissures mark the divisions of the lung into separate lobes.*

In cross section, the spongy alveolar tissue and the bronchi (main airways) and bronchioles (smaller airways) are obvious.

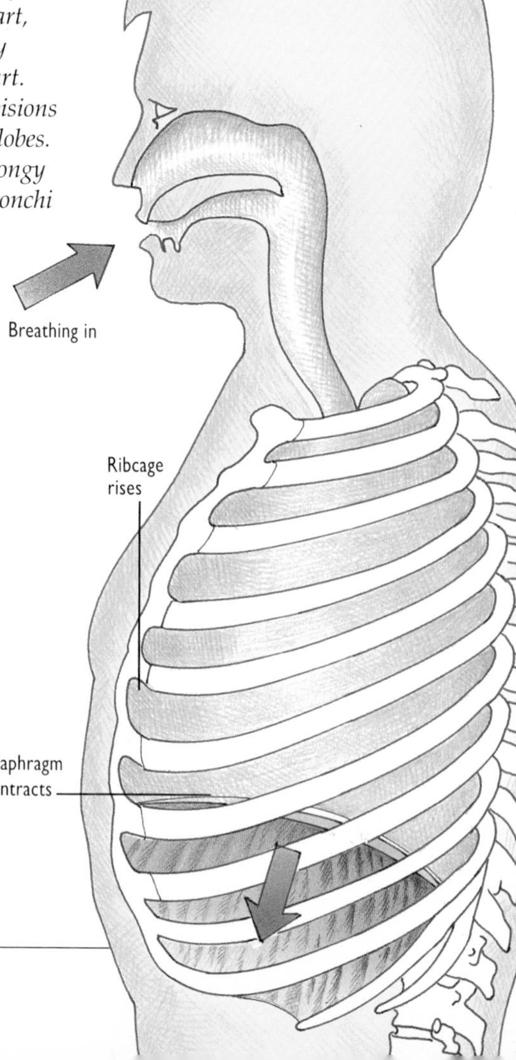

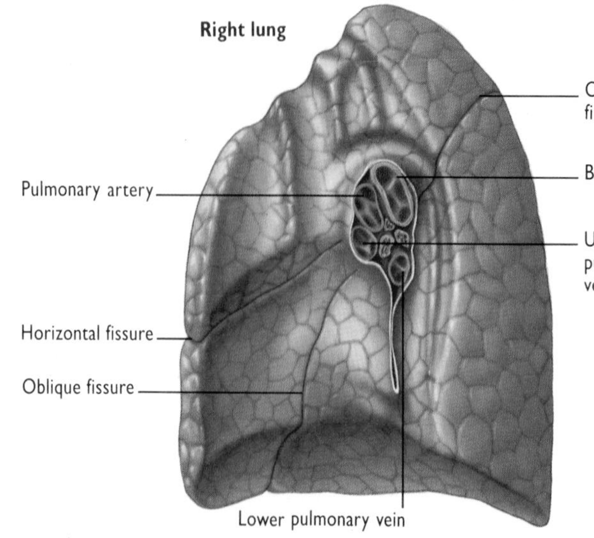

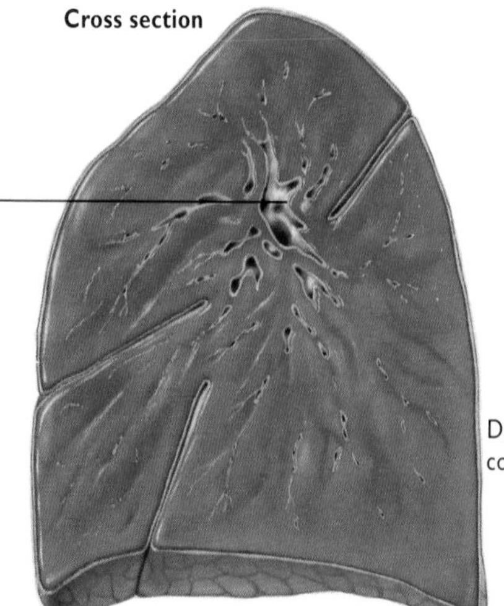

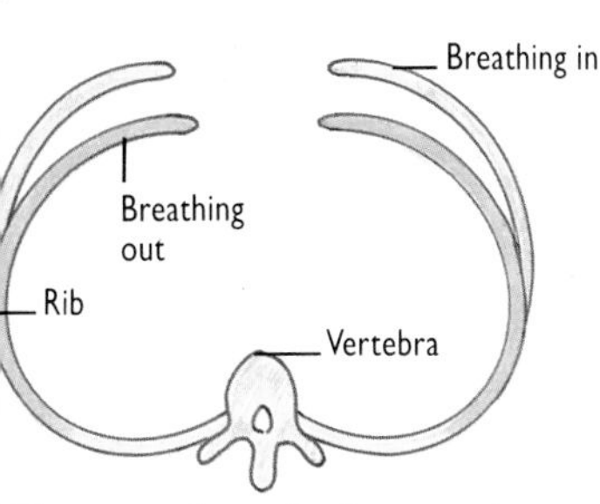

Vhen you breathe in, muscles aise and expand the rib cage, while he diaphragm, a domed muscular heet under the lungs, flattens. These novements increase the volume of he pleural cavity, the space in vhich the lungs sit, and air rushes n to equalize pressure. With the everse movements, the volume lecreases and air is pushed out.

BREATHING FACT FILE

The lungs are not identical – the right has three lobes; the left, which is smaller than the right, has two.

Ave. weight of left lung	20 ounces (565 g)
Ave. weight of right lung	22 ounces (625 g)
Total surface area of lungs	645–755 sq. feet (60–70 m^2)
Ave. no. of breaths a minute	16
Breathing rate (adult man) max.	300 quarts/min (300 liters/min)
resting	8 quarts/min (8 liters/min)
No. of alveoli per lung	More than 350 million
Diameter of alveoli	$^1/_{1,000}$ inch (25 micrometers)
Thickness of alveoli walls	$^1/_{6,000}$ inch (4 micrometers)
Oxygen	inhaled air 21% by volume
	exhaled air 16% by volume
Carbon dioxide	inhaled air 0.04% by volume
	exhaled air 4% by volume

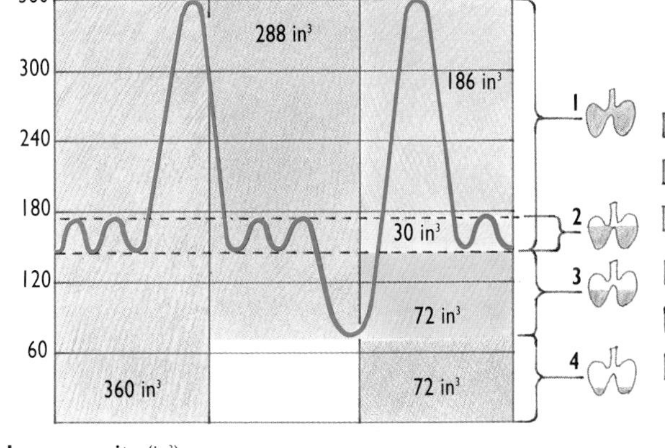

Lung capacity (in³)

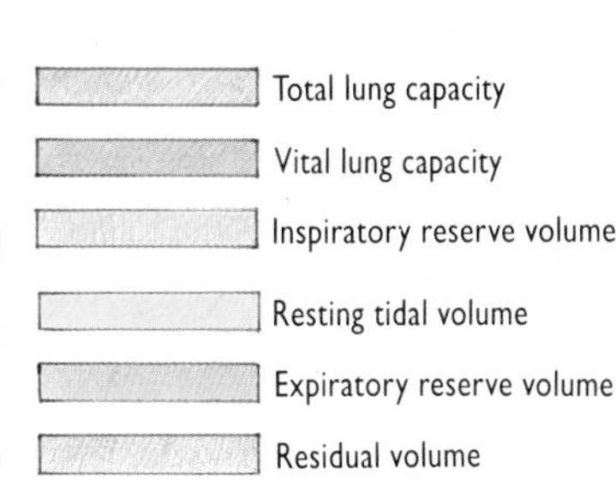

Resting tidal volume *is the amount of air breathed in and out during light breathing* (**2**). *The extra volume that can be filled with the deepest breath is the inspiratory reserve volume; together they add up to the inspiratory capacity* (**1**). *The extra volume breathed out by the strongest possible exhalation is the expiratory reserve volume* (**3**). *Air left in the airways after the strongest exhalation is the residual volume* (**4**). *The vital lung capacity is the amount of air that can be taken in by the strongest inhalation after the strongest exhalation. Vital lung capacity added to residual volume gives the total lung capacity.*

Diaphragm relaxes

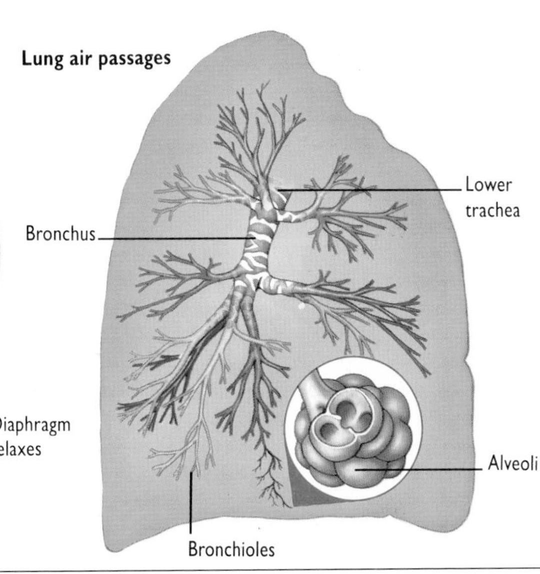

Inside a lung, *the airway divides and redivides like the branches of a tree or the tributaries of a river system. Each lung has 10 segments, each served by individual bronchi and bronchioles. Segments are further divided into lobules, each of which is served by a single tiny bronchiole – the final division in the system.*

In each individual lobule are hundreds of alveoli (***inset***). *It is in the alveoli that inhaled air comes into contact with tiny blood vessels containing deoxygenated blood. Gases are exchanged between the air and the blood, which picks up oxygen and gives out carbon dioxide.*

See also

ENERGY
▶ Fueling the body 94/95

▶ The gas exchange 110/111

▶ Cells at work 114/115

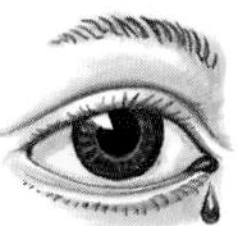

SUPPORT AND MOVEMENT
▶ Holding it together 20/21

▶ Smooth operators 30/31

CONTROL AND SENSATION
▶ Steady as you go 34/35

▶ The automatic pilot 48/49

CIRCULATION, MAINTENANCE, AND DEFENSE
▶ Transportation systems 118/119

▶ Repelling invaders 144/145

REPRODUCTION AND GROWTH
▶ The newborn baby 176/177

The gas exchange

Deep in the lungs, tiny molecules move to and fro between air from outside and blood from within.

Each lung is an absorbent sponge, and breathing pulls air into the spaces of that sponge with every inhalation. The tiniest air-filled spaces are known as alveoli, and in them oxygen passes into the bloodstream for circulation around the body, and waste carbon dioxide leaves the blood to be discharged in exhaled air.

There are about 700 million alveoli in both lungs, and their combined surface area is some 84 sq. yards (70 m^2) – not far short of the surface area of a tennis court. Gases such as oxygen and carbon dioxide move across this huge area by diffusion, which takes place because substances tend to move from a region where they are highly concentrated to one where they are relatively less concentrated. Since the oxygen concentration in the air of the alveoli is higher than that in the deoxygenated blood in the capillaries around the alveoli, oxygen moves into the capillaries from the air. Similarly, capillary blood contains more carbon dioxide than inhaled air, so this waste gas from the respiration (energy production) of body tissues passes from the blood into the alveolar air spaces to be breathed out.

The alveoli themselves are clustered around the closed inner endings of the thinnest branches of the airways (the terminal bronchioles) like minute bunches of grapes. Each individual alveolus "grape" is only 1/1,000 inch (25 micrometers) in diameter, so 40 would have to be placed side by side to make a bunch 1/25 inch (1 mm) across.

***Flour in a sifter** stays there until it is shaken. Then the fine flour particles fall through the much wider holes in the sifter. This movement is rather like the diffusion of oxygen across the wall of an alveolus in the lungs.*

Oxygen molecules are so small that they can move easily through the alveolar wall. The energy source that moves them is the vibration energy that all molecules have; this is like the shaking of the sifter which sends the flour through the holes under the influence of gravity. Gravitational force can be likened to the concentration difference that allows oxygen to diffuse from air to blood.

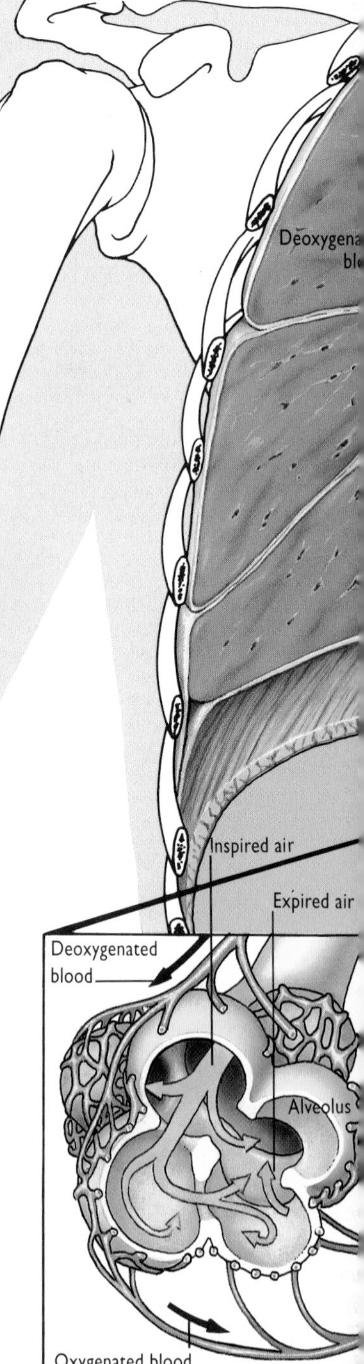

***The lungs** have a spongy texture and are shot through with branching airways – bronchi and bronchioles – and blood vessels.*

Deoxygenated blood on the right side of the heart is pushed by the right ventricle through the two pulmonary arteries and along a series of fine arterioles until it reaches the capillaries around an alveolus. There it picks up oxygen and returns to the left side of the heart via a network of veins that become increasingly wide and ultimately form the pulmonary vein. The oxygen is then pumped around the body from the left ventricle.

***Alveoli occur in bunches (inset).** Each bunch is fed by its own tiny airway, or bronchiolus, and the cluster of air sacs is surrounded by a mesh of branching capillaries. As you breathe, air moves in and out of the central spaces, and gas exchanges take place across the alveolar wall: blood takes oxygen from the air and returns carbon dioxide to the air for exhalation.*

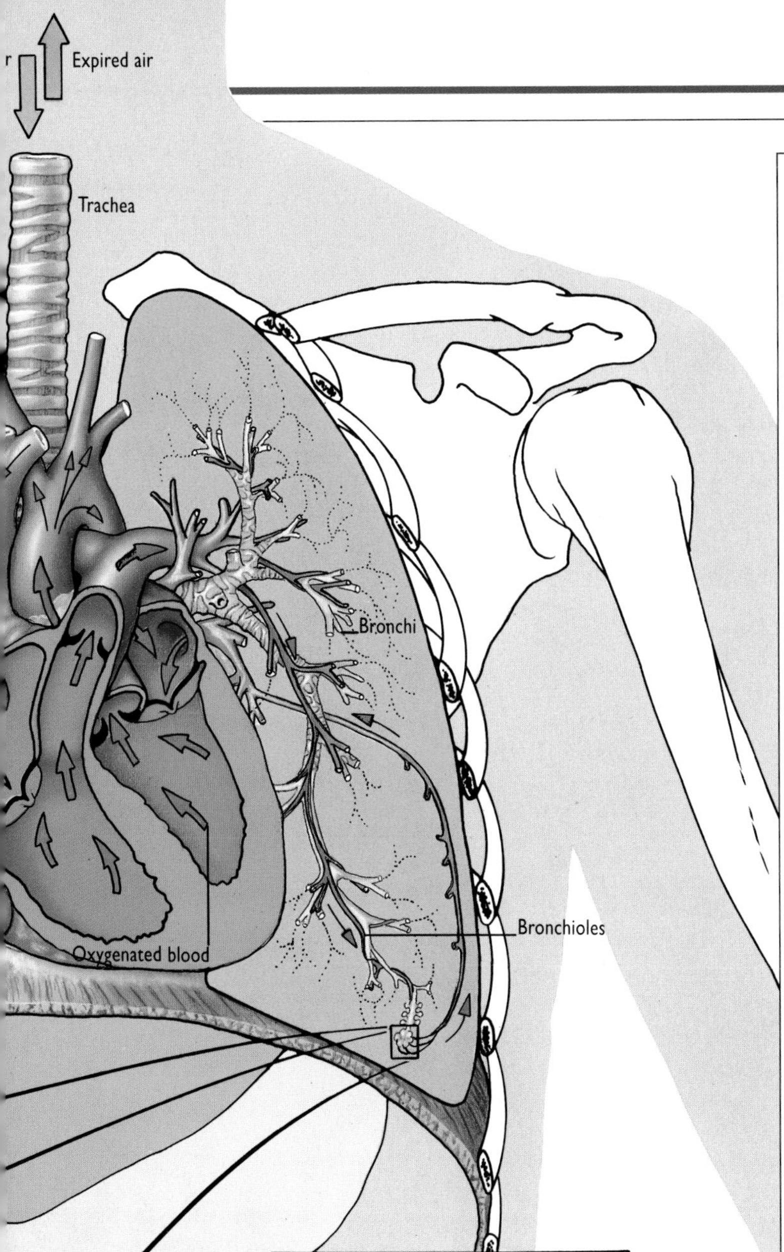

The total ***thickness*** *of the barrier between alveolar air and the blood n the capillaries is nly about 1/6,000 inch 4 micrometers), about alf the diameter of a ingle red blood cell. he extreme thinness of his barrier means that as diffusion in both irections is rapid nd efficient.*

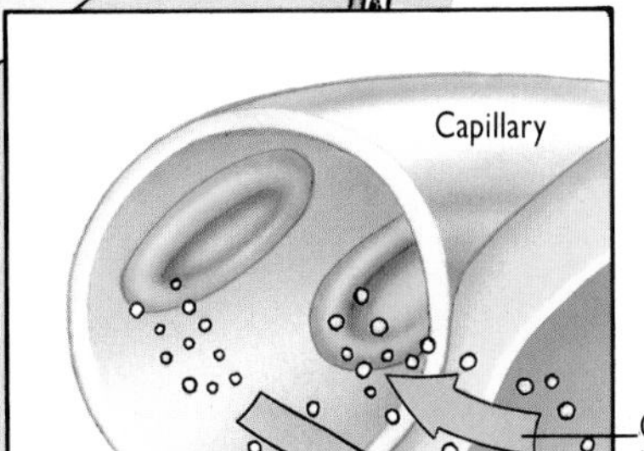

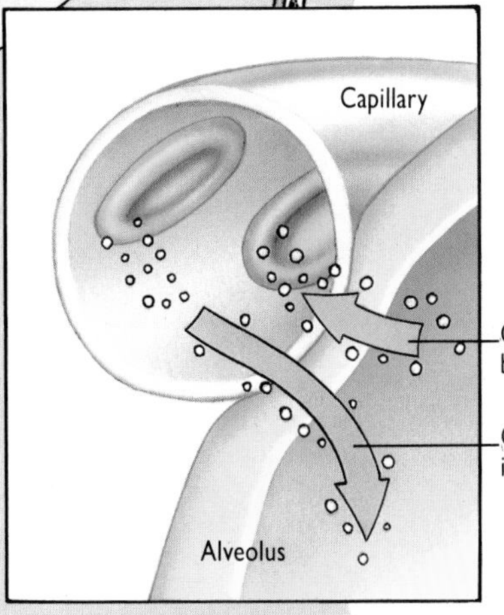

CROSSING THE BARRIER

Although the barrier that oxygen and carbon dioxide have to cross during gas exchange in the lungs is extremely thin, it is nevertheless a complex, multilayered sandwich of tissue. The "bread" consists of two flattened layers of thin cells: one slice of alveolar cells, the other slice of cells that make up the capillary wall. A thin filling of basement membrane fibers separates the two cell layers. To an oxygen molecule, however, these layers seem insubstantial; and the gas molecules, dissolved in a watery liquid, move across them by diffusion.

Oxygen diffusing into the blood *from an alveolus first meets a biodetergent known as lung surfactant that is secreted by cells on the inner wall of the alveolus. This reduces the surface tension inside each alveolus so that it can easily be inflated by inhaled air.*

The oxygen passes through a membrane into the alveolar cell and out through the other side. It then crosses the basement membrane before traversing a cell in the wall of the capillary. There it meets up with a red blood cell and oxygenates it.

Carbon dioxide moving from the blood into the air in the alveolus passes across the same barriers by diffusion, but in the opposite direction.

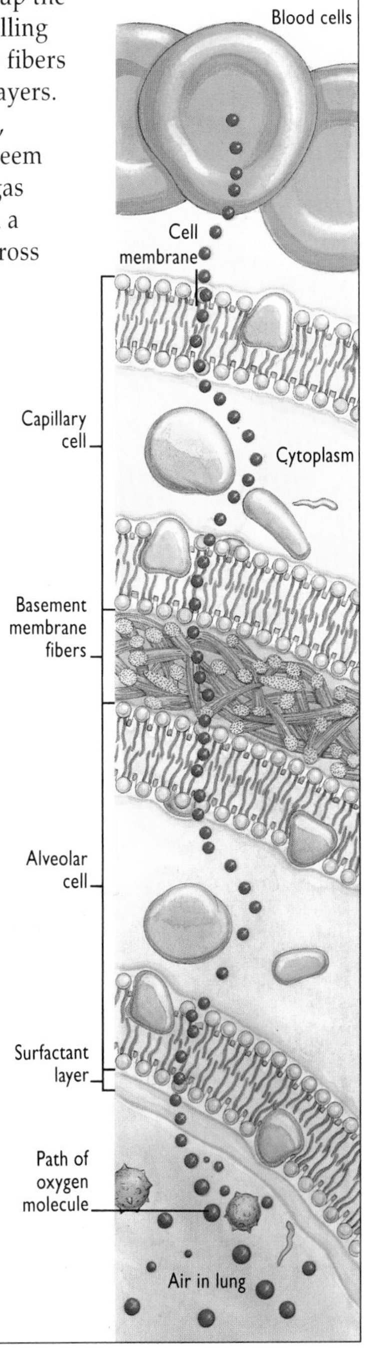

See also

ENERGY
▶ Fueling the body 94/95

▶ A deep breath 108/109

▶ The cell and energy 112/113

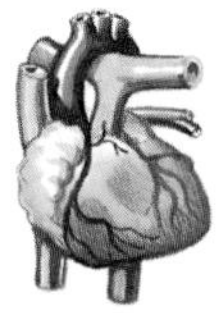

SUPPORT AND MOVEMENT
▶ Holding it together 20/21

CONTROL AND SENSATION
▶ Steady as you go 34/35

CIRCULATION, MAINTENANCE, AND DEFENSE
▶ The living pump 120/121

▶ Growing heart 122/123

▶ In circulation 124/125

▶ Blood: supplying the body 126/127

REPRODUCTION AND GROWTH
▶ Being born 174/175

The cell and energy

Inside a typical cell are various sub-units, each with its own vital function. All of them need energy.

Every part of a person's body is made of microscopic cells or their products. There are trillions of cells in total, and they vary enormously in shape, dimensions, and function. But all of them, except for a few specialized types, have essentially the same organization. The basic plan is a volume of semifluid matter, or cytoplasm, held in shape by a cell membrane. Within the cytoplasm is the controlling nucleus. It contains the set of 46 chromosomes that carry the DNA molecules that hold the genes responsible for passing inherited characteristics from one generation to the next.

The cytoplasm and the nucleus house the specialized "tools" of the cell's living machinery. Some hold the cell in shape; others are manufacturing sites that make the products the cell needs. Others are power plants that provide energy: in mitochondria – the main energy generators – the energy-rich molecule ATP is formed.

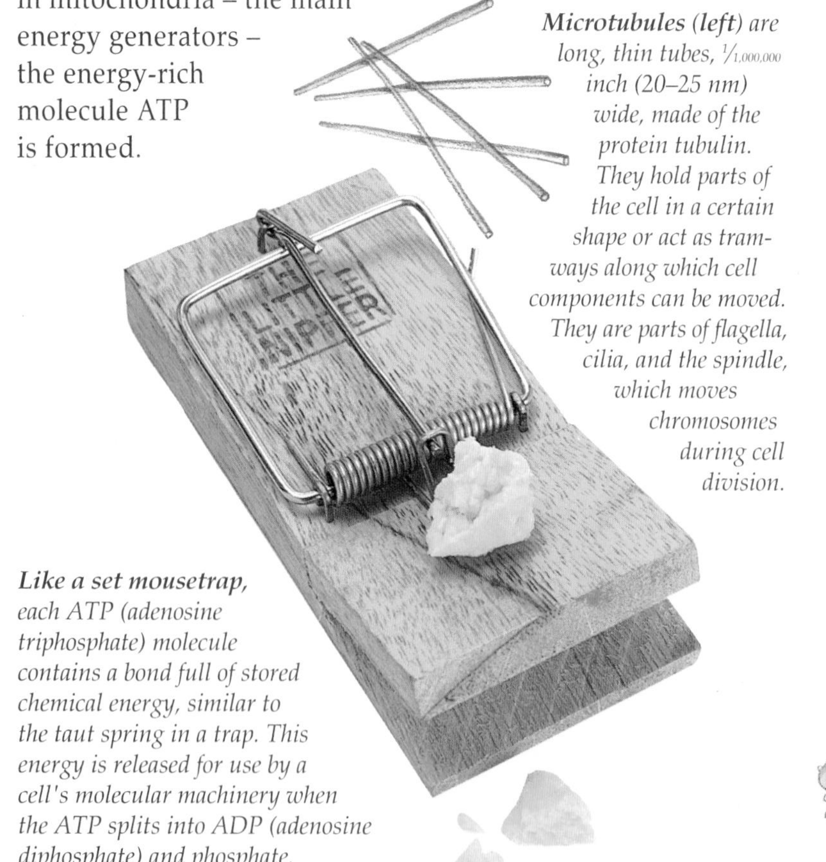

Microtubules (left) *are long, thin tubes, 1/1,000,000 inch (20–25 nm) wide, made of the protein tubulin. They hold parts of the cell in a certain shape or act as tramways along which cell components can be moved. They are parts of flagella, cilia, and the spindle, which moves chromosomes during cell division.*

Like a set mousetrap, *each ATP (adenosine triphosphate) molecule contains a bond full of stored chemical energy, similar to the taut spring in a trap. This energy is released for use by a cell's molecular machinery when the ATP splits into ADP (adenosine diphosphate) and phosphate.*

Cell respiration, *or the generation of energy-rich compounds, entails charging up ADP molecules to make ATP molecules. This takes place in a series of chemical reactions, mostly in the mitochondria. There are three distinct phases: glycolysis, the Krebs cycle, and the electron transportation system. Together they break down glucose molecules (each containing six carbon atoms) into carbon dioxide and water, producing more than 30 ATPs per glucose molecule.*

Glycolysis, which takes place in the cytoplasm, breaks down the 6-carbon glucose into two 3-carbon fragments, releasing a small number of ATPs. The main ATP production shifts to the mitochondria, where, in an altered form, the carbon fragments are fed into a cycle of reactions, the Krebs cycle. It makes carbon dioxide and joins hydrogen atoms onto transfer molecules NAD and FAD. The hydrogen drives the ATP-making machinery of the electron transportation system on the knobs on the cristae in the mitochondria. Once ATP is made, the hydrogen atoms link with oxygen to form water.

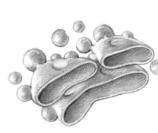

Glucose molecule with six carbon atoms

Glycolysis

CO_2

Krebs cycle

FAD

The endoplasmic reticulum (below) *consists of folded sheets of membrane. It is found in the cytoplasm (the region inside the cell membrane, but outside the nucleus) of almost all cells. In places its membranes may merge with those of the envelope around the nucleus and with the cell membrane.*

Much of the reticulum carries rounded granules, or ribosomes, made of RNA and protein. Ribosomes are the protein-making machinery of the cell and they translate messenger RNA (mRNA), from the nucleus, into the proteins for which the RNA carries the code. Newly made protein is held initially in transfer vesicles – small bags of reticulum membrane.

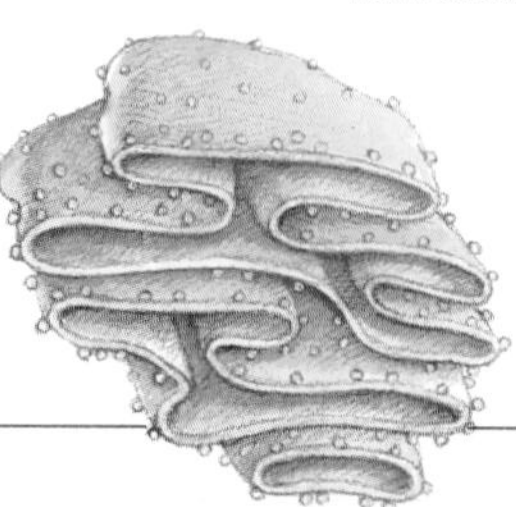

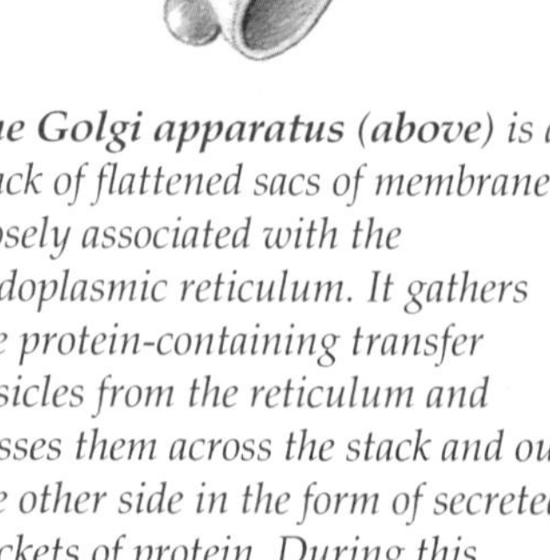

The Golgi apparatus (above) *is a stack of flattened sacs of membrane closely associated with the endoplasmic reticulum. It gathers the protein-containing transfer vesicles from the reticulum and passes them across the stack and out the other side in the form of secreted packets of protein. During this transfer, the Golgi sacs add sugars to the proteins to convert them into glycoproteins.*

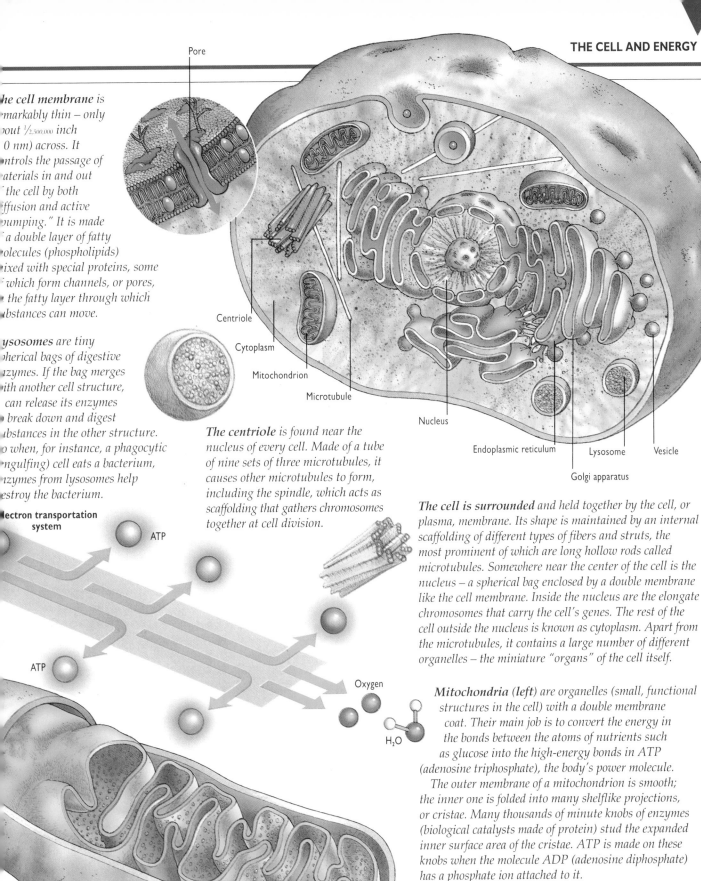

he cell membrane is
markably thin – only
out 1/2,500,000 inch
0 nm) across. It
ntrols the passage of
aterials in and out
the cell by both
ffusion and active
umping." It is made
a double layer of fatty
olecules (phospholipids)
ixed with special proteins, some
which form channels, or pores,
the fatty layer through which
bstances can move.

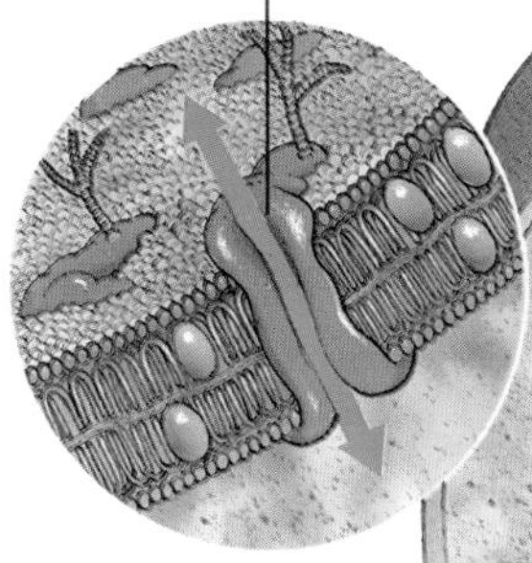

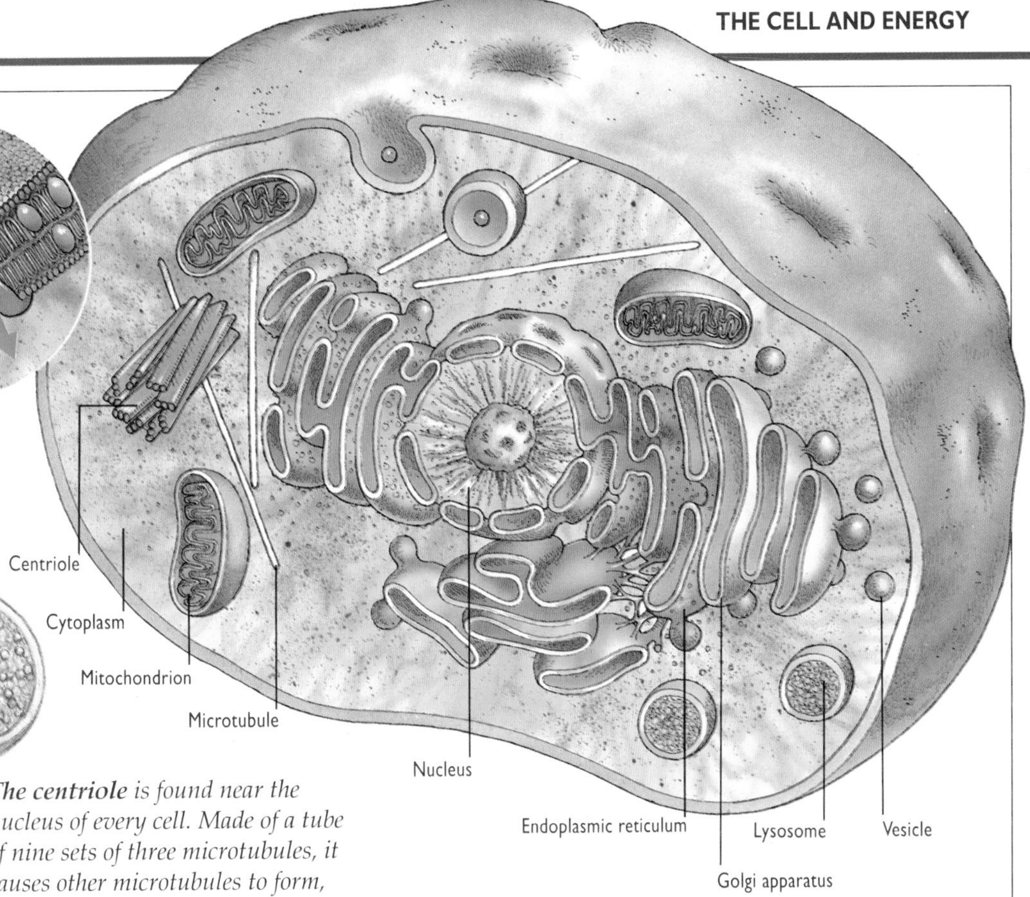

ysosomes are tiny
herical bags of digestive
zymes. If the bag merges
ith another cell structure,
can release its enzymes
break down and digest
bstances in the other structure.
o when, for instance, a phagocytic
ngulfing) cell eats a bacterium,
zymes from lysosomes help
estroy the bacterium.

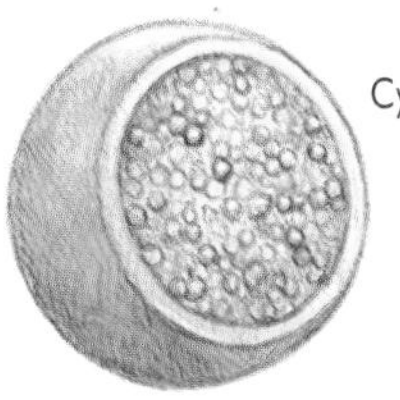

The centriole is found near the nucleus of every cell. Made of a tube of nine sets of three microtubules, it causes other microtubules to form, including the spindle, which acts as scaffolding that gathers chromosomes together at cell division.

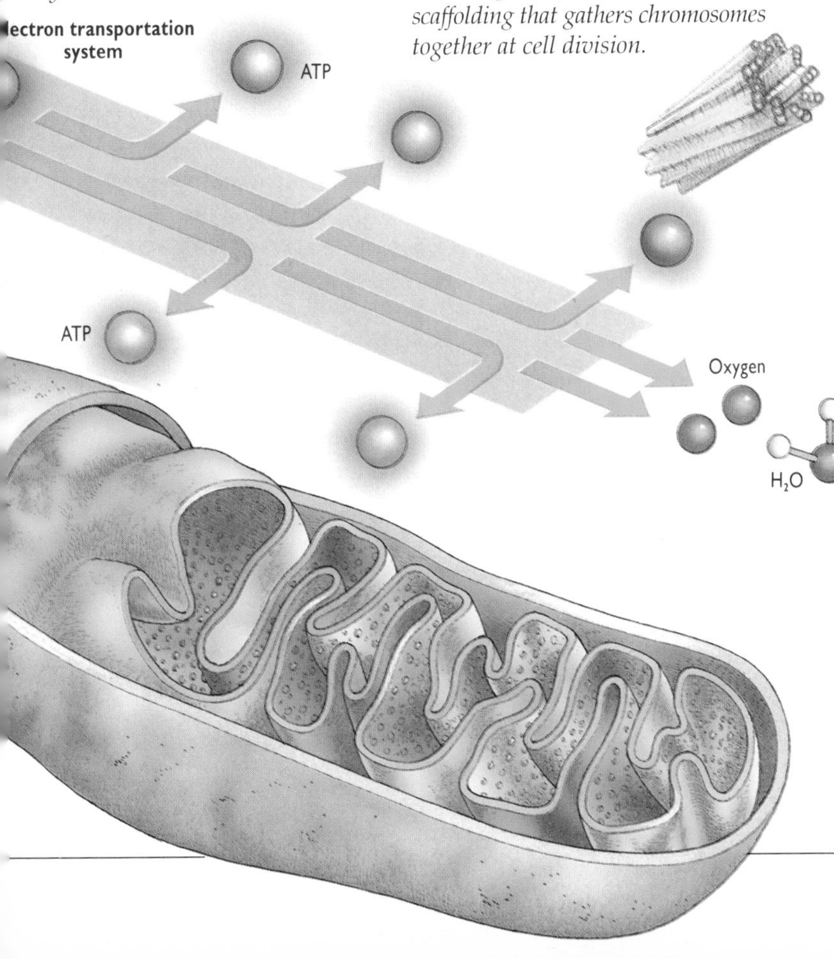

The cell is surrounded and held together by the cell, or plasma, membrane. Its shape is maintained by an internal scaffolding of different types of fibers and struts, the most prominent of which are long hollow rods called microtubules. Somewhere near the center of the cell is the nucleus – a spherical bag enclosed by a double membrane like the cell membrane. Inside the nucleus are the elongate chromosomes that carry the cell's genes. The rest of the cell outside the nucleus is known as cytoplasm. Apart from the microtubules, it contains a large number of different organelles – the miniature "organs" of the cell itself.

Mitochondria (left) are organelles (small, functional structures in the cell) with a double membrane coat. Their main job is to convert the energy in the bonds between the atoms of nutrients such as glucose into the high-energy bonds in ATP (adenosine triphosphate), the body's power molecule.

The outer membrane of a mitochondrion is smooth; the inner one is folded into many shelflike projections, or cristae. Many thousands of minute knobs of enzymes (biological catalysts made of protein) stud the expanded inner surface area of the cristae. ATP is made on these knobs when the molecule ADP (adenosine diphosphate) has a phosphate ion attached to it.

See also

ENERGY
▶ The speed of life 92/93

▶ Cells at work 114/115

SUPPORT AND MOVEMENT
▶ The living framework 16/17

▶ Muscles at work 26/27

CONTROL AND SENSATION
▶ Key chemicals 42/43

CIRCULATION, MAINTENANCE, AND DEFENSE
▶ Routine replacement 138/139

▶ Waste disposal 130/131

REPRODUCTION AND GROWTH
▶ Language of life 158/159

▶ Building bodies 160/161

▶ The sexual advantage 162/163

Cells at work

The majority of the body's cells have identical genes, yet each cell type has its own specific structure and shape.

All the different cell types use energy for different reasons to perform their major functions. Yet almost all, whether skin cells, nerve cells, or muscle cells, use the same energy-rich molecules, such as ATP (adenosine triphosphate), during energy exchange. Although the various cells' reasons for using energy are diverse, there are three major patterns of energy use common to many of them.

The first of these is in the building up of complex molecules from simpler ones – a secretory cell, for instance, uses most energy in manufacturing its secretory products from chemical sub-units and then exporting them from the cell. In the pancreas there are two main types of secretory cells: one type makes the hormone insulin and releases it into the blood, the other synthesizes digestive enzymes which are passed into the duodenum.

The second pattern of energy use concentrates on producing changes in cell shape; inside muscle cells, highly ordered arrays of actin and myosin fibers slide past one another to produce muscle contraction. The myosin molecules need an energy source of ATP to achieve this. The third is in the pumping of materials across cell membranes, which may involve taking up useful materials such as nutrients into a cell, pushing out unwanted ones, or in the case of excitable cells like nerves, maintaining an electrical potential difference between the inside and the outside of the cell. For example, a cell in the wall of the proximal convoluted tubule of a kidney nephron expends great amounts of energy pumping sodium ions from the newly formed urine back into the bloodstream to maintain the correct level of sodium in the blood.

By using energy *to "pump" charged particles, or ions, from the inside to the outside of the cell membrane, and vice versa, many types of cells generate tiny differences in voltage, or potentials. Nerve cells, for instance, have ion pumps that create a potential across their membranes so they are ready to pass on nerve impulses. An impulse is a sudden change in the distribution of ions between the inner and outer surfaces of the cell's membrane. After an impulse has passed, ions are pumped across the membrane to restore the potential.*

When a cell needs *to take up a useful substance such as a nutrient from its surroundings, it has two alternatives: it can do it without use of energy, by diffusion; or in an energy-using way by active transportation. If the substance the cell needs is at a higher concentration outside the cell than it is inside, and if the cell membrane is permeable to it, the substance moves in by diffusion. But with the concentrations reversed, the cell must use energy to work a molecular pump in its membrane to pull the desired molecule in.*

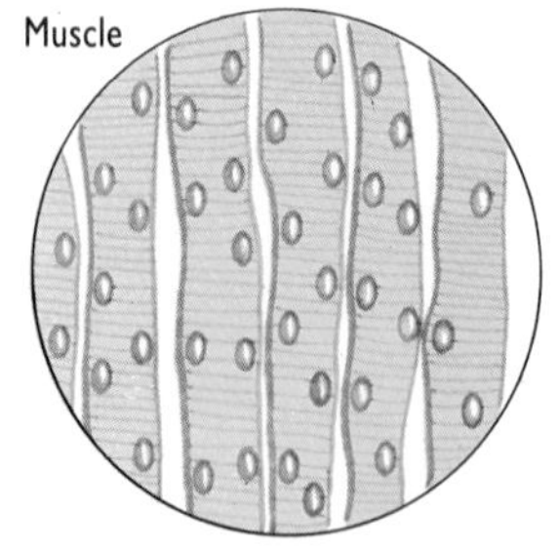

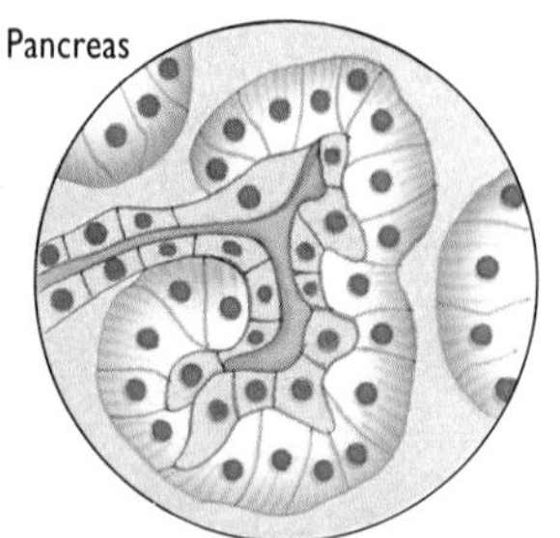

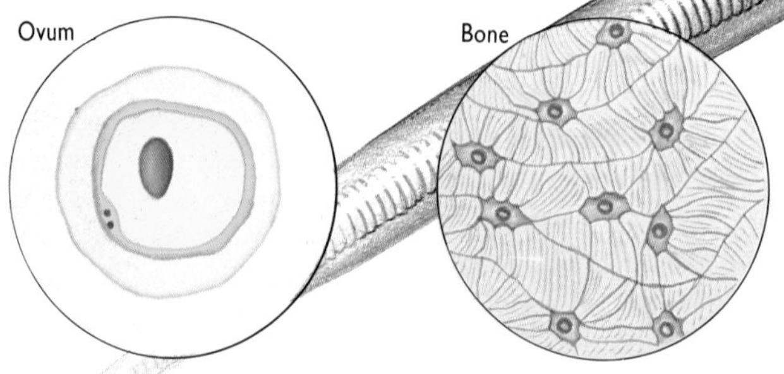

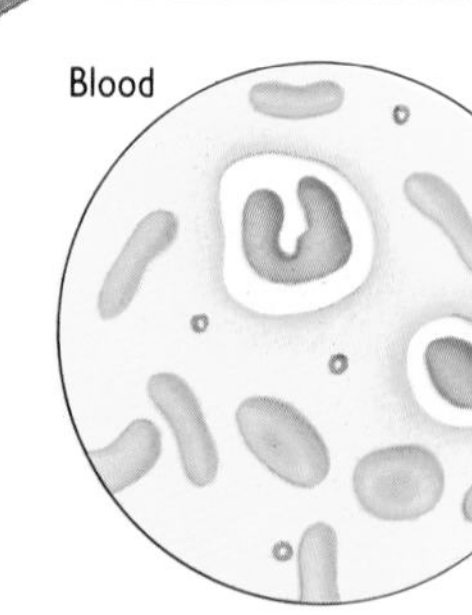

THE RANGE OF CELLS – THE GENOME IN ACTION

Almost every single cell among the billions in a single human body has the same genome – the same set of genes. The only obvious exceptions to this, at least in a healthy body, are the sex cells – eggs and sperm – each of which contains only a random 50 percent of a person's total genome. But if all the cells have the same genes, why are there so many different sorts of cells in the body? How, for instance, can the same genes form products as disparate as elongated, yard-long nerve cells, free-floating rounded white blood cells only the minutest fraction of an inch across, and star-shaped bone cells embedded in the solid mineral matrix of bone in tight-fitting chambers? The answer is to be found in the unfolding pattern of cell differentiation during the

***human sperm burns up energy on its mission** to nd an egg and fuse with it to fertilize it. To do this, the perm must swim through the cervical mucus at the top f the vagina, through the fluids in the uterus, then along Fallopian tube. The minute sperm cell thus has to swim everal inches through the female reproductive tract, hich it achieves by beating its tail, or flagellum.*

Enzymes in the sperm's mitochondria break down ugar fragments to form ATP, which is then used to enerate shape changes – and thus movement – in the agellum. ATP-powered bending movements in the laments of the tail drive the sperm forward in a ead-first direction. The base of the region ontaining the mitochondria – the owerhouse of the cell – is arked by a structure alled the annulus.

Nucleus

Annulus

Mitochondria

nts

***Cells use energy** to generate shape changes. The most obvious examples of this are the movements generated by muscles where individual cells contract in concert. Other cells that use energy to change shape are those with cilia and flagella – cell appendages that wave to and fro either to move the cell (flagella) or move liquid past the cell (cilia). In all cases, ATP is used as the immediate energy source.*

***To manufacture** a complex molecule, a cell needs an energy source. So when, for instance, a protein is built by linking its individual component amino acids together with peptide bonds, this requires an input of chemical energy. Only reactions that break chemical bonds and form simpler molecules out of complex ones release energy rather than taking it in – energy from such reactions powers protein building.*

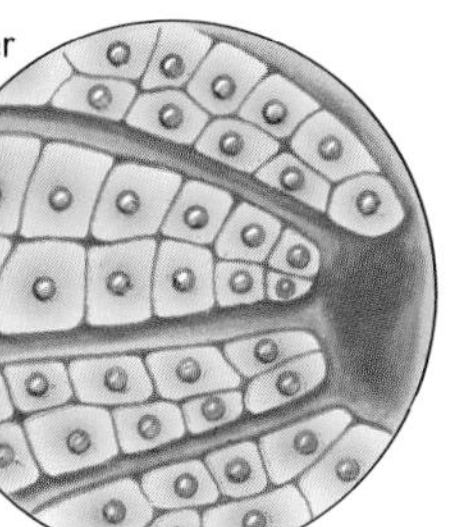

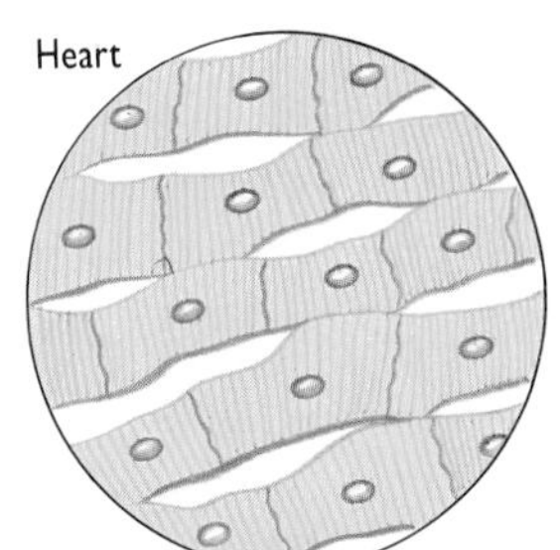

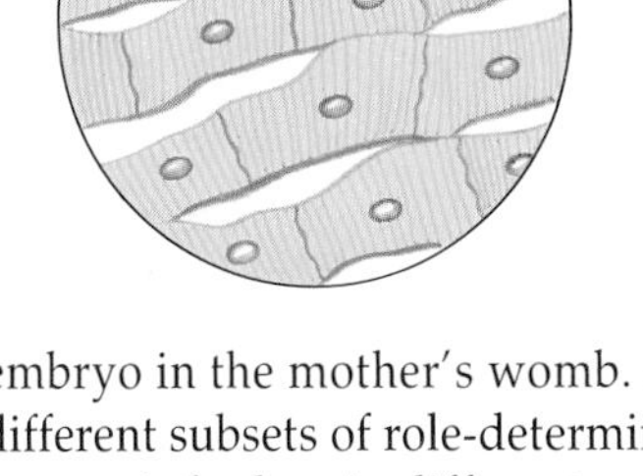

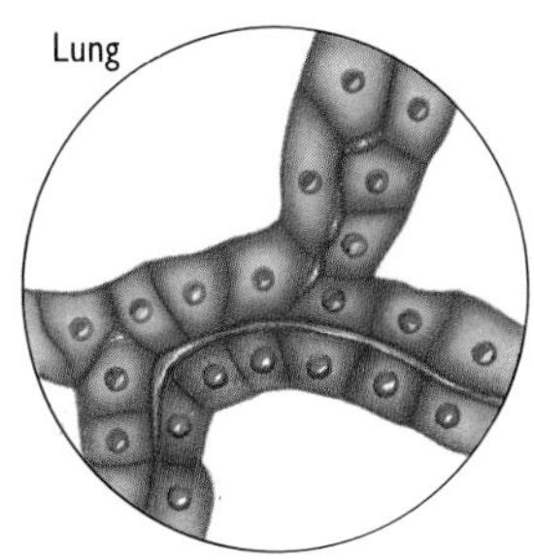

Kidney

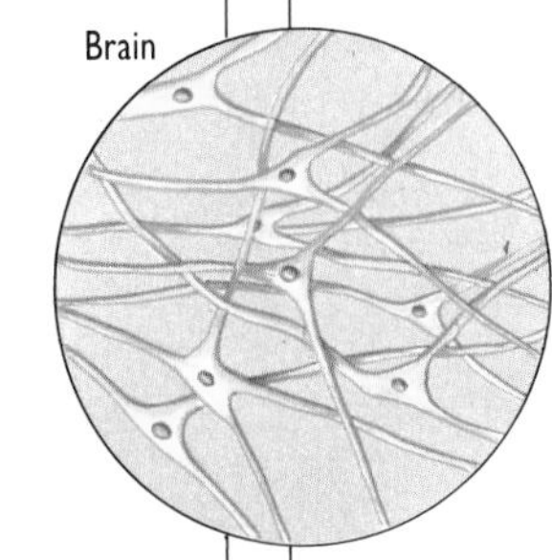

evelopment of an early embryo in the mother's womb. As evelopment progresses, different subsets of role-determining enes in the total genome are switched on in different groups f cells, turning them into different cell types. Thus there are eart muscle cells that spontaneously and rhythmically ontract and lining cells of the alveoli of the lungs which form the respiratory surface for gas exchange. This story of different end points emerging from the same genome is no more surprising than the fact that a cabbage white butterfly caterpillar, the pupa it turns into, and the butterfly that emerges from the pupa all have the same genes – same genes, different gene sets in operation.

See also

ENERGY
▶ The cell and energy 112/113

SUPPORT AND MOVEMENT
▶ The living framework 16/17

▶ Muscles at work 26/27

▶ The heart's special muscle 28/29

CONTROL AND SENSATION
▶ Down the wire 46/47

CIRCULATION, MAINTENANCE, AND DEFENSE
▶ Blood: supplying the body 126/127

▶ Inside the kidney 134/135

REPRODUCTION AND GROWTH
▶ Language of life 158/159

▶ The growing plan 170/171

The Cow-Pock — or — the Wonderful Effects of the New Inoculation! — Vide. the Publications of ye Anti-Vaccine Society.

SEIKO

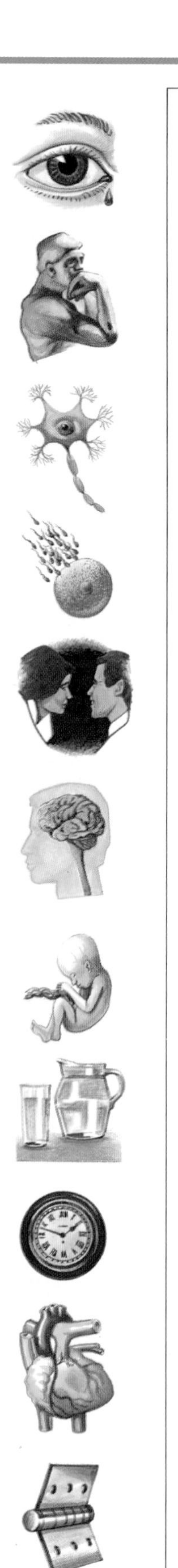

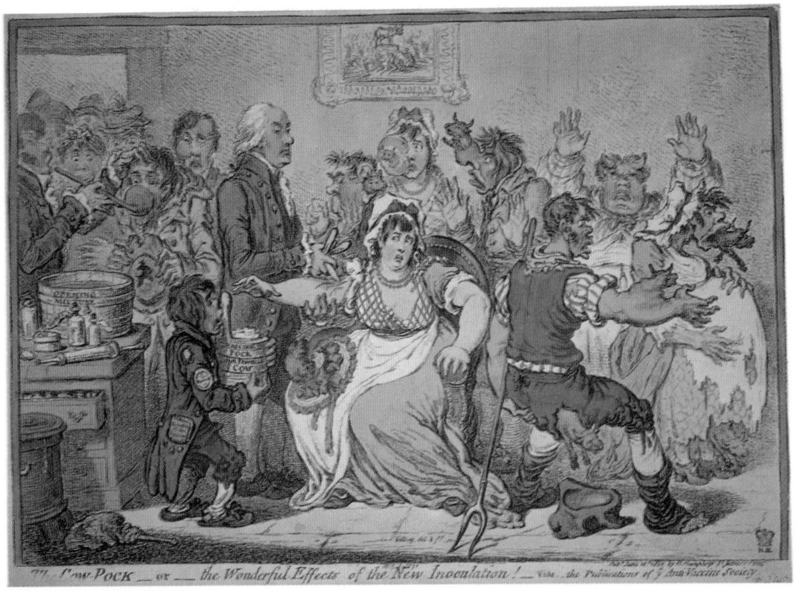

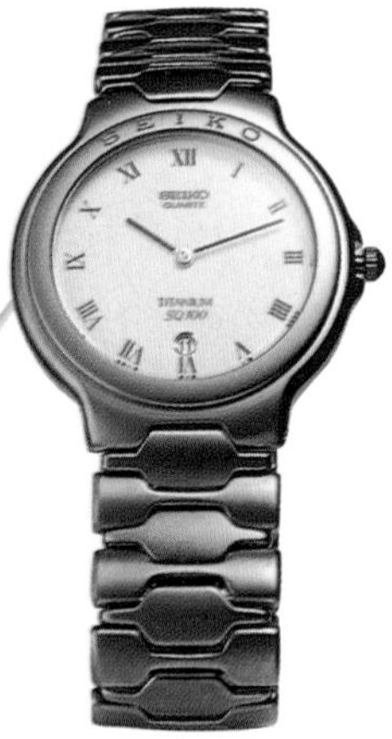

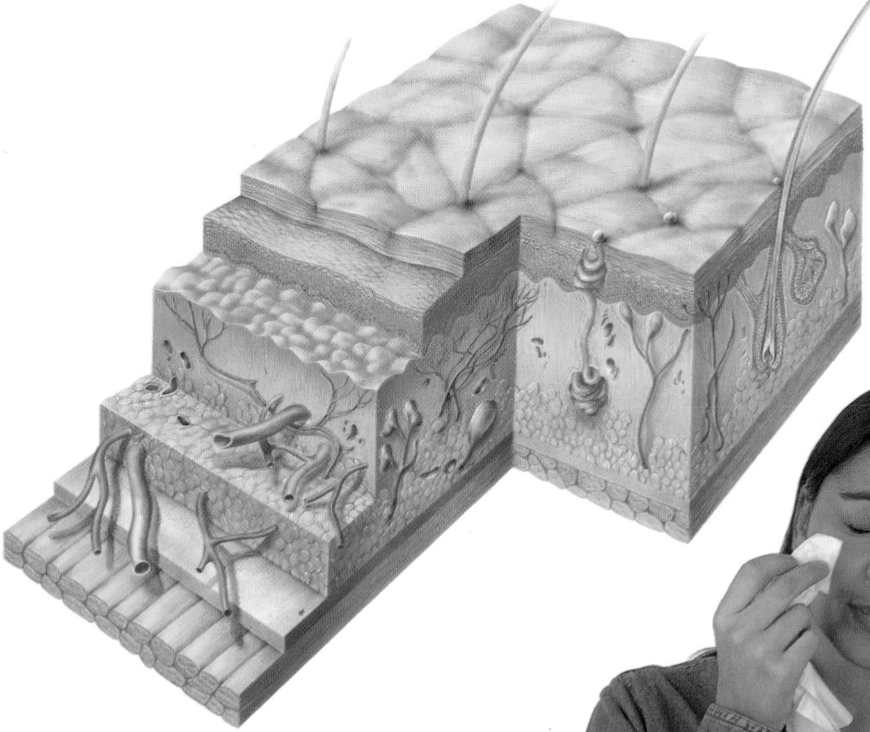

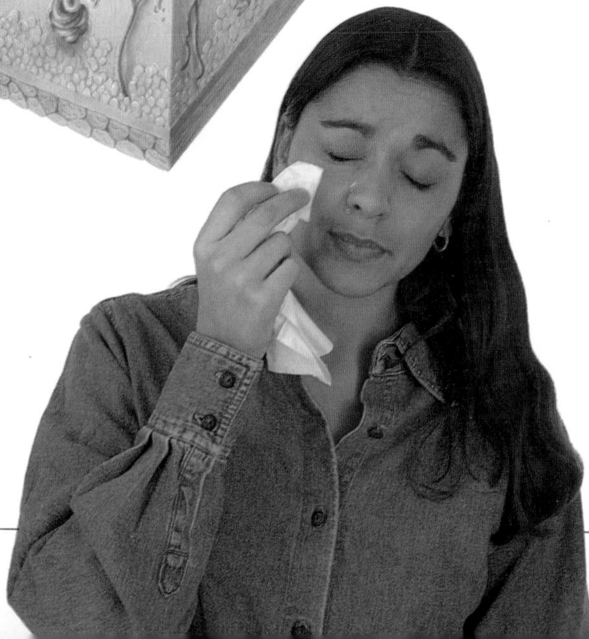

Circulation, Maintenance, and Defense

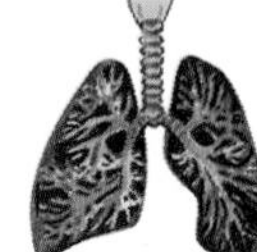

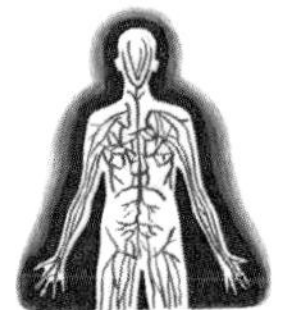

The blood circulating in veins and arteries is not the only transportation system in the body. It is, perhaps, the most important, but there are other types of transportation at work, including systems that mop up fluid between tissues and those that send wastes out via the body's disposal routes.

Other systems maintain the structure of the body, routinely replacing and repairing cells that die due to wear and tear. And still others are concerned with its defense; they exist at both macro and micro levels – from the skin barrier to the tiny hunter-killer cells borne in the blood. And they can deal with almost any invading germ, reacting fast to mount a devastating counterattack.

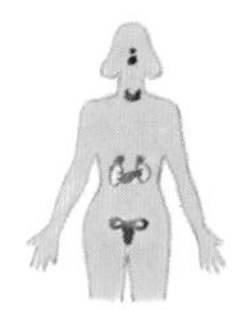

Left (clockwise from top): *anti-vaccination propaganda; titanium – safe inside; killing tears; skin, the outer barrier.* **This page (right)**: *the microscopic marvels of the immune system;* **(left)** *a bacterium – know the enemy.*

Transportation systems

Much of the activity in the body depends on moving materials around it, as well as in and out.

If an animal is small enough – only a fraction of an inch across – it can depend on purely passive physical processes to solve its internal transportation problems. The main process is diffusion, in which molecules – which are always in motion, jostling each other about – move, on average, from a region where they are in high concentration to areas of lower concentration until the concentrations even out. Over the distances in tiny animals, diffusion works well enough to let oxygen in and carbon dioxide out without the need for a circulatory system. And nutrients and other key molecules can move into the creature – and around inside it – by diffusion alone.

Any large animal, though, requires specialized transportation systems to move materials and cells around when simple diffusion cannot do the job. These systems can be closed loops, as in the circulatory system which allows blood to flow continuously around the body through the same tube; unidirectional, as in the urinary system; or two-way, as happens with the in-and-out movement of air in our lungs.

Whatever the direction of the movement or the type of channel, there are only a few driving forces for the transportation of bulk materials. The main mechanisms are muscle contraction, secretion pressure, the beating of cilia (tiny hairs), and the force of gravity.

Breathing and the movement of food through the digestive tract use muscle action. The production of a secretion in a gland forces the secretion to move out from the gland like toothpaste being squeezed from a tube. For instance, the sebaceous (oily) secretions in skin are squeezed onto the skin and hair by the pressure of secretions behind them.

The waving to and fro of microscopic hairs on cell surfaces, or ciliary beating, can move liquids and objects. The mucus lining the airways of the lungs is moved along like a conveyor belt by cilia, taking airborne pollution such as dust particles out of the lungs. Similarly, rows of ciliated cells in a woman's Fallopian tube move an egg along the tube to where it can be fertilized. Gravity is a driving force for internal transportation. For example, part of the flow of blood back to the heart through veins in the head and neck region depends on gravity.

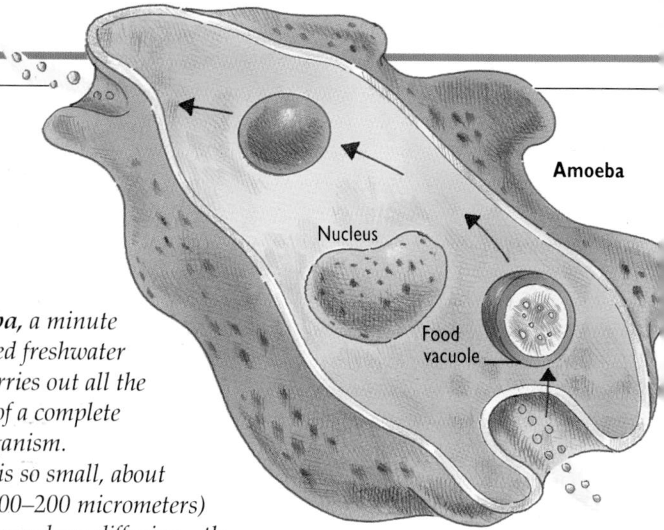

An amoeba, *a minute single-celled freshwater animal, carries out all the functions of a complete animal organism.*

Since it is so small, about 1/150 inch (100–200 micrometers) across, it depends on diffusion – the passive movement of molecules from areas of high to low concentration – to take oxygen from the surrounding water into the cell and move carbon dioxide out. Food is taken in by tucking the cell membrane around particles, to form food vacuoles. They move around the cell while the food is digested. Nutrients then diffuse from the vacuole into all parts of the cell.

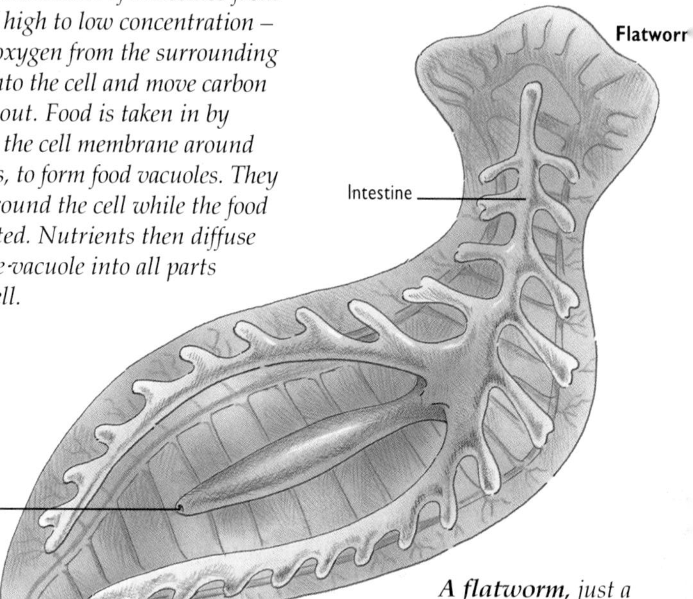

A flatworm, *just a fraction of an inch long, has no specialized respiratory or blood transportation systems. The necessary gas exchanges and nutrient movements through the body occur by diffusion. It does have a digestive tract, though, with a single opening that has to operate as both a mouth and an anus.*

An earthworm *is several inches long and has a more complex internal setup than a flatworm. For instance, it has a range of transportation systems, in particular a unidirectional digestive tract with a mouth at one end, an intestine in the middle, and an anus at the other. It also has a true circulatory system with multiple pumping hearts to move blood around the body.*

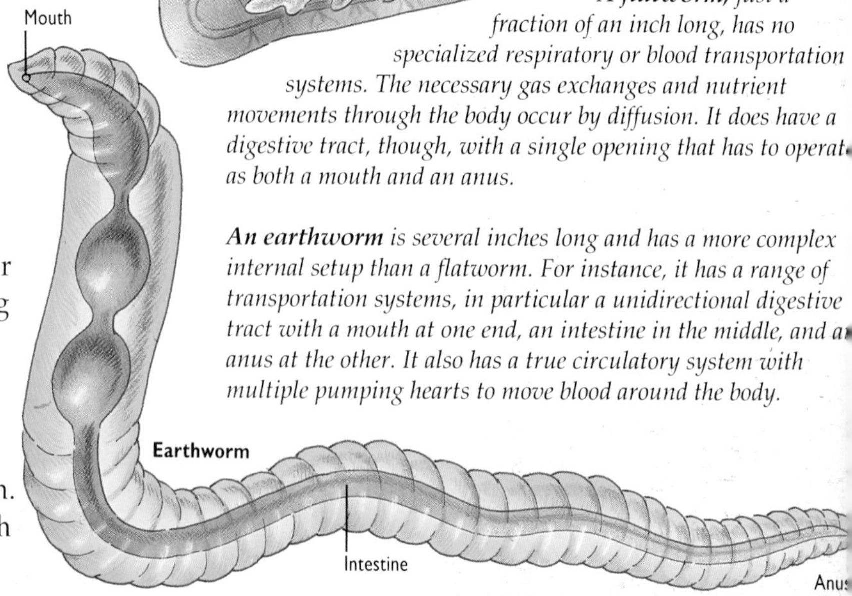

WAVING PUMPS

The sea anemone belongs to the major division of animals known as the coelenterates – invertebrates (spineless creatures) with saclike bodies and just one opening to the gut. It has no circulatory system, so depends on diffusion for the exchange of gases, such as oxygen and carbon dioxide, and the absorption of nutrients. The molecules diffuse through easily since its body wall is extremely thin.

To keep its shape, the sea anemone relies on a system similar to the transportation system in the human lungs. When it takes in water and closes its opening, grooves full of cilia (tiny waving hairlike cell extensions) down its length pump water into the anemone's gut space to bulk the animal up into its normal shape.

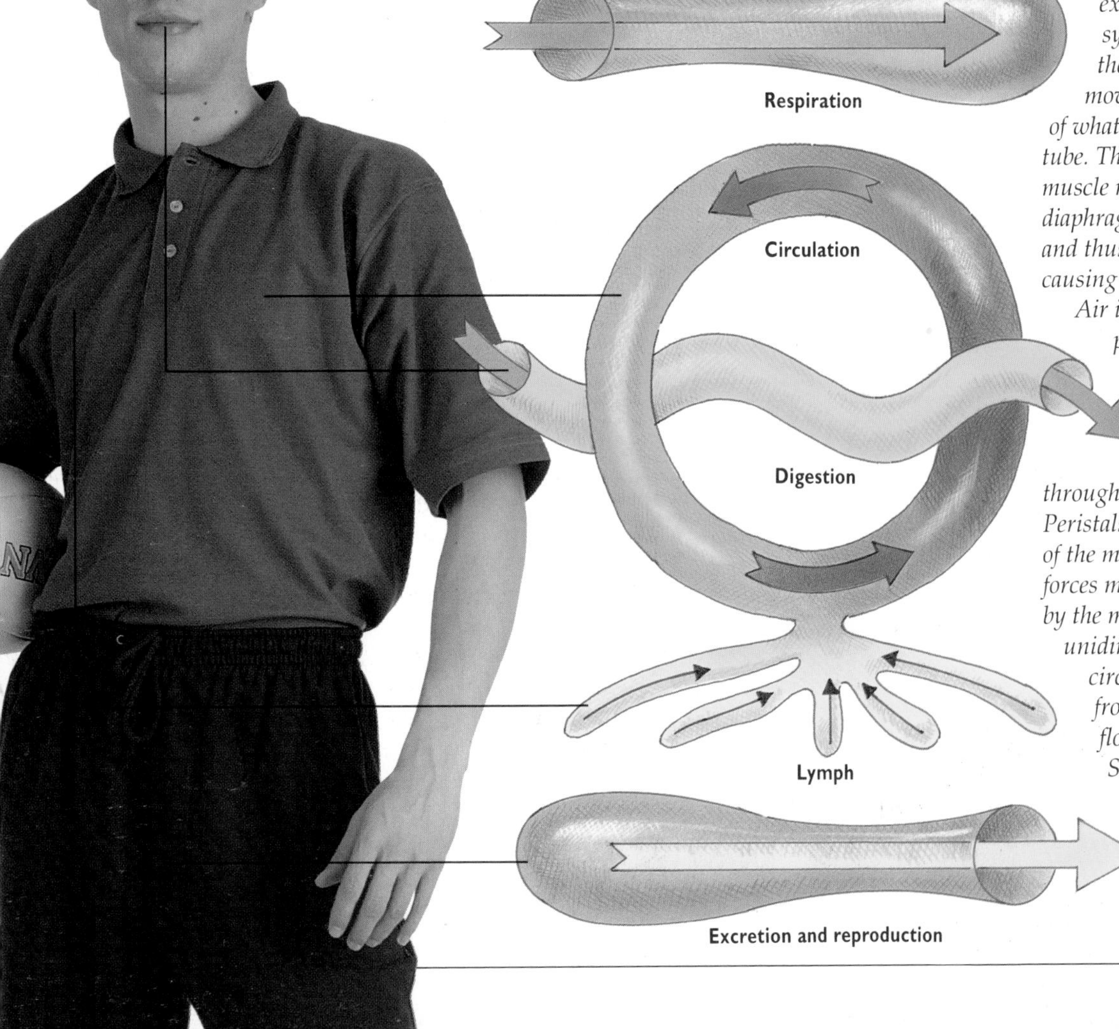

Four types of flow patterns *exist in our transportation systems. Breathing shows the to and fro, or oscillatory, movements of air in and out of what is, in effect, a closed-end tube. The air is pumped in by muscle movements that lower the diaphragm, increase lung volume, and thus reduce internal pressure, causing air from outside to rush in. Air is forced out along the same path when the diaphragm moves up again. The digestive tract shows the unidirectional movement of food through an open-ended system. Peristalsis – sequential contraction of the muscles lining the tract – forces matter along. Blood, pumped by the muscular heart, moves unidirectionally around a closed, circular system. Lymph fluid from the lymphatic system flows into the blood. Secretions such as excreted sweat, urine, and reproductive products move in a single outward direction along their own specialized ducts.*

See also

CIRCULATION, MAINTENANCE, AND DEFENSE

- The living pump 120/121
- Growing heart 122/123
- In circulation 124/125
- Blood: supplying the body 126/127
- The body's drain 128/129
- Waste disposal 130/131
- The water balance 132/133
- Inside the kidney 134/135

ENERGY

- A deep breath 108/109
- Cells at work 114/115

REPRODUCTION AND GROWTH

- Being born 174/175

The living pump

Beating steadily in your chest, your heart sends blood coursing to all parts of your body.

A pump about the size of a clenched fist, the heart sends blood around the body's circulatory system – a closed, tubular system of arteries, veins, and capillaries. It is like a hot-water central heating system: the piping and radiators correspond to the blood vessels, while the water pump, which drives hot water through the pipes, is the equivalent of the heart. All living tissue has to be supplied with blood – it delivers nutrients and oxygen and takes away waste products. If blood flow to a part of the body stops – even for just a few minutes – that region will be seriously damaged and may die.

To force blood through the blood vessels, the whole heart contracts, squeezing blood out of its internal chambers. The heart is, in fact, a double pump with two pairs of pumping chambers. Specialized flap valves in the internal spaces of the organ mean that each side of the heart allows blood to move in only one direction.

Each side of this double pump has its own role. The right side pumps oxygen-depleted blood from the rest of the body to the lungs, where it gains a new oxygen supply. The reoxygenated blood then moves to the left side of the heart, from where it is pumped back to the remainder of the body. Part of the output from the left side supplies the heart muscle itself via the coronary arteries.

The four chambers of the heart, two atria and two ventricles, are built of muscle – cardiac muscle. It is a specialized type of muscle in which individual barrel-shaped muscle-cell sections

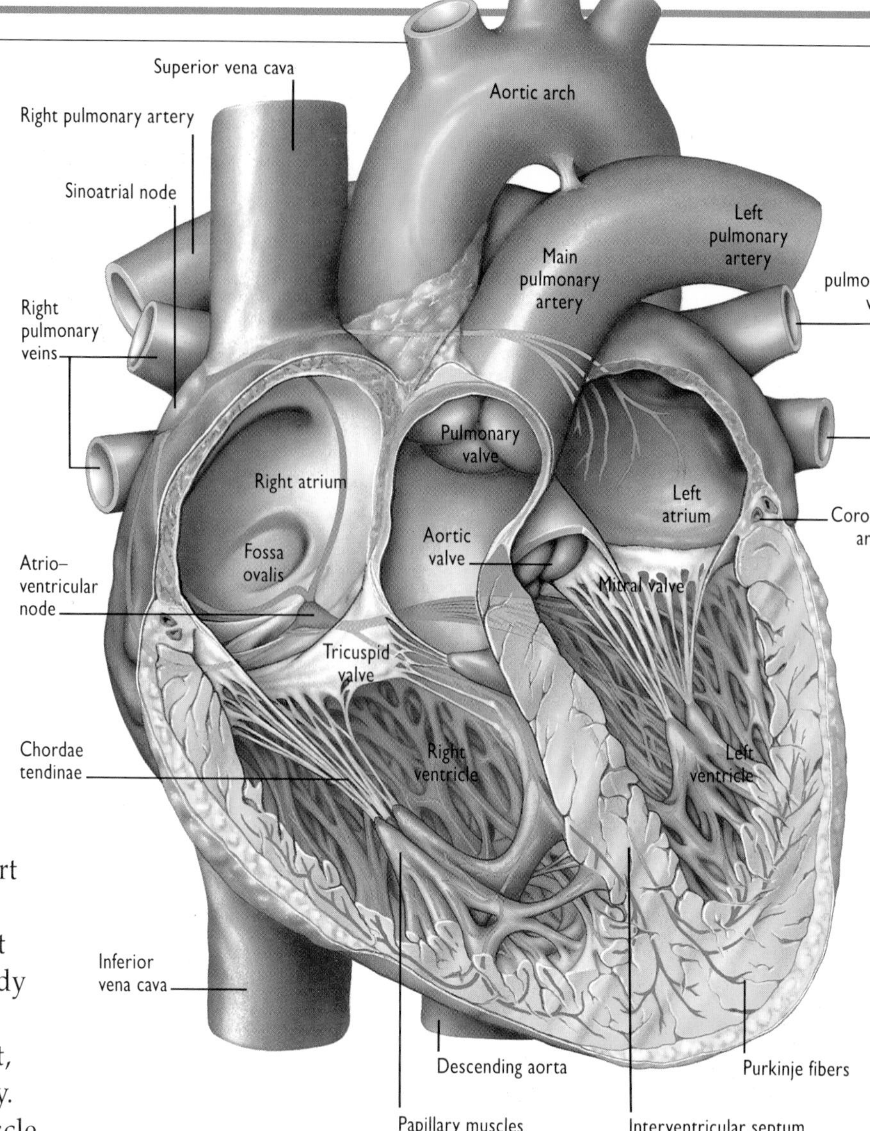

***The muscle** that makes up the left ventricle of the heart, which includes the interventricular septum (the wall separating the two lower chambers), is thicker than that of the right – the left ventricle has to be stronger since it pumps at higher pressure. Blood coming from the left ventricle is fed via the aorta – the body's main artery – to every part of the body, which demands high-pressure output. Blood from the right ventricle only has to go via the pulmonary arteries to the lungs.*

The sinoatrial node, atrioventricular node, and Purkinje fibers are part of the heart's own pacemaking and internal nerve signal transmitting apparatus.

***Like a foot pump,** the heart contains one-way valves to make sure flow is unidirectional. But unlike the heart, a foot pump does not have built-in muscles to help it force out its contents.*

re linked together end to end at junctions known as ntercalated disks. These tie the cells together and ensure hat there is synchronized contraction of heart muscle.

Certain regions of cardiac muscle are modified to form a ype of living pacemaker, which spontaneously beats out the hythm of the heart's regular contractions and transfers that empo to the rest of the heart muscle. These pacemaker regions re connected to the body's nervous system to make sure that he speed, strength, and capacity of the heart's pumping is natched to the body's moment-to-moment needs. So when ou run upstairs, your heart rate is increased to push more lood both to your muscles, which require glucose and xygen, and to the lungs to pick up oxygen more quickly.

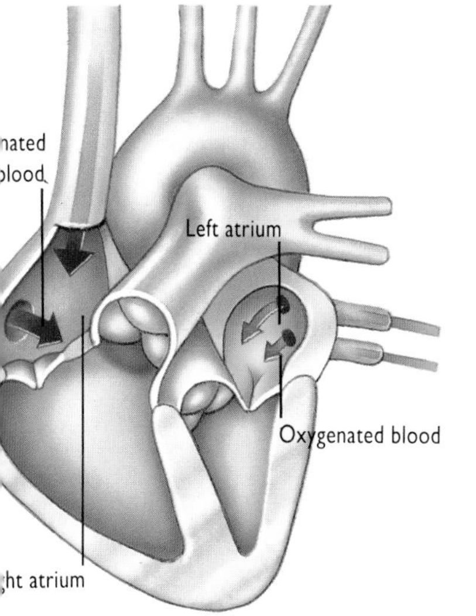

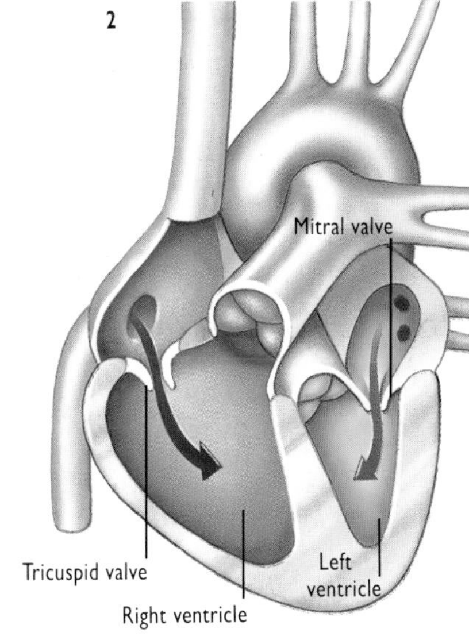

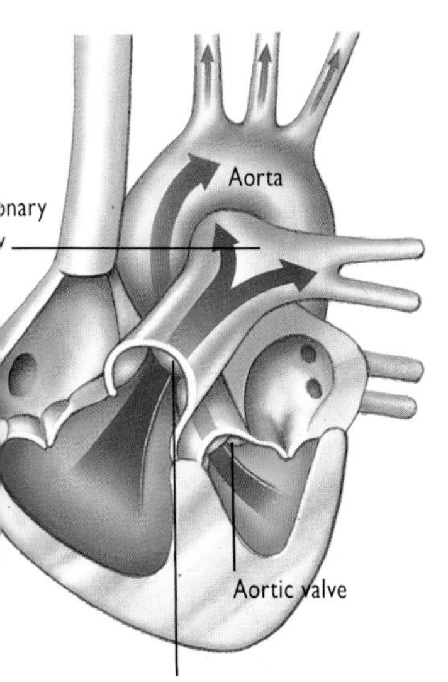

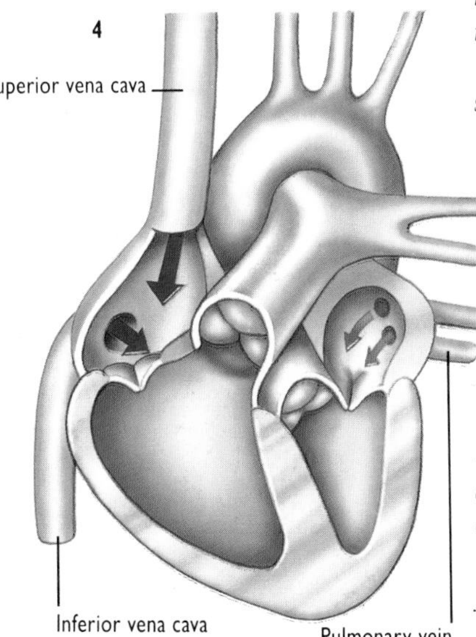

HEART FACT FILE

The heartbeat can be monitored by sounds (with a stethoscope). An unusual heart sound is called a murmur and is often caused by a valve leaking.

Ave. size of heart		4¾ x 3½ inches (12 x 9 cm)
Weight of heart		8–14 ounces (250–390 g)
No. of valves		4
Ave. beats per	minute	70
	day	100,000
	lifetime	2.5 billion
Ave. blood pumped per	beat	2½ ounces (75 ml) at rest
	day	1,980 gallons (7,500 liters)
	year	660,500 gallons (2.5 million liters)
Blood sent at rest to	heart	8½ ounces/min (250 ml/min)
	muscles	1¼ quarts/min (1,200 ml/min)
Ventricular pressure	left	120 mmHg on contraction
	right	20 mmHg on contraction

***Each beat of the heart** (contraction or systole) is a carefully orchestrated sequence of events. But even when the heart is not undergoing contraction, there is still plenty happening. In its resting phase (diastole), the two smaller chambers, the atria, are filling with blood* (**1**)*. The right atrium fills with "spent" deoxygenated blood returning from the body, the left with freshly oxygenated blood coming back from the lungs. A beat starts when muscle in the atria contracts from the top down, forcing blood into the main chambers, or ventricles* (**2**)*. Blood from the right atrium is squeezed into the right ventricle through the interconnecting tricuspid valve, which is forced open by the pressure in the atrium. Similarly, blood in the left atrium is forced through the mitral valve into the left ventricle.*

The contraction continues in the ventricles. The muscle starts to contract from the bottom, so blood in them is squeezed upward (**3**)*, but the two interconnecting one-way valves close so blood does not flow back into the atria. Blood from the right ventricle moves under pressure into the pulmonary artery through the pulmonary valve, and on to the lungs where it is oxygenated. Likewise, blood from the left ventricle is forced through the aortic valve into the aorta.*

When the power stroke finishes, the aortic valve and pulmonary valve slam shut to stop blood from flowing back into the ventricles. The heart enters its resting phase again when all four valves are closed. Meanwhile, the cycle starts again: the right atrium fills with blood from the superior and inferior venae cavae, and the left atrium fills with blood from the pulmonary veins (**4**).

See also

CIRCULATION, MAINTENANCE, AND DEFENSE
▶ Transportation systems 118/119
▶ Growing heart 122/123
▶ In circulation 124/125
▶ Blood: supplying the body 126/127

SUPPORT AND MOVEMENT
▶ The heart's special muscle 28/29

CONTROL AND SENSATION
▶ Control systems 36/37
▶ The chemicals of control 38/39

ENERGY
▶ Fueling the body 94/95
▶ Cells at work 114/115

REPRODUCTION AND GROWTH
▶ Being born 174/175

Growing heart

The first fully functioning system to form in a fetus is the heart and circulation.

By just 30 days after an egg has been fertilized by a sperm, there is a working blood-transporting system in the baby-to-be. It takes oxygen and nutrients to all parts of the fetus's 1/12-inch (0.2-cm) long body.

The final fully formed heart is a four-chambered muscular pump for moving blood around the body. Its right side receives deoxygenated blood back from the body and pushes it into the lungs where it is oxygenated. Oxygenated blood from the lungs then passes to the left side of the heart, from where it is pumped around the body to supply the tissues.

A heart with almost all of the adult components is in place about eight weeks after fertilization. There is one key difference, however – since the fetus is in the uterus, its lungs do not contain any air and so cannot oxygenate blood as they do after birth. The fetus's blood is oxygenated instead in the placenta and passes from there to the right atrium of the heart. Some of this blood passes to the right ventricle and then on to the lungs to provide them with oxygen and nutrients for growth and development.

Most of the oxygenated blood either shunts from the right to the left atrium through a hole between the two atria – a special fetal feature known as the oval window – or from the pulmonary arteries into the aorta, the main artery that comes from the left ventricle, through the ductus arteriosus, again specific to the fetus. Both these shunts enable oxygenated blood to be pushed around the developing fetus's body by the left ventricle, which is, in the fetus and in later life, the heart's most powerful chamber.

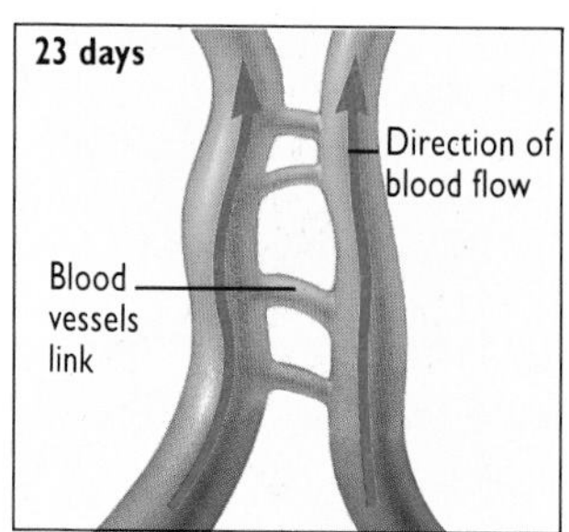

In the extraordinarily tiny embryo *23 days after it has been fertilized (shown* ***below*** *at actual size), two minute and primitive blood vessels have grown adjacent to each other. Cross linkages between these two vessels start to form. This is the first stage in the construction of what will become the heart.*

From this point on, over a period of about four weeks, this primitive structure develops to become a heart that is fully formed, albeit in miniature.

Actual size at 23 days

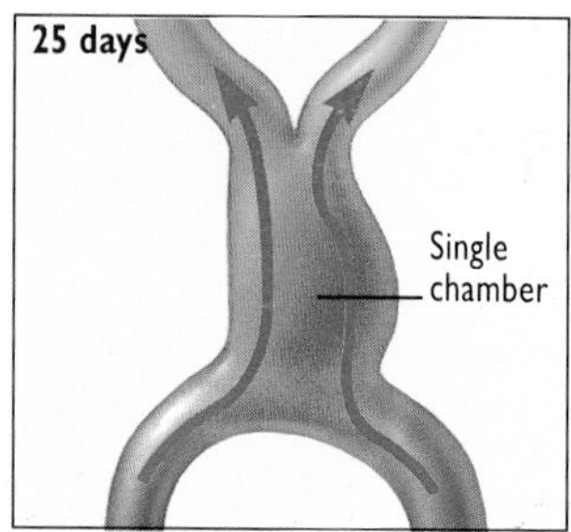

Shortly after the blood vessels form links, *they draw together and fuse to make a single chamber. This occurs about 25 days after fertilization.*

The blood vessels and the single-chambered "heart" that they form are surrounded by a sheath of muscle. This sheath will later grow into the muscular walls of the heart which, when they contract, will tirelessly pump blood around the circulatory system.

Actual size at 25 days

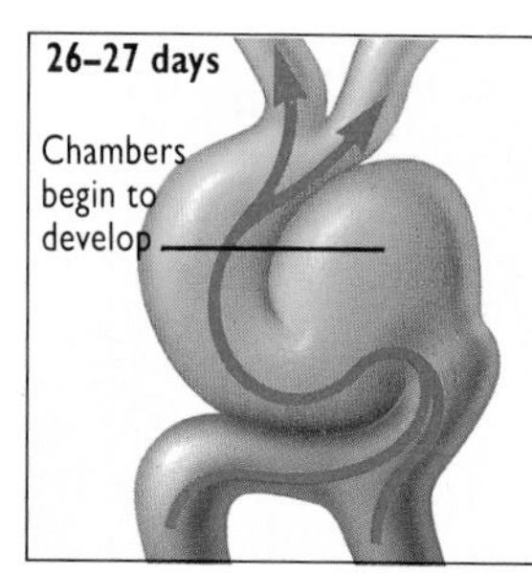

The single chamber, *made by the fusing of the two vessels, now bulges and twists, forming an S-shaped object that is less than 1/25 inc (0.1 cm) long. This marks the start of the process of forming more chambers.*

Even at this early stage, the proto-heart beats, or pulses. But since there are no valves in the heart at this time, the blood ebbs and flows and does not have the one-way flow of the fully developed organ.

Actual size at 26–27 days

WHEN THE GAPS DON'T CLOSE

There are three defects that occasionally occur in newborn babies' hearts. In all of them, deoxygenated blood is mixed with oxygenated blood and sent around the arterial system. Since the oxygen content of the blood is lowered, an infant with one of these defects may have a bluish tinge, because deoxygenated blood is blue and oxygenated is red. If the gaps fail to close on their own, all three conditions can be cured with surgery.

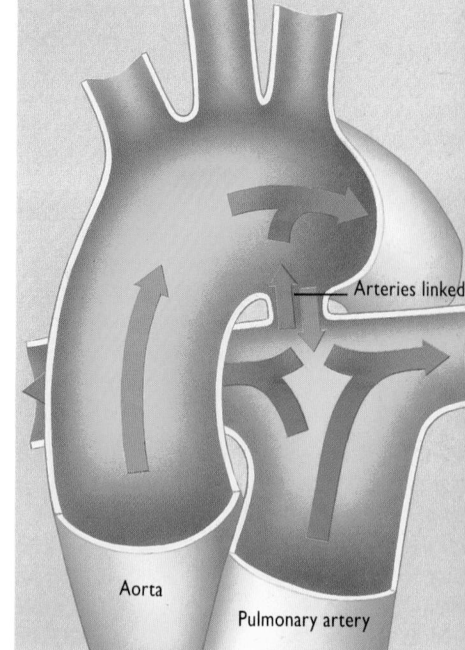

In the defect known as patent ductus arteriosus, *the channel that allows blood to bypass the lungs in the fetus does not close fully after the baby is born. This leaves a link between the aorta and the pulmonary artery and allows deoxygenated blood from the pulmonary artery to mix with oxygenated blood in the aorta.*

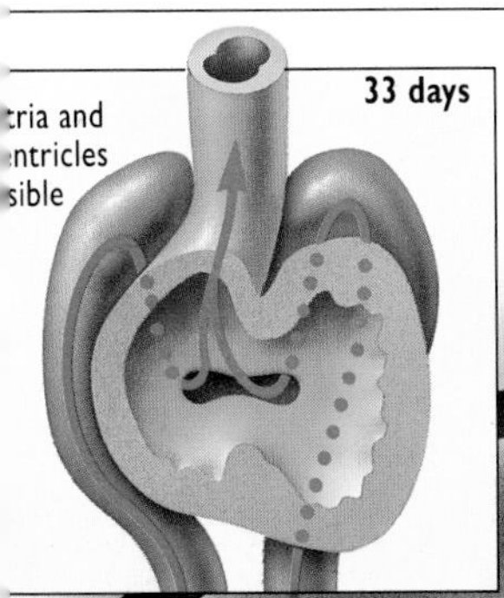

33 days

tria and ntricles sible

Actual size at 33 days

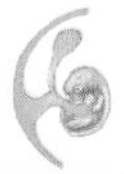

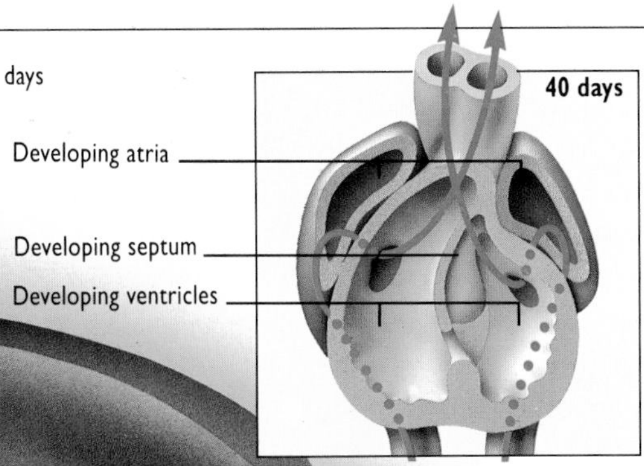

Developing atria

Developing septum

Developing ventricles

40 days

Actual size at 40 days

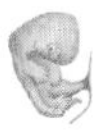

After about 33 days the atrioventricular canal is gradually split by ridges of tissue. By this time the proto-atrium has divided in two, making the left and right atria (**far left**).

By approximately 40 days, *or six weeks, a ridge or partition – the septum – begins to divide the right and left ventricles. At six or seven weeks, valves form that will ensure one-way flow of blood.*

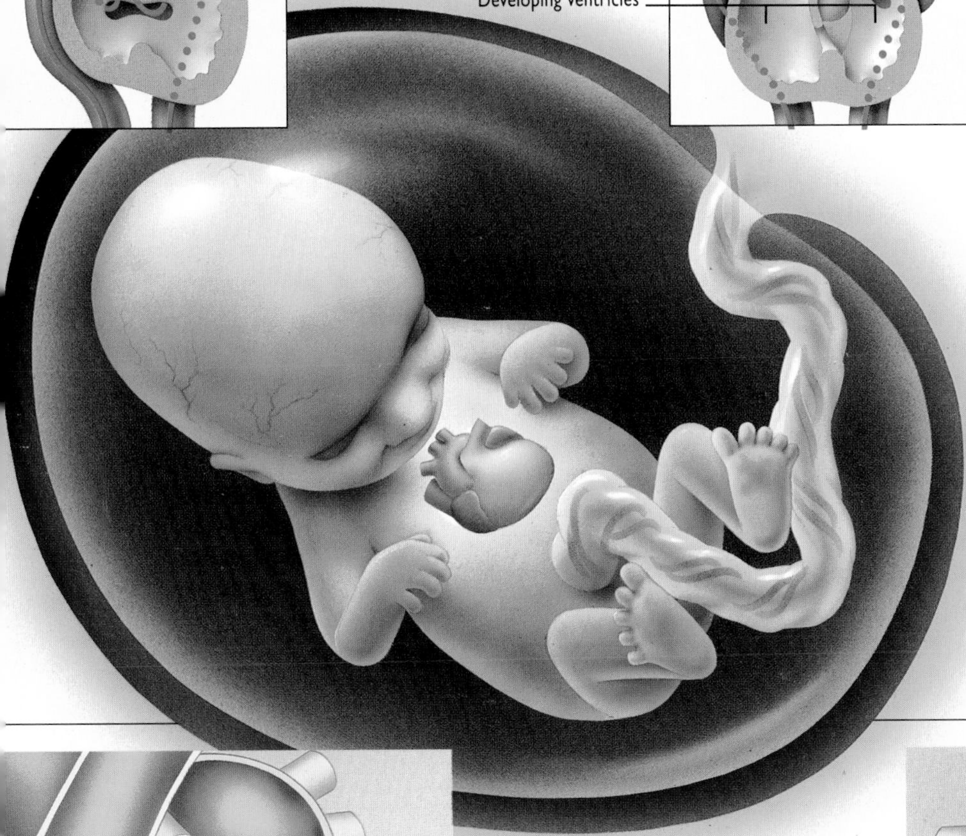

In a fetus at 55 days, *about eight weeks, the heart has four chambers working as twin, side-by-side pumps, and has a one-way flow. It now has the features of an adult heart, yet the fetus is barely 1 inch (2.3 cm) in length. One of the advantages of one-way flow is a greater delivery of blood to all parts of the fetus's rapidly developing body.*

55 days

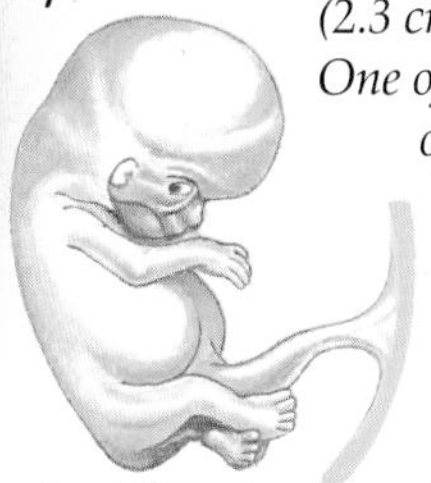

1 inch (2.3 cm)

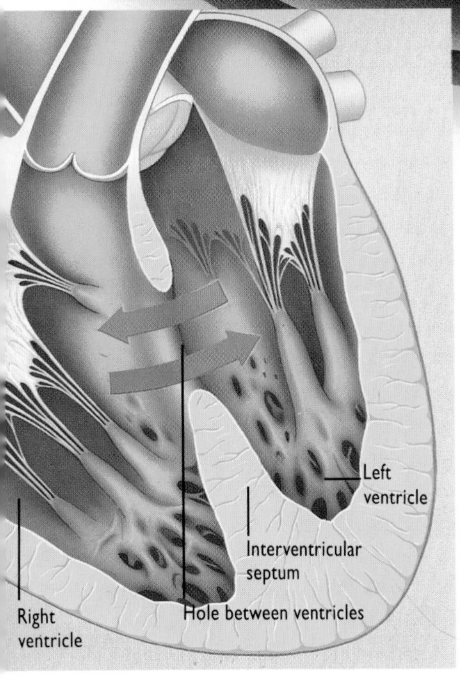

A "hole in the heart" (left)*, or ventricular septal defect, is the most common congenital heart disease. It is caused by failure of the interventricular septum, or partition, to grow sufficiently, leaving a hole between the ventricles. If it persists after birth, it means that deoxygenated blood from the right ventricle passes into the left chamber and is pumped around the body, starving the body of oxygen.*

Left ventricle

Interventricular septum

Right ventricle

Hole between ventricles

If the partition between the two atria *– the so-called oval window – fails to close, an atrial septal defect* (***right***) *occurs. Again, oxygenated and deoxygenated blood mix before being pumped around the infant's body.*

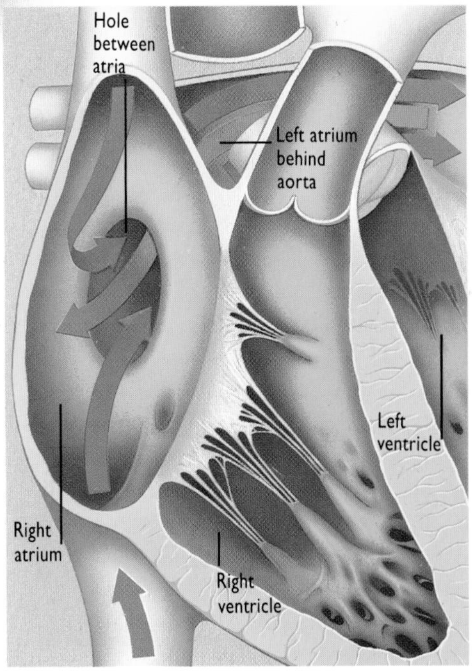

Hole between atria

Left atrium behind aorta

Left ventricle

Right atrium

Right ventricle

See also

CIRCULATION, MAINTENANCE, AND DEFENSE
▶ Transportation systems 118/119

▶ The living pump 120/121

▶ In circulation 124/125

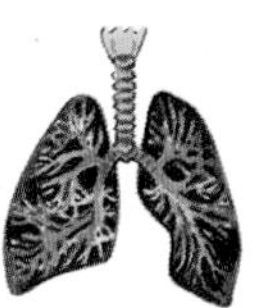

SUPPORT AND MOVEMENT
▶ The heart's special muscle 28/29

ENERGY
▶ The gas exchange 110/111

REPRODUCTION AND GROWTH
▶ The first month 168/169

▶ The growing plan 170/171

▶ Baby in waiting 172/173

▶ Being born 174/175

In circulation

Arteries deliver blood to the capillaries, where it does its work, then veins return it to the heart for recycling.

Blood truly circulates – it passes continuously around the body, returning to its starting point again and again. In one complete tour of the system, blood goes in turn around each of the body's two circulation tracks: from the heart, through the lungs and back to the heart; and from the heart to the rest of the body and back.

In both tracks thick-walled arteries carry blood under relatively high pressure away from the heart, and thinner-walled veins carry blood under much lower pressure back again. Linking the arterial and venous blood vessels are networks of increasingly fine arteries and veins (arterioles and venules) which eventually join up in a mesh of the thinnest blood vessels – capillaries. These are so narrow that red blood cells, which are only 1/3,600 inch (0.0007 cm) across, have to pass through them in single file. The capillary walls consist of a single, ultrathin layer of tissue made of flattened cells. The thinness of this layer allows wastes, nutrients, and gases, such as oxygen and carbon dioxide, to pass between the cells in the body's tissues and the blood in the capillaries. The capillary networks of all the organs are the exchange zones of the blood system – the rest of the system merely distributes the blood.

Surface veins in the skin of this exercising body builder stand out because they are inflated by the large volume of blood returning to the heart from his muscles.

CIRCULATION FACT FILE

The largest blood vessels are more than 3,000 times the width of the smallest blood vessels.

Diameter largest vein or artery	1–1 1/5 inch (2.5–3 cm)
smallest capillary	1/3,200 inch (0.0008 cm)
Percentage of blood in	
veins	75
arteries	20
capillaries	5
Amount of blood in circulation	
ave. man	5–6 quarts (5–6 liters)
ave. woman	4–5 quarts (4–5 liters)
Time taken for blood to circulate through	
lungs	4–8 seconds
body	25–30 seconds

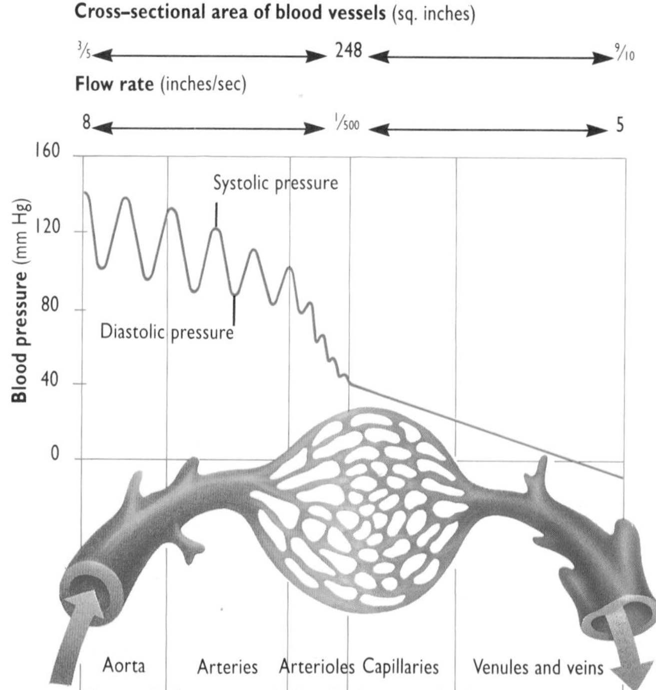

Blood pressure rises and falls in the arteries according to the contractions (systoles) and relaxations (diastoles) of the muscles of the heart's left ventricle, which force the blood around the body. Overall, this pulsating pressure drops with increasing distance from the heart until there are no variations in the capillaries and veins. Just before the veins return to the heart, pressure can drop so low that the blood must be sucked in by the atria. Blood thus moves fastest in the arteries and slowest in the capillaries, picking up speed again in the veins.

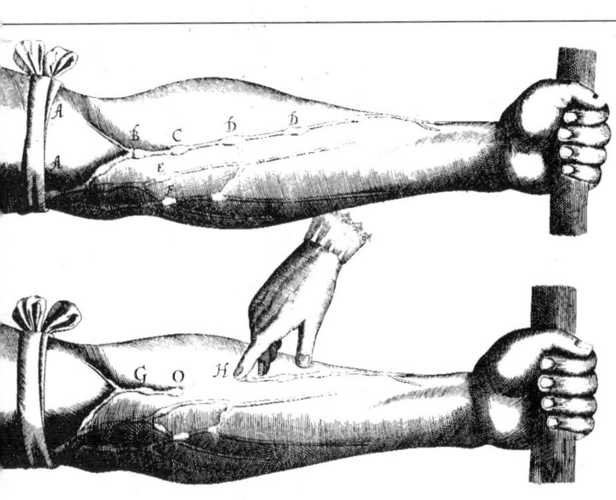

HARVEY AND THE VALVES

In 1628 British physician William Harvey (1578–1657) published *On the Motions of the Heart and Blood,* a small book in which he cleared up the hitherto mysterious nature of the flow of blood. He showed beyond doubt that blood circulated. These lithographs from the book show the effects of non-return valves in veins just beneath the skin in the arm.

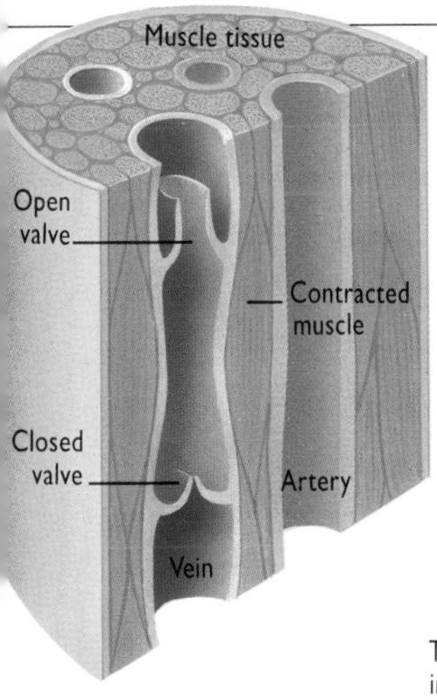

Muscle tissue
Open valve
Contracted muscle
Closed valve
Artery
Vein

The incredible highway of blood vessels *carries blood to and from almost every part of the body. Only specialized regions, for instance the cornea of the eye, the enamel of the teeth, and the dead outer layers of the skin, hairs, and finger- and toenails, are without blood vessels.*

Superior vena cava
Carotid artery
Jugular vein
Brachial artery
Cephalic vein
Inferior vena cava
Aorta
Femoral artery
Femoral vein

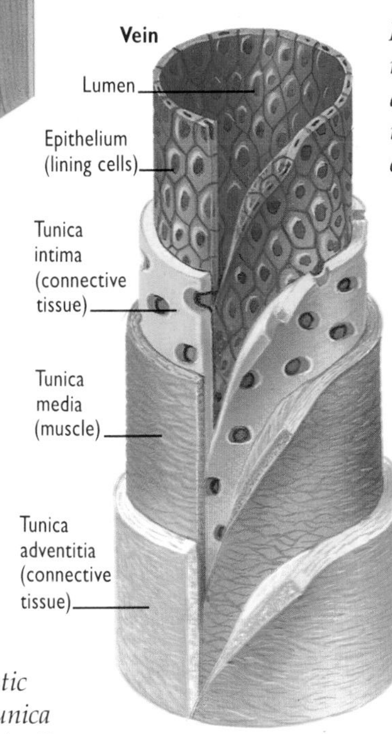

Many veins and arteries *run through muscles (**left above**). With low pressure in some veins, muscle contractions help to massage blood along the veins, aided by one-way valves which guarantee that the blood flows only in the correct direction.*

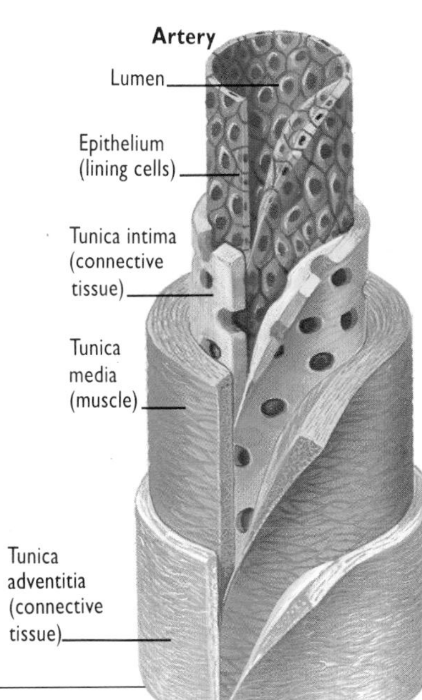

'eins and arteries *have ructures that are xtremely similar. Both ave an internal lining f flattened cells known s the epithelium. This is ıside the tunica intima, hich is made of onnective tissue and lastic fibers, while utside that is the tunica ıedia, which consists of mooth muscle cells and elastic ibers. The outer layer, the tunica dventitia, is made mainly of collagen nd elastic fibers. The main difference between veins nd arteries is that the tunica media is thicker in arteries; hese thicker walls are "designed" to cope with the higher ressures of blood from the heart.*

In many parts of the arterial network, the smooth ıuscles of the tunica media can contract under the ontrol of the nervous system to alter blood pressure.

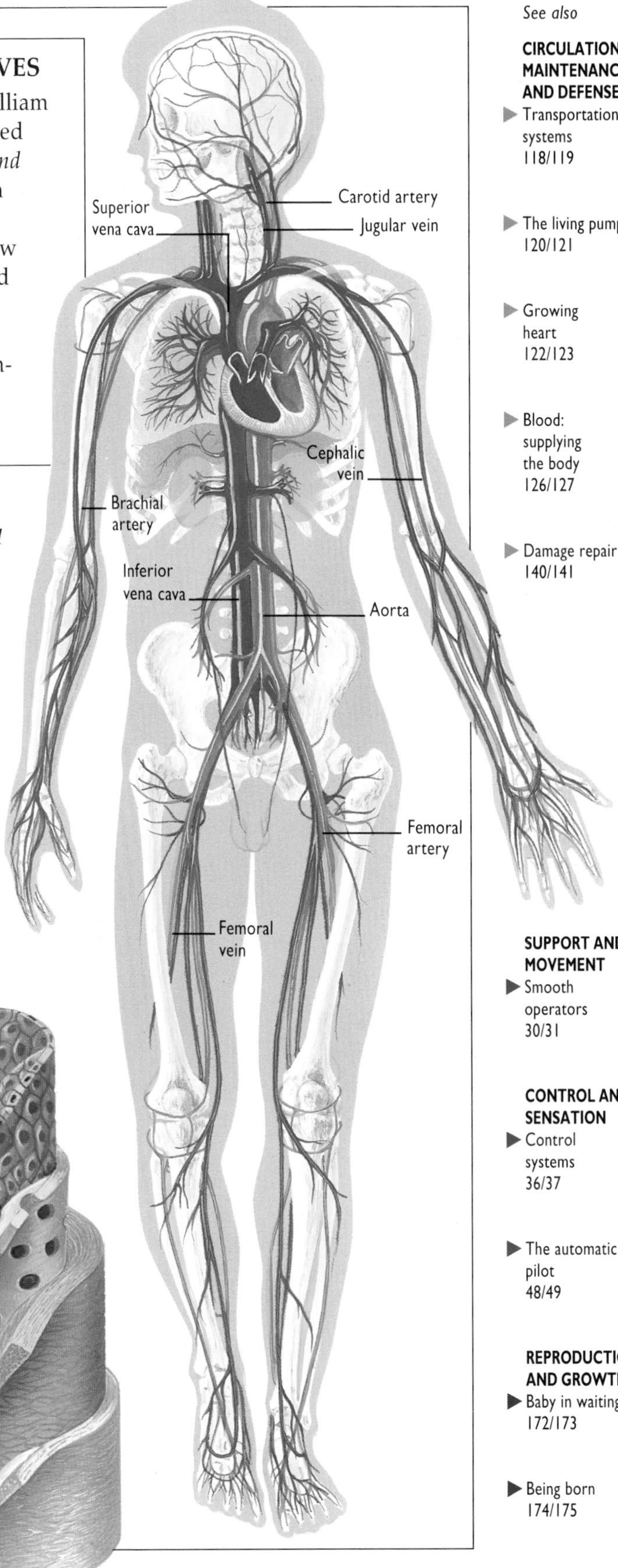

See also

CIRCULATION, MAINTENANCE, AND DEFENSE
- Transportation systems 118/119
- The living pump 120/121
- Growing heart 122/123
- Blood: supplying the body 126/127
- Damage repair 140/141

SUPPORT AND MOVEMENT
- Smooth operators 30/31

CONTROL AND SENSATION
- Control systems 36/37
- The automatic pilot 48/49

REPRODUCTION AND GROWTH
- Baby in waiting 172/173
- Being born 174/175

Blood: supplying the body

The whole body depends on a constant supply of blood to sustain, nourish, and protect its tissues.

Blood is the central transportation facility of the body and is as important to human activity as road, rail, and airline systems are to the workings of a modern industrialized society. Our blood is like a rich soup of many ingredients and, like a soup, the most plentiful ingredient is water. In blood the watery background, or plasma, has materials dissolved in it; it also has a wide range of free-floating blood cells suspended in it. Most of these are oxygen-transporting red blood cells, or erythrocytes. The others are white blood cells, or leucocytes, of many types, most of which have functions to do with body defense, repair, and immunity.

The blood is, in fact, a liquid tissue that, like any other type of tissue, has a variety of roles. For instance, its respiratory job is to carry oxygen to all body tissues. The oxygen is bound to hemoglobin, an iron-containing protein in the red blood cells. At the tissues, the hemoglobin gives up its vital gas in regions where oxygen concentration has become lowered. The waste product carbon dioxide, mainly in the form of bicarbonate, is carried by the blood away from respiring body cells for transporting to the lungs and removal in exhaled air. Blood also has nutrients dissolved in it, including sugars, amino acids, fatty acids, fats, minerals, and vitamins, and is a vehicle for transporting information in the form of hormones.

__Blood cells__ are continually made to replace those that die in the normal course of events. Red cells, platelets, and some white cells – all made in bone marrow – originate from one type of cell __(left)__, but differentiate and follow separate paths, some with many intermediary cell types.

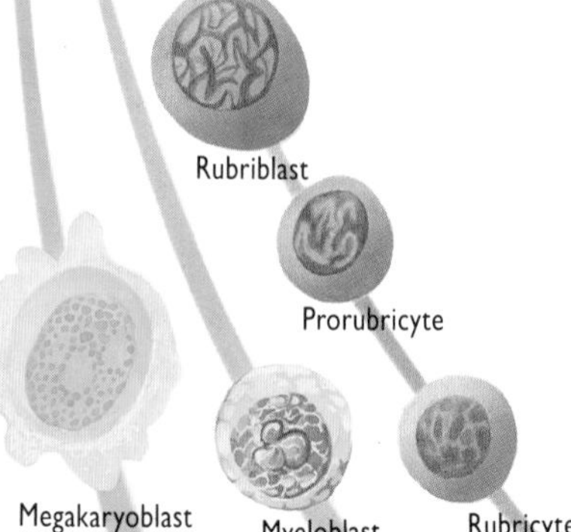

__The line__ that makes red blood cells (erythrocytes) has six steps. Along the way the cells repeatedly divide to increase number.

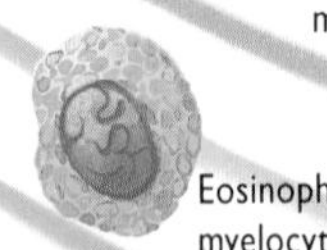

__One development line__ makes huge metamegakaryocyte cells. These cells fragment to give vast numbers of tiny platelets which have a crucial role in blood clotting.

BLOOD FACT FILE

Blood in arteries gets its redness from oxygenated hemoglobin in red cells; deoxygenated hemoglobin gives blood in veins a bluish color. Blood from a cut vein looks red because as soon as blood contacts air, hemoglobin takes in oxygen and becomes red.

Volume of blood	man	5–6 quarts (5–6 liters)
	woman	4–5 quarts (4–5 liters)
No. of red blood cells in body		25 trillion (25×10^{12})
Red cell dimensions*	width	$^1/_{3,600}$ inch (7 micrometers)
	thickness	$^1/_{12,500}$ inch (2 micrometers)
Lifetime of red cell		80–120 days
Ratio of red to white blood cells		1,000:1
pH of blood		7.35–7.45

*About 30 red blood cells placed side by side would stretch across the period at the end of this sentence.

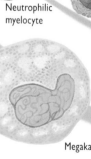

Blood is half water and nearly half red cells. Other matter suspended or dissolved in blood, including white cells and platelets, makes up $^1/_{20}$.

Water 50%
Plasma substances 4%
White blood cells and platelets 1%
Red blood cells 45%

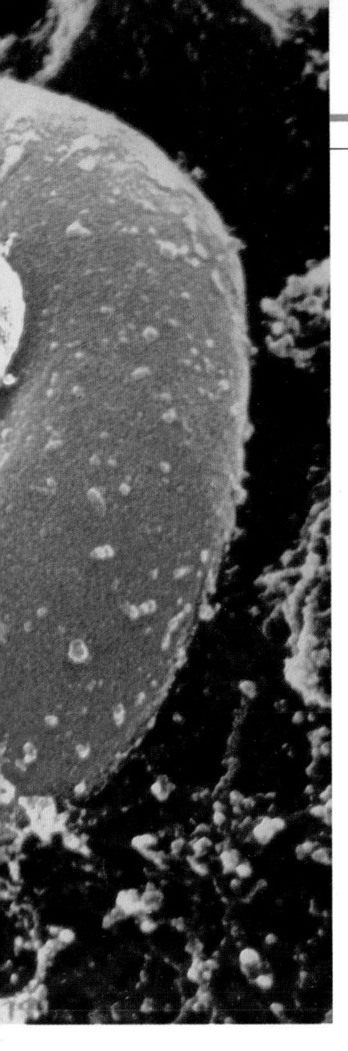

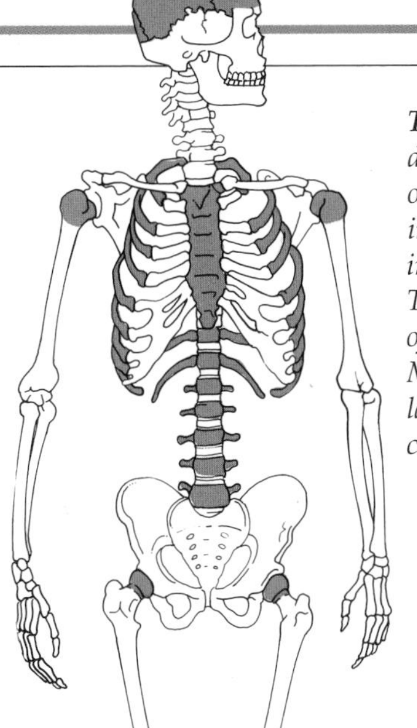

The white cells *made in lymph system tissue develop into lymphocytes (B-cells and T-cells) or monocytes. Lymphocytes are key players in immune responses, the body's defense against invading foreign substances or organisms. The B-cells, for instance, are the origin of cells that secrete antibodies. Monocytes can transform into large phagocytic (engulfing) cells called macrophages.*

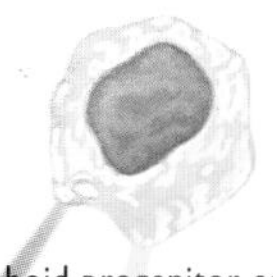

Lymphoid progenitor cell

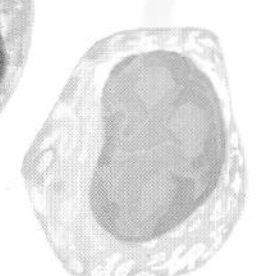

Lymphoblast

Monoblast

In lymph tissue

Lymphocyte

Lymphocyte

B–cell lymphocyte

Promonocyte

In blood

T–cell lymphocyte

Monocyte

In adult life *red blood cell production (erythropoiesis) takes place in the bone marrow of certain bones (in red, left). This happens particularly in the skull, backbone, ribs, breastbone (sternum), and the upper ends of the humerus (upper arm bone) and femur (thigh bone).*

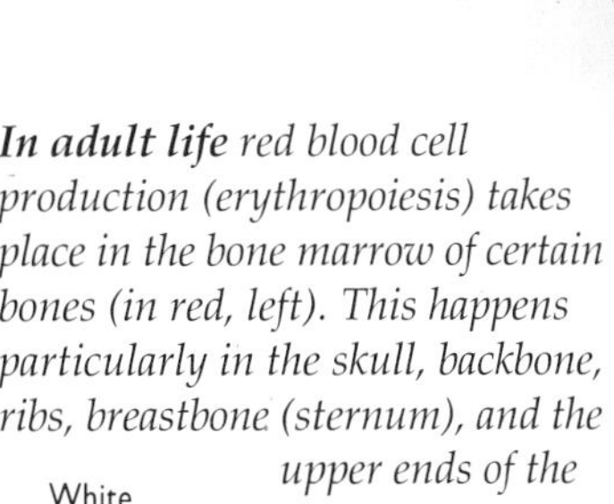

Red blood cells

Erythrocyte

White blood cells

Basophil

Reticulocyte

Eosinophil

Neutrophil

Basophilic band cell

...ocyte

Eosinophilic metamyelocyte

Eosinophilic band cell

Neutrophilic band cell

Metamegakaryocyte

red blood cell, *shaped like a ...ughnut with no hole, is adapted ... have maximum surface area ... gas exchange, specifically ...e oxygen it transports ...ound the body. The ...llow object on the ...ll is a platelet.*

The white blood cells *produced in bone marrow come in three forms: neutrophils, eosinophils, and basophils.*

In bone marrow

In blood

Platelets

Thrombocytes

Neutrophils *can attack bacteria. Eosinophils and basophils have complex functions that are linked with the allergy response and inflammation.*

Blood cells are formed *either in the bone marrow or in the tissues of the lymphatic system, such as the lymph nodes, tonsils, and spleen. Red blood cells (erythrocytes), white cells known as granulocytes (neutrophils, basophils, and eosinophils), and platelets are all made in bone marrow; other white cells are made in lymph tissue.*

In each of the two sites, an early cell type undergoes multiple sets of cell division to make hundreds, then millions, of daughter cells. As they multiply, different sets of these cells become progressively differentiated into the various specialized cell types that are ultimately found in the blood.

See also

CIRCULATION, MAINTENANCE, AND DEFENSE

▶ Transportation systems 118/119

▶ The living pump 120/121

▶ In circulation 124/125

▶ The water balance 132/133

▶ Inside the kidney 134/135

▶ Damage repair 140/141

CONTROL AND SENSATION

▶ Control systems 36/37

ENERGY

▶ The gas exchange 110/111

▶ The cell and energy 112/113

▶ Cells at work 114/115

REPRODUCTION AND GROWTH

▶ Baby in waiting 172/173

▶ Being born 174/175

The body's drain

Not only does the lymphatic system drain excess fluid from tissues, it is also in the first line of immune defense.

A drainage ditch in an agricultural system is as vital as the irrigation channel that brings water to the crops. The drain takes waste water from the land to prevent waterlogging. In the human body the lymphatic system acts, among other things, as the drainage canal network, removing excess fluid from all the body tissues and returning it to the blood system. To do this it has collecting vessels throughout the body, and this means that as a system it is excellently equipped to monitor problems in the areas that it drains. It also functions, therefore, as a rapid response section of the body's immune system.

There is always fluid – known as interstitial, or extracellular, fluid – in and around the body's tissues and thus between the cells that the body is made of. It leaks out of the blood capillaries and passes in and out of the cells as they perform their metabolic processes. There are other substances in the fluid as well as water – dissolved chemicals, molecules, and stray items such as disease-causing microorganisms that have found

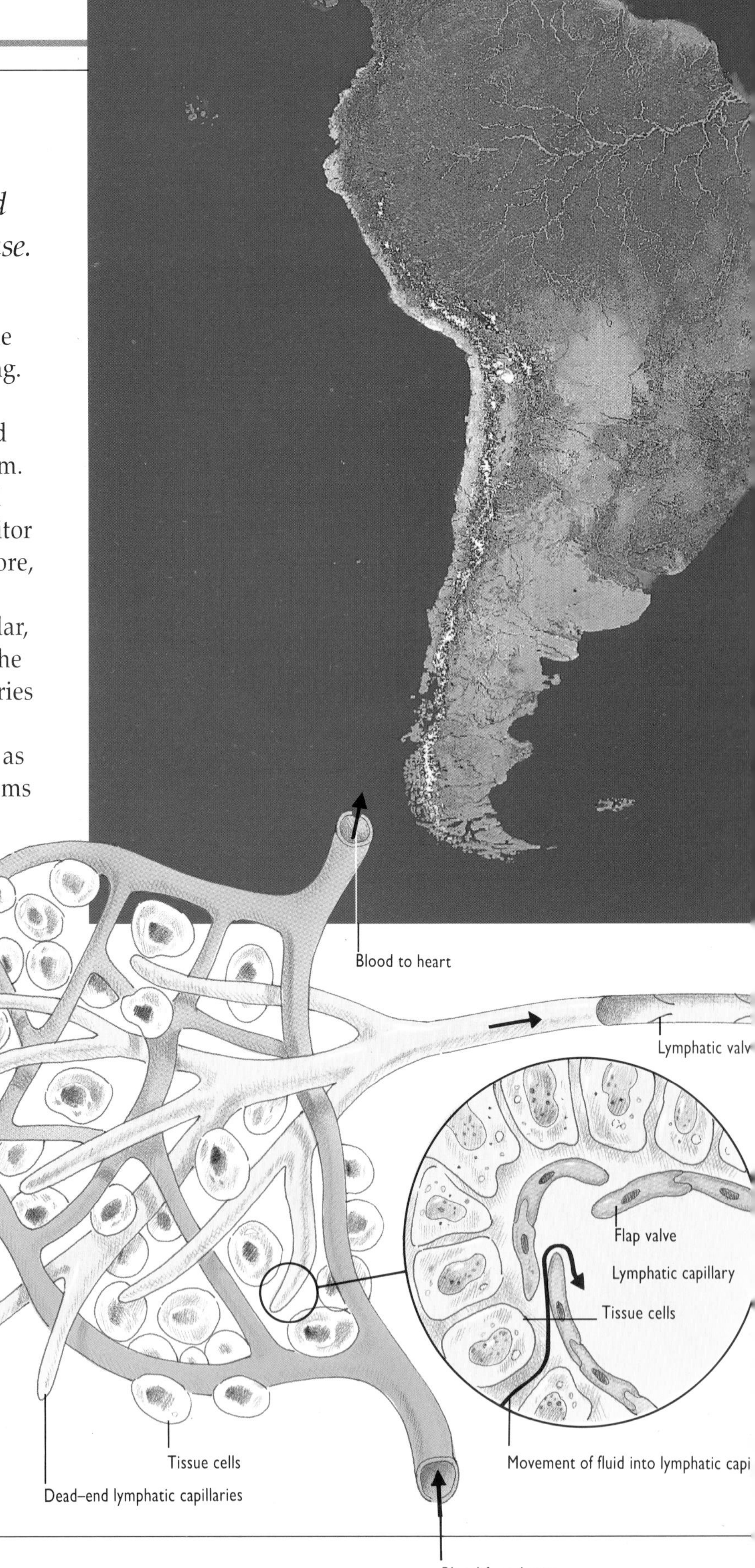

Dead-end lymphatic capillaries collect fluid *from between the cells. The fluid enters through one-way valves, called flap valves, that line the lymph capillaries. Inside the capillaries more flap valves provide a one-way flow of fluid away from the tissues through the lymphatic system. Larger flap valves in the lymphatic ducts maintain the direction of flow into filtering lymph nodes. From there the lymph returns to the blood system through the thoracic duct, which joins the left subclavian vein.*

LYMPHATIC SYSTEM FACT FILE

Unlike blood, which is moved by the heart, lymph is not pushed along by a special pump. Movements of surrounding organs push it through the system.

Amount of lymph fluid in body Percentage of body weight	1–2 quarts (1–2 liters) 1–3
Largest lymphoid body Size of spleen	spleen 4¾ inches (12 cm) across
No. of lymphocytes in body	2 trillion (2×10^{12})
Max. rate of increase of lymph flow during exercise	15 times

their way into the body, perhaps through a cut.

The lymphatic system starts as a series of dead-end tubes – lymphatic capillaries – close to the blood capillaries in tissues. Extracellular fluid passes into the lymphatic capillaries, which eventually join together to form larger ducts, or lymphatics. At intervals these pass through "way stations," the lymph nodes, where foreign material, including invading microorganisms drained from the tissues, is filtered out.

There are also larger lymphatic organs, including the tonsils, thymus gland, and spleen. The fluid moving through them, now known as lymph, comes into contact with large numbers of the immune system's white blood cells, which are on the lookout for matter foreign to the body. Finally, lymph is returned to the blood system where a large lymphatic vessel – the thoracic duct – joins the left subclavian vein taking blood to the heart.

***Just as the myriad tributaries of the Amazon** come together as they drain a large part of South America, so the lymphatic capillaries of the lymphatic system merge in a branching network to drain the body tissues of excess extracellular fluid.*

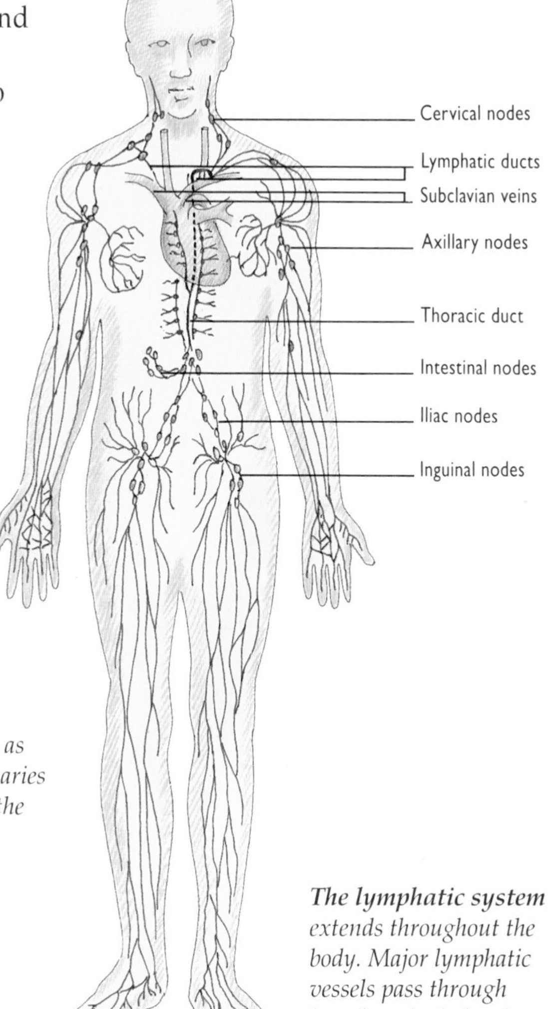

***The lymphatic system** extends throughout the body. Major lymphatic vessels pass through lymph nodes before lymph fluid is returned to the blood for circulation.*

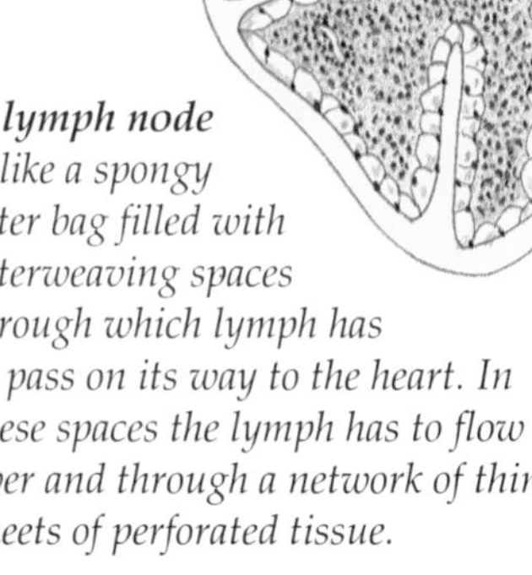

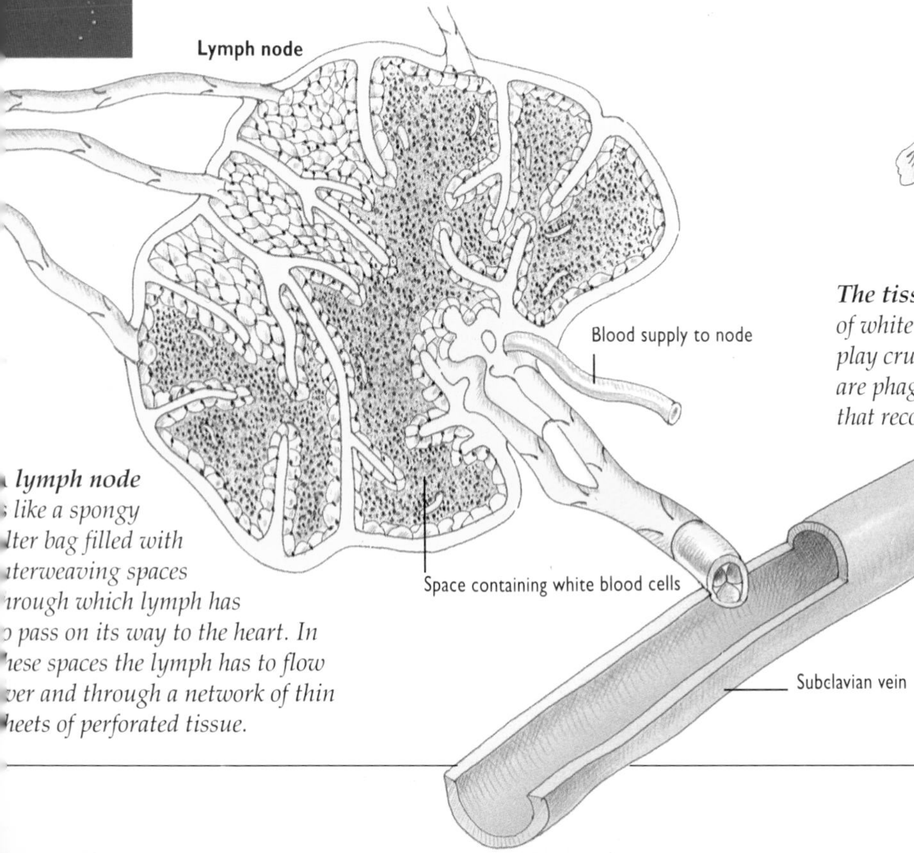

***A lymph node** is like a spongy filter bag filled with interweaving spaces through which lymph has to pass on its way to the heart. In these spaces the lymph has to flow over and through a network of thin sheets of perforated tissue.*

***The tissue sheets** in lymph nodes have large numbers of white blood cells on them. These include the cells that play crucial roles in the body's immune defenses. There are phagocytes that engulf and destroy, and lymphocytes that recognize invading matter and prime the immune system for action.*

This screen of white blood cells – acting like lookouts for the immune system – effectively filters the lymph of pathogens (dangerous microorganisms) and monitors the fluid for foreign material of all sorts. This filtering-out explains why, when infection starts, the lymph nodes are often swollen with the activity of the immune system swinging into action.

See also

CIRCULATION, MAINTENANCE, AND DEFENSE
- In circulation 124/125
- Blood: supplying the body 126/127
- Waste disposal 130/131
- The water balance 132/133
- Inside the kidney 134/135
- Sentinels of immunity 146/147
- Helpers and killers 148/149

ENERGY
- The food processor 100/101
- Absorbing stuff 104/105

Waste disposal

Toxins and wastes, wherever they come from, must be dealt with by the body.

One basic process – energy generation in cells – creates waste water and carbon dioxide. Water is sweated, breathed, urinated, and excreted in feces. Carbon dioxide can be breathed out or combined with other chemicals and passed out in urine. A by-product of another basic process – digestion – is roughage, the parts of food including the cell walls of plant material that cannot be digested. It is passed out of the body as feces.

Another type of waste is ammonia, which is highly toxic. The liver changes it into the less poisonous waste urea, which exits the body in urine. Ammonia conversion is an example of the vital process known as detoxification, which happens mostly in the liver. Batteries of enzymes in liver cells turn toxic chemicals into less toxic ones that are easy to remove, usually in urine or bile. The toxins made harmless in this way are produced by the body itself or are poisons: pollutants and drugs taken in from outside.

On most marine oil-producing platforms, *waste gases from the oil production process are burned off as a flare at the end of a long exhaust pipe away from the rig. All complex manufacturing processes – including those inside the body – produce wastes that have to be disposed of.*

Benzene *(C_6H_6) – a volatile minor part of gasoline – is highly toxic and if inhaled is a health hazard. If small amounts enter the body, they are rapidly detoxified by liver cells in a two-stage process. First, enzymes called mixed function oxidases change the benzene to phenol.*

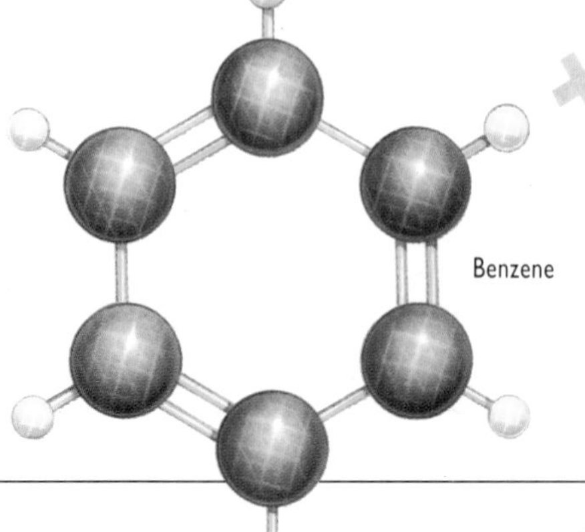

Oxygen

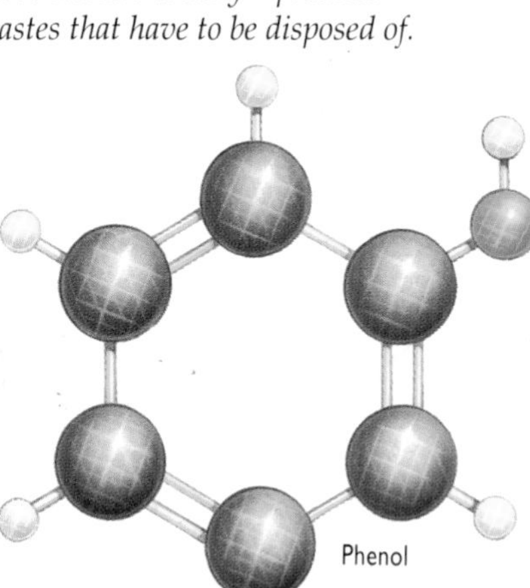

Phenol *is benzene to which an oxygen atom is added to give it a hydroxyl (-OH) group.*

Second, *another enzyme, glucuronidase, adds glucuronic acid – a sugar acid – to the hydroxyl group.*

Glucuronic acid

There are five major sources of body waste products: *toxins from outside; internally produced chemicals such as hormones that are no longer needed; food; nitrogen-containing waste left over from protein re-use; and energy production. After processing, the resulting products are eliminated from the body along one of the five disposal routes. Indigestible fiber, for instance, moves through the intestine to make the bulk of solid feces. Water exits in all outputs.*

Source	Process
Toxins	Detoxification in liver
Hormones	Metabolism
Food	Digestion
Nitrogenous wastes	Deamination and urea p
Energy production	Cellular respiratio

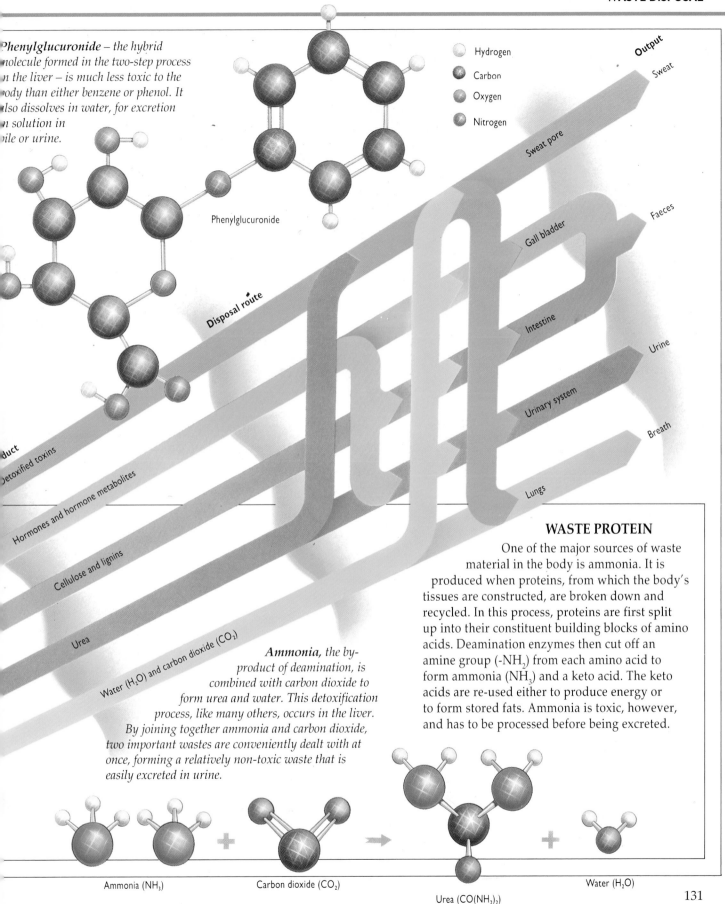

Phenylglucuronide – the hybrid molecule formed in the two-step process in the liver – is much less toxic to the body than either benzene or phenol. It also dissolves in water, for excretion in solution in bile or urine.

Hydrogen
Carbon
Oxygen
Nitrogen

Phenylglucuronide

Output
Sweat
Sweat pore
Faeces
Gall bladder
Intestine
Urine
Urinary system
Breath
Lungs

Disposal route

duct
Detoxified toxins
Hormones and hormone metabolites
Cellulose and lignins
Urea
Water (H_2O) and carbon dioxide (CO_2)

WASTE PROTEIN

One of the major sources of waste material in the body is ammonia. It is produced when proteins, from which the body's tissues are constructed, are broken down and recycled. In this process, proteins are first split up into their constituent building blocks of amino acids. Deamination enzymes then cut off an amine group (-NH_2) from each amino acid to form ammonia (NH_3) and a keto acid. The keto acids are re-used either to produce energy or to form stored fats. Ammonia is toxic, however, and has to be processed before being excreted.

***Ammonia,** the by-product of deamination, is combined with carbon dioxide to form urea and water. This detoxification process, like many others, occurs in the liver. By joining together ammonia and carbon dioxide, two important wastes are conveniently dealt with at once, forming a relatively non-toxic waste that is easily excreted in urine.*

Ammonia (NH_3)
Carbon dioxide (CO_2)
Urea ($CO(NH_2)_2$)
Water (H_2O)

See also

CIRCULATION, MAINTENANCE, AND DEFENSE
- The water balance 132/133
- Inside the kidney 134/135
- Outer defenses 142/143

SUPPORT AND MOVEMENT
- Smooth operators 30/31

CONTROL AND SENSATION
- The chemicals of control 38/39

ENERGY
- You are what you eat 96/97
- The energy storehouse 106/107
- The gas exchange 110/111
- The cell and energy 112/113

The water balance

Our bodies are almost two-thirds water. So where is all that liquid and how are fluid levels controlled?

Each and every cell in the human body contains water and is bathed in water. Indeed, there is more water in the body than any other substance. And since water transports all the sustaining substances that the body's cells require, it is no wonder that for a healthy existence the total volume of water in the body and the concentrations of materials dissolved in it have to be kept within extremely tight limits.

At the heart of the physiological machinery that manages the body's water balance are the kidneys. They work by first filtering water and a wide range of soluble molecules from the blood and then reabsorbing useful molecules and rejecting unwanted waste materials, which are expelled from the body in solution in urine. They are so important that untreated kidney failure leads to death within days. Modern medicine provides a solution to failure in the form of dialysis – the filtration of the blood by artificial kidneys – or kidney transplant.

Urine is formed in the kidneys by filtration of plasma – the liquid in which blood cells are suspended. This entails the removal of large quantities of water and all soluble materials from it. Useful substances and water are then selectively returned to the blood, while unwanted materials, such as urea (a waste product of the breakdown of proteins), stay in the filtrate. In 24 hours the kidneys filter out 40 gallons (150 liters) of water from the blood but return an average of 99 percent of it, keeping only 3 pints (1.5 liters) to form a day's worth of urine.

By contrast, of the 2 ounces (50 g) of urea filtered out daily by the kidneys, almost half is excreted dissolved in the urine. There are many other soluble substances in the blood, all of which pass through the kidneys. Each is returned to the blood or excreted in urine in a particular proportion. Glucose, for instance, is usually all reabsorbed, except in a disease such as diabetes, when some is lost.

Some 60 percent of our body weight *is water – only 40 percent is made up of solid material. In an average-sized adult man, that 60 percent of water makes up about 84 pints (40 liters). The water inside cells, or intracellular water, makes up about two-thirds of the total body water. All the remaining water, about 32 pints (15 liters) in an average person, is extracellular. It is found in a variety of body fluids. Most is present in the so-called interstitial fluid – the liquid medium which immediately surrounds all cells. Most of the rest is in lymph fluid in the lymphatic vessels, liquid in brain and spinal spaces (cerebrospinal fluid), and blood plasma.*

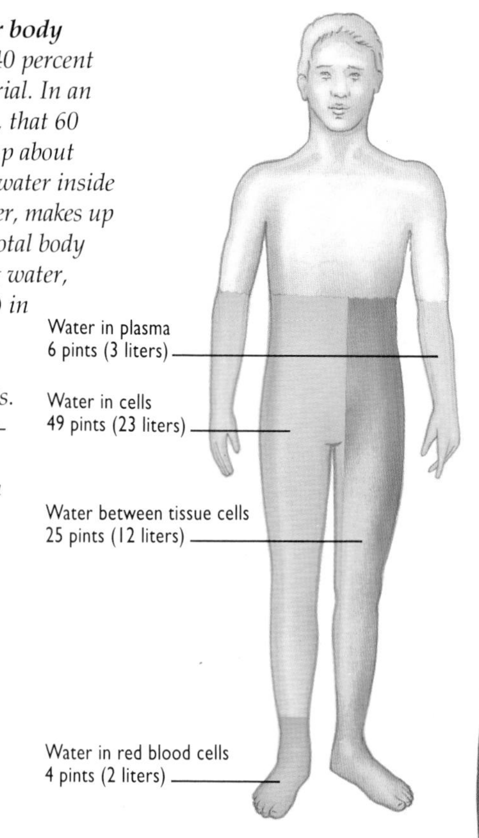

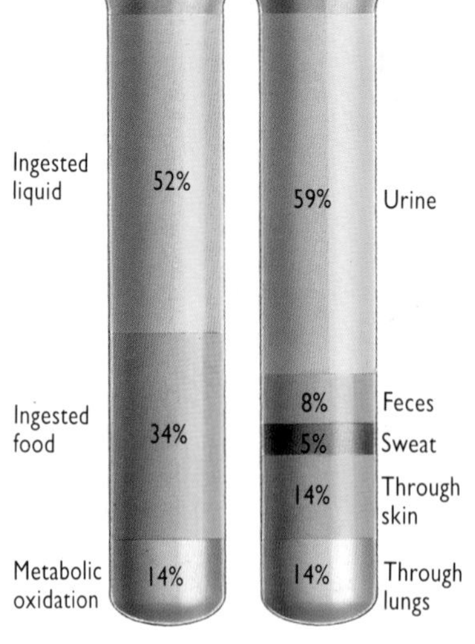

The amount of water *taken in plus that made internally by the conversion of glucose (the body's fuel source) to energy matches that lost by various means. The average daily intake, and thus output, is about 5¾ pints (2.7 liters).*

KIDNEYS FACT FILE

Blood enters the kidneys at 20–25 percent of the rate at which it leaves the heart – they receive proportionately more blood than any other organ.

Dimensions	length	4⅓ inches (11 cm)
	width	2⅓ inches (6 cm)
	thickness	1⅕ inches (3 cm)
Weight of kidney		5 ounces (140 g)
No. of nephrons per kidney		1 million
Total length of nephron tubules		50 miles (80 km)
Kidney blood flow per hour		19 gallons (72 liters)
Pressure of blood entering kidney		75 mmHg
Time to filter total blood plasma		30 minutes
Length of	ureter	12 inches (30 cm)
	urethra (man)	8 inches (20 cm)
	urethra (woman)	1½ inches (4 cm)
Water content of	body	60%
	blood (most)	83%
	fat (least)	10%
Solids excreted in urine		2 ounces (50 g) per day
Ave. urine passed on urination		11–14 ounces (300–400 ml)
pH of urine		4.8–7.8

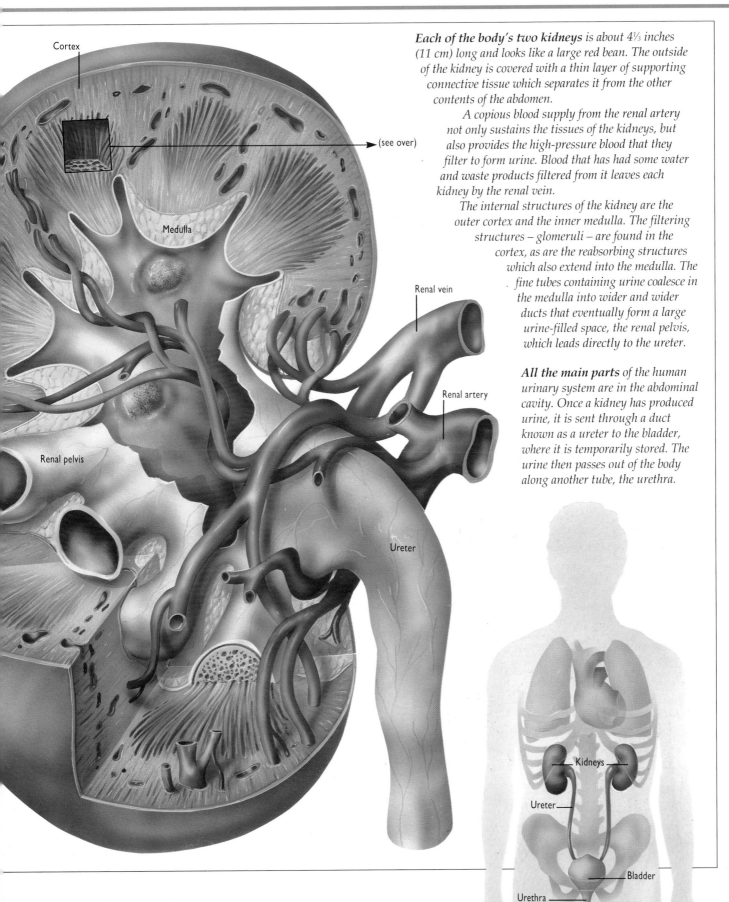

Cortex

(see over)

Medulla

Renal vein

Renal artery

Renal pelvis

Ureter

__Each of the body's two kidneys__ is about 4⅓ inches (11 cm) long and looks like a large red bean. The outside of the kidney is covered with a thin layer of supporting connective tissue which separates it from the other contents of the abdomen.

A copious blood supply from the renal artery not only sustains the tissues of the kidneys, but also provides the high-pressure blood that they filter to form urine. Blood that has had some water and waste products filtered from it leaves each kidney by the renal vein.

The internal structures of the kidney are the outer cortex and the inner medulla. The filtering structures – glomeruli – are found in the cortex, as are the reabsorbing structures which also extend into the medulla. The fine tubes containing urine coalesce in the medulla into wider and wider ducts that eventually form a large urine-filled space, the renal pelvis, which leads directly to the ureter.

__All the main parts__ of the human urinary system are in the abdominal cavity. Once a kidney has produced urine, it is sent through a duct known as a ureter to the bladder, where it is temporarily stored. The urine then passes out of the body along another tube, the urethra.

Kidneys

Ureter

Bladder

Urethra

See also

CIRCULATION, MAINTENANCE, AND DEFENSE

- ▶ Transportation systems 118/119
- ▶ In circulation 124/125
- ▶ Blood: supplying the body 126/127
- ▶ The body's drain 128/129
- ▶ Waste disposal 130/131
- ▶ Inside the kidney 134/135

CONTROL AND SENSATION

- ▶ Steady as you go 34/35
- ▶ The chemicals of control 38/39
- ▶ The automatic pilot 48/49

ENERGY

- ▶ Absorbing stuff 104/105
- ▶ Cells at work 114/115

Inside the kidney

Urine is the end product of a sophisticated process involving millions of tiny filtration units.

Although each kidney looks like a single entity, it consists of more than a million micro-organs, or nephrons. Each one produces its own tiny output of urine to yield a combined output of about 1½ quarts (1.5 liters) per day. Urine production is a convoluted process, reflected in the structure of the nephron.

Blood under high pressure enters a nephron through a glomerulus, a tangled knot of capillaries fed by one of the renal arteries. The plasma of the blood is filtered through the specialized wall of the glomerular capillary and the inner lining of the pocket in which the glomerulus sits. This pocket is known as a Bowman's capsule. Almost all the water and dissolved substances, such as glucose and sodium, found in plasma pass from the glomerulus into the Bowman's capsule, and from there into the nephron tubule.

The tubule and its contents take a highly complex, meandering track through the kidney. Most, if not all, of the substances that are useful to the body, for example glucose, are reabsorbed back into the plasma via the intricate network of capillaries that hug the sides of the tubule. This is achieved in a number of ways, including passive diffusion (where molecules in solution move from regions of high to low concentration), osmosis (where water passes through a semipermeable membrane from a weak solution to a strong solution), and active transportation of molecules across the tubule membrane. By the end, all that remains in the tubule is urine – a mixture of waste products such as urea and some salts and water. This passes into the renal pelvis and ultimately, via the ureter, to the bladder.

Reabsorption of salt and water is under complex hormonal control. A key factor that can influence the rate of water and salt loss by the kidneys is the state of the body's water balance. For instance, if you become dehydrated in a hot climate, hormonal changes ensure that water reabsorption rates in the tubule are increased so you lose less fluid.

Detail of kidney (from previous page)

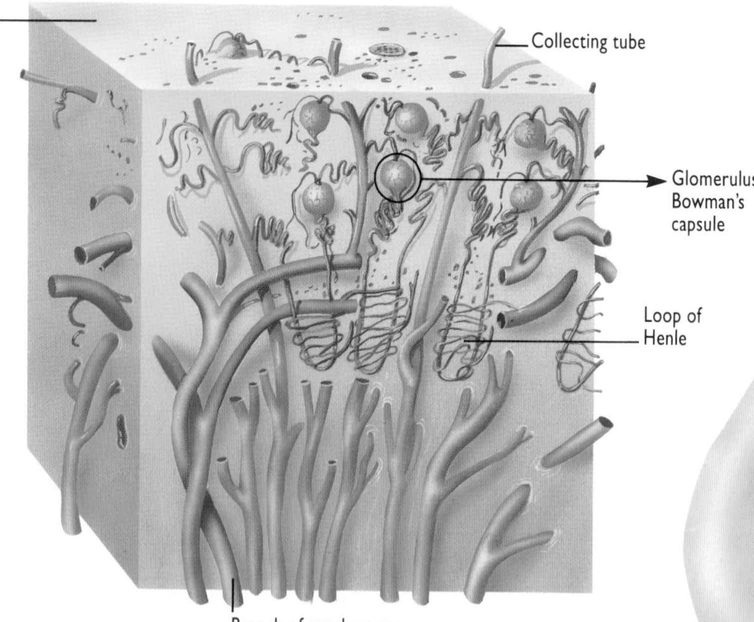

Within kidney tissue *a glomerulus – enclosed in a Bowman's capsule and supplied by a branch of the renal artery – is found at the outer end of each nephron. The nephron tubule leaves the glomerulus and dips down – in the loop of Henle – into the medulla, where it is surrounded by a mesh of fine blood vessels. The tubule finally joins with the collecting tube which carries urine to the renal pelvis.*

Collecting tube

Grounds are filtered *from coffee to make a palatable drink. In the same way, the glomerulus and Bowman's capsule filter out larger molecules and the blood cells, allowing only small molecules – including water – to pass into the nephron tubule.*

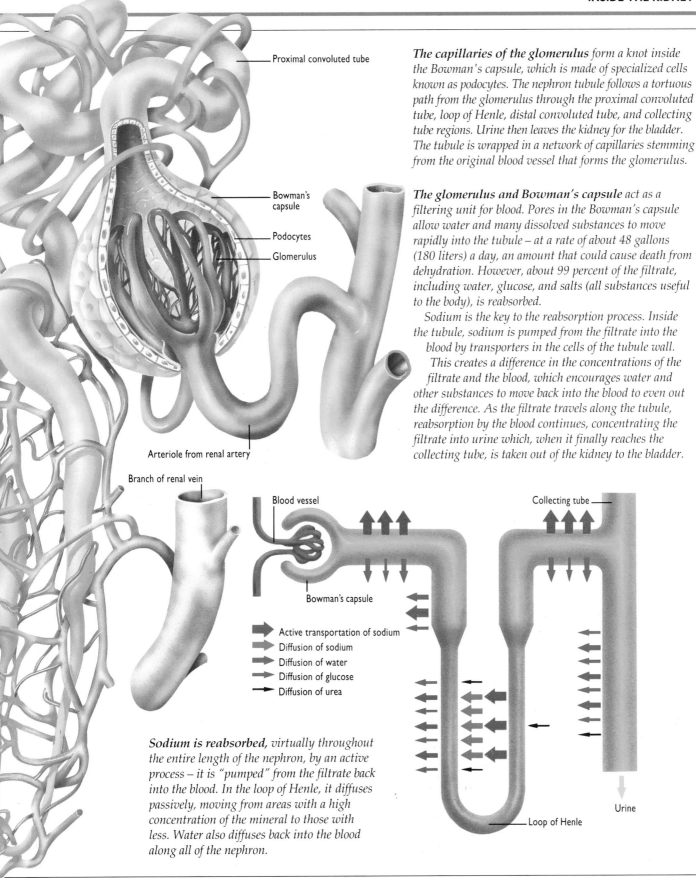

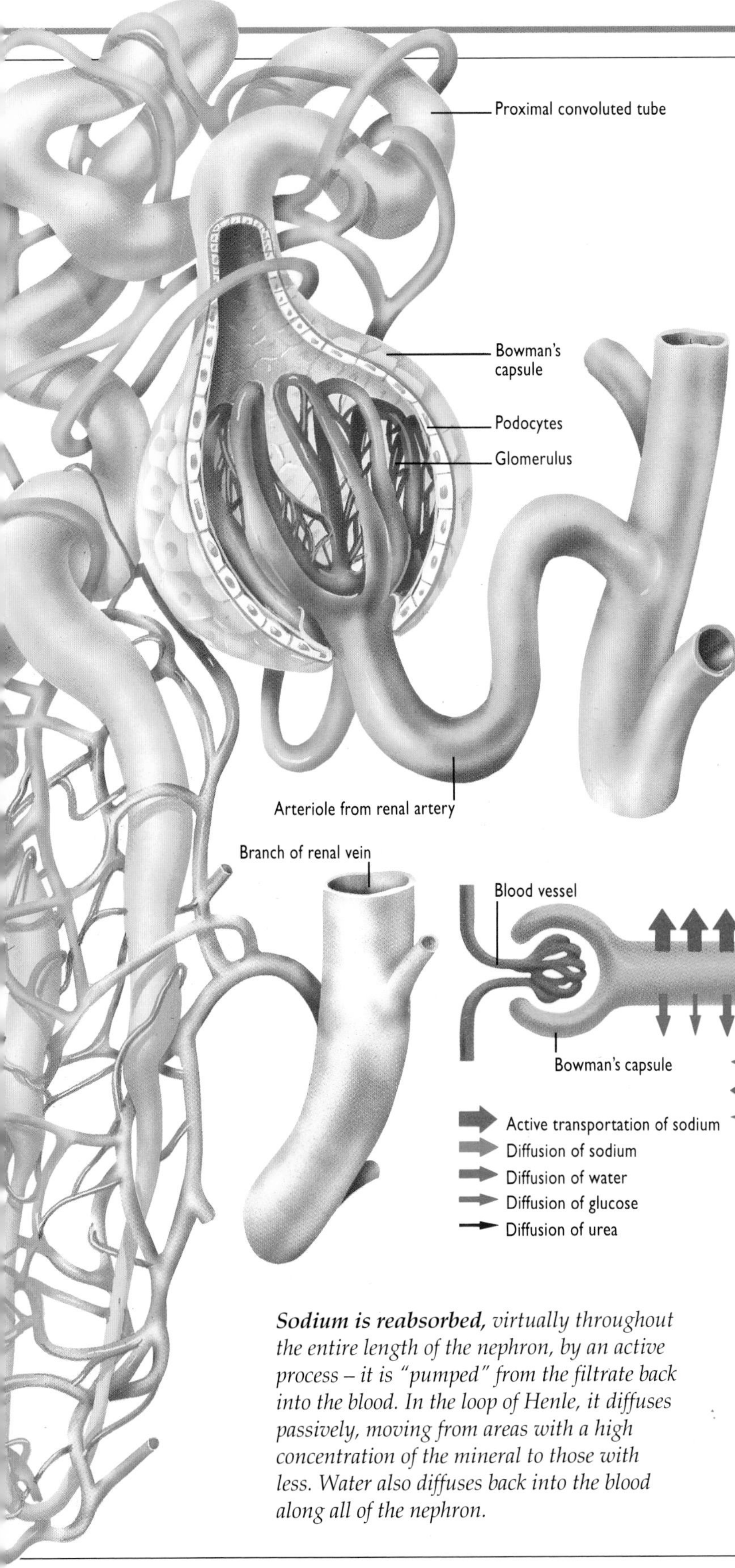

Proximal convoluted tube

Bowman's capsule

Podocytes

Glomerulus

Arteriole from renal artery

Branch of renal vein

The capillaries of the glomerulus *form a knot inside the Bowman's capsule, which is made of specialized cells known as podocytes. The nephron tubule follows a tortuous path from the glomerulus through the proximal convoluted tube, loop of Henle, distal convoluted tube, and collecting tube regions. Urine then leaves the kidney for the bladder. The tubule is wrapped in a network of capillaries stemming from the original blood vessel that forms the glomerulus.*

The glomerulus and Bowman's capsule *act as a filtering unit for blood. Pores in the Bowman's capsule allow water and many dissolved substances to move rapidly into the tubule – at a rate of about 48 gallons (180 liters) a day, an amount that could cause death from dehydration. However, about 99 percent of the filtrate, including water, glucose, and salts (all substances useful to the body), is reabsorbed.*

Sodium is the key to the reabsorption process. Inside the tubule, sodium is pumped from the filtrate into the blood by transporters in the cells of the tubule wall. This creates a difference in the concentrations of the filtrate and the blood, which encourages water and other substances to move back into the blood to even out the difference. As the filtrate travels along the tubule, reabsorption by the blood continues, concentrating the filtrate into urine which, when it finally reaches the collecting tube, is taken out of the kidney to the bladder.

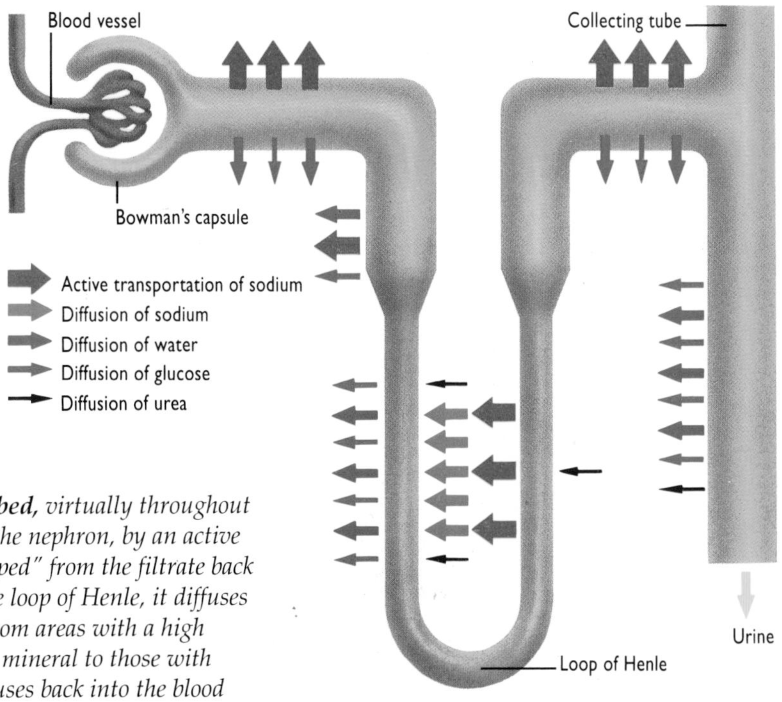

Blood vessel

Bowman's capsule

Collecting tube

Active transportation of sodium

Diffusion of sodium

Diffusion of water

Diffusion of glucose

Diffusion of urea

Loop of Henle

Urine

Sodium is reabsorbed, *virtually throughout the entire length of the nephron, by an active process – it is "pumped" from the filtrate back into the blood. In the loop of Henle, it diffuses passively, moving from areas with a high concentration of the mineral to those with less. Water also diffuses back into the blood along all of the nephron.*

See also

CIRCULATION, MAINTENANCE, AND DEFENSE

▶ Transportation systems 118/119

▶ In circulation 124/125

▶ Blood: supplying the body 126/127

▶ The water balance 132/133

CONTROL AND SENSATION

▶ Control systems 36/37

▶ The chemicals of control 38/39

▶ Key chemicals 42/43

oop of Henle

Maintaining the system

The body maintains, repairs, and protects itself with an astounding array of systems – and directing them all is the brain.

A house has to be intrinsically strong and resilient to withstand the ravages of weather, decay, subsidence, woodworm, termites, and burglars. The physical robustness of the house – its frame, roof, bricks, cement, doors, and windows – provides the first line of defense against these potential problems. The body has equivalent defenses to protect it against attack by viruses, bacteria, fungal infections, and poisons. These range from the protective outer covering of the skin to complex physiological and chemical defenses inside the body.

Wear and tear mean that different parts of the house also need repair and replacement. The body likewise requires running repairs, almost all of which are carried out by cells that have the capacity to divide, producing new cells to replace the old ones. In addition, damaged organs such as broken bones and cut blood vessels can also be mended.

Deliberate attacks on a house and its contents provide the most severe tests. An attempted break-in by an intruder requires sophisticated defenses because the strategy and cunning of the invader must be overcome. It is much the same when the body has to confront attacks by disease-causing organisms, whether they are viruses and bacteria that are invisible to the naked eye or large parasites such as tapeworms. Since harmful invaders (pathogens) have active mechanisms to help them breach the body's layers of protection, countering infection requires precise and flexible defensive weaponry. The cells and antibodies of the immune system are the physiological equivalents of surveillance cameras, burglar alarms, and detectives with long memories.

BRAINS INSIDE THE SYSTEM

In addition to the body's brilliant array of automated defense, repair, and maintenance systems, there is another dimension to the maintenance of the human system – human brain power and problem-solving abilities. This means that defense against invading disease agents also has an overriding and unique human dimension, equivalent in some ways to the architect whose skills and intellect plan a house in the first place. Thus, through health education, we can change our behavior to minimize disease risks. We can purify water before drinking it, sleep under mosquito nets to avoid infection by the bite of malaria-infected insects, and make sure that foods are cooked appropriately to kill germs in them. Beyond such preventative measures, we have also developed vaccines to protect us from infections and drugs to eliminate them if they have managed to establish themselves.

Like the walls of a house, *the skin provides the outermost physical protective barrier. Tough and flexible, with a surface built from dead skin cells packed with keratin (a fibrous structural protein), the skin can stop invasion by disease-causing microbes. Its crucial role is seen in severe burn injuries, which are almost always fatal because it is impossible to stop bacterial infection without the outer skin layers.*

The lining *of the mouth, throat, lungs, and reproductive system is constructed from layers of cells that secrete large amounts of watery fluid known as mucus. The mucus traps invading microbes, and the antibacterial enzyme lysozyme in the mucus helps to keep microorganisms under control. From the lungs, the mucus, containing foreign particles, is moved outward toward the throat as a cleaning mechanism.*

ike a house, *the human body must be maintained and efended. Even a house as brilliantly designed as allingwater (**above**), built in 1936 at Bear Run, ennsylvania, and designed by American architect Frank loyd Wright (1867–1959), will survive only if it is looked fter and protected.*

)EFENDING THE BODY

'he symbols (**right and above**) identify functions of ne body's defense system. Like a house, the body as an in-built passive defense system – walls, a oof, and damp-proofing to keep weather and ntruders out – shown here by the wall and brick ymbols. It also has repairers, analogous to painters, laziers, and builders (paint can symbols), and an ctive defense system, like a security alarm with a irect line linked to the police (alarm bell symbols).

A static defense system *waiting passively in the stomach is high acidity caused by secretion of hydrochloric acid by the stomach lining. Conditions are such that almost all disease-causing organisms taken in accidentally with food or drink are killed in the stomach by the acid.*

The complement system *– a complex mixture of proteins in the liquid part of the blood (plasma) – is normally in a non-activated state. But when these proteins are "switched on" by contact with bacteria, they can destroy invading bacterial cells.*

Damage repair systems *have to deal with the failure of major structural components. For instance, if bones are put under enough strain, they crack and fracture. When this happens new cells are formed in a blood clot at the broken bone ends. These cells transform into bone cells that knit the bone together again.*

Regular maintenance *involves replacing worn out or dead cells. In the skin, for instance, the dead outer layers constantly fall off. They are replaced by new cells that were formed in the living lower layer, where rapidly dividing cells produce a constant supply of replacement cells. A similar process occurs in the hair follicle in each hair "root" so hairs constantly grow at their bases.*

Repairing the wall *of a damaged blood vessel as quickly as possible is important so that the flow of blood can be stemmed. This is done first with a quick-setting temporary patch made from platelets. A longer term repair is then made with fibrin – an insoluble blood protein. Finally, new cell growth makes a permanent repair in the vessel wall.*

Active defenses *that react to invaders include antibodies – specialized proteins of the immune system which are secreted by white blood cells. They recognize and stick to foreign molecules which have invaded the body. This "brands" the invading molecules as targets for attack. Once an invader has been recognized, more and more antibodies are made to counter it. When the invader has been defeated, the immune system is ready for it the next time it invades.*

The immune system's *phagocytic cells "eat" and digest invading microbes. Once the microbes have been identified by antibodies, the phagocytes, including cells such as macrophages, consume the invaders, digest them, and in the process kill them.*

Some of the body's white blood cells *(lymphocytes) can recognize and react to cancer cells in the body as well as to body cells which have been infected with an invading virus. These lymphocytes are called natural killer cells – they attach to altered body cells and kill them by secreting deadly chemicals which make the cell's contents leak out.*

See also

CIRCULATION, MAINTENANCE, AND DEFENSE

- Blood: supplying the body 126/127
- Routine replacement 138/139
- Damage repair 140/141
- Outer defenses 142/143
- Repelling invaders 144/145
- Sentinels of immunity 146/147
- Helpers and killers 148/149
- Knowing me, knowing you 150/151

SUPPORT AND MOVEMENT

- The living framework 16/17

CONTROL AND SENSATION

- Problem solving 82/83

ENERGY

- The food processor 100/101

Routine replacement

The body is a community of countless cells which, like any community, sees both death and birth.

When a piece of machinery breaks down or wears out, someone, usually a skilled engineer, has to diagnose the problem and fix the machine. By contrast, the human body does not need some outside agency – except in unusual circumstances – to come along and mend it or replace a worn-out part. Our bodies are self-repairing and self-replenishing. In fact, almost all the living parts of the body are in a state of constant flux. It has been estimated that during each hour almost 200 billion cells die – but in a healthy body these dying cells are simultaneously replaced by new ones. So although individual cells become ineffectual and then die, the body replaces its worn-out parts seamlessly and continuously.

So where do these new cells come from? They are formed by the division of previously existing cells into pairs of daughter cells. This division, or splitting apart, is a phase of the cell cycle. In this continuous cycle of changes, both to the cell nucleus and to the cytoplasm region outside the nucleus, the chromosomes – containing the gene-carrying DNA molecules – first copy themselves so the nucleus has a double set. Then, in a process known as mitosis, the two sets of chromosomes are

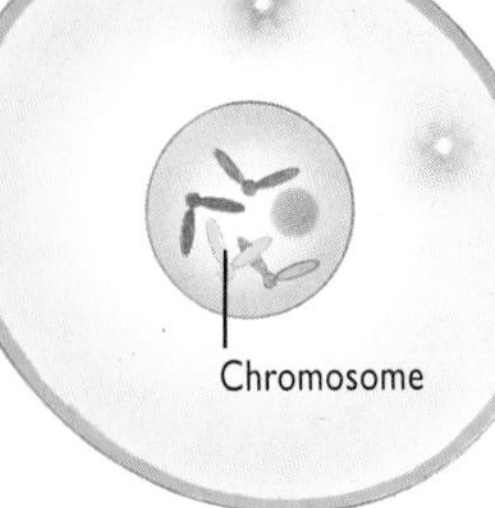

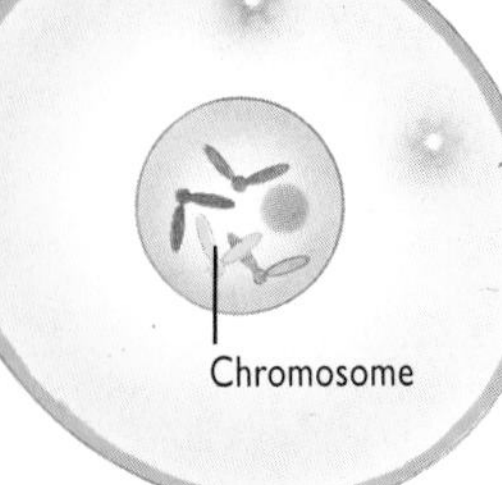

Cell turnover *in the body is like the seemingly constant appearance of a waterfall. A fall looks the same from moment to moment even though at every successive instant it is made of different droplets.*

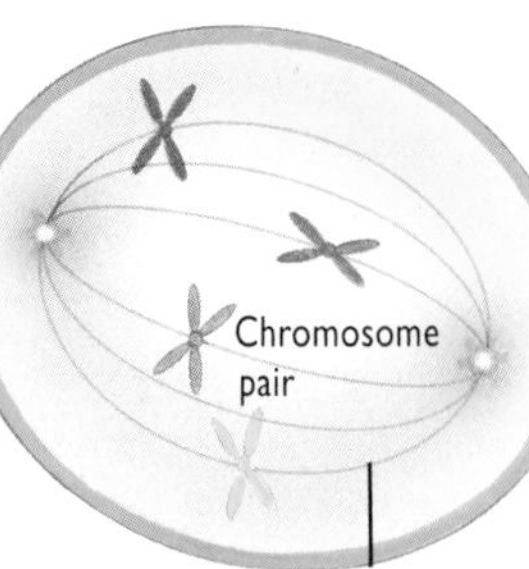

When a cell divides *to copy, or replicate, itself, the nucleus divides in two, and each new nucleus has an identical set of the cell's genetic information, or genes. This division of the nucleus is called mitosis. Before this happens, in the interphase stage (**above**), the genetic material copies itself. In early prophase, the genetic material becomes condensed to form short pairs of chromosomes.*

Metaphase

In late prophase, *ch attach to the fibers of apparatus that is forr organelles, or centri nuclear envelope this time. In the metaphase, the line up around equator zone of the spindle.*

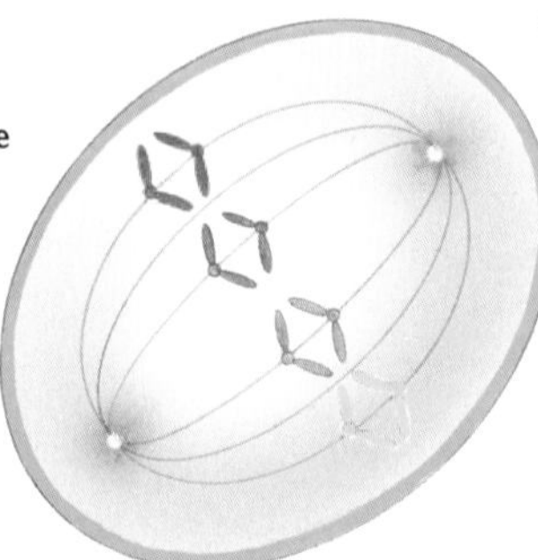

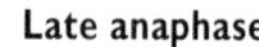
Late anaphase

THE LIFE SPAN OF DIFFERENT CELLS

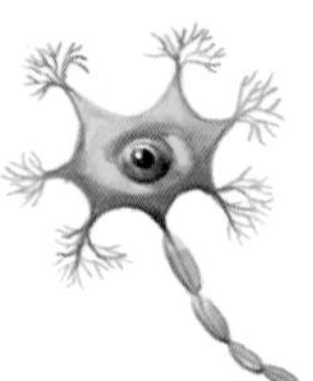

Nerve cells are the body's longest cells – they can have a complex, elongated, and branched shape. Once a nerve cell attains this shape during development in the womb or in early childhood, it cannot divide further. This means that some nerve cells live for as long as the person they are part of. Another implication of this is that once a cell of the nervous system dies, it is not replaced, whether in the brain itself or elsewhere in the body.

Liver cells, or hepatocytes, in the lobules that make up the bulk of the liver lobes are metabolically complex cells that often carry out many processes simultaneously. So great is their ability to divide – and therefore multiply – that even if an entire liver lobe is removed by surgery, the remaining liver cells can slowly replace the lost liver tissue. The approximate life span of a liver cell is 500 days.

carefully and accurately separated from one another in a perfectly orchestrated set of microscopic movements which carry identical sets of chromosomes to opposite ends of the cell. After the newly reproduced chromosomes have separated like this, they form two new nuclei and then the cell itself splits in two – each half with an identical set of genes in its nucleus.

In this way, one turn of the cell cycle transforms one cell into two. Carefully regulated multiplication of this sort can exactly compensate for the dying and dead cells in a healthy body. This balanced loss and birth of cells means that although a person's body does not seem to change, it is made of new cells day by day. So even though all the cells in the skin of your face can be replaced, your face still retains its characteristic looks.

The life of an individual cell can end in one of two ways: the cell can divide, forming two identical daughter cells; or it may become diseased or worn out, so that it dies before it is able to duplicate itself. Different cell types have characteristically varying life spans linked either to the rapidity with which they divide or to the time during which they are able to work properly. Some cells cannot replicate themselves, notably the neurons in the brain and nervous system.

Interphase

Late telophase

arly telophase

During anaphase and telophase, the two chromosome sets move to opposite ends of the cell, driven by the action of fibers in the spindle. At the end of telophase, two new nuclei have formed and the cell itself splits in half with a new nucleus in each daughter cell.

New red and some white blood cells are formed by the actively dividing cells that are found in the bone marrow of many large bones. Their phenomenally active rate of cell division means that bone marrow cells have one of the shortest life spans of any human cell. They divide every 10 hours or so – about 900 times a year. About 200 billion red blood cells are made a day.

Once formed, red and white blood cells move from the bone marrow into the bloodstream. Red blood cells, which do not possess nuclei and therefore cannot undergo mitosis, live for about 120 days in the blood before dying. Nucleated white blood cells last for an average of only 13 days before dying or dividing.

See also

CIRCULATION, MAINTENANCE, AND DEFENSE
▶ Blood: supplying the body 126/127

▶ Maintaining the system 136/137

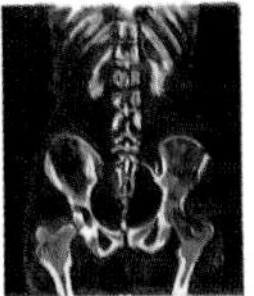

SUPPORT AND MOVEMENT
▶ The living framework 16/17

▶ Muscles at work 26/27

CONTROL AND SENSATION
▶ Down the wire 46/47

ENERGY
▶ The energy storehouse 106/107

▶ The cell and energy 112/113

▶ Cells at work 114/115

REPRODUCTION AND GROWTH
▶ Language of life 158/159

▶ Cycle of life 164/165

Damage repair

If you are unfortunate enough to get hurt, you can rely on your body to look after you.

The human body has self-regulating methods for replacing the cells in tissues that wear out and die. New skin cells, hair cells, and blood cells are made in a ceaseless and precisely controlled fashion to take the place of those that are lost in cellular aging. In addition to this process of continuous repair and replacement, the body has dramatic capacities to repair itself when more catastrophic types of damage occur.

These large-scale abilities for emergency repair are clearly shown in one of the most crucial organ systems in the body, the blood-carrying circulatory system. When blood vessels are ruptured, "first-aid" responses happen almost instantaneously to minimize the immediate risks. Then a step-by-step, longer-term set of changes begins to repair the damage.

Since the whole circulatory system is pressurized, a hole in the wall of a blood vessel is potentially fatal. Unless the hole is rapidly sealed, the internal pressure will mean that blood continues to flow from the opening. The body has a three-phase set of emergency responses to deal with this "red alert" situation. First, the muscles in the blood vessel wall contract to minimize the hole size. Second, platelets in the blood stick together to make a soft clot over the hole. Third, blood clotting, or coagulation, produces a much more solid patch over the breakage.

Damage is also repaired in other parts of the body. Broken bones can grow back together – and be as strong as before the break. Dislocated or sprained joints are immobilized by fluid in the tissues and held in place while healing takes place. Even severed nerves can join up to restore feeling and control of muscles to a limited extent. And in the brain, the network of neurons can rewire itself so that damage can be minimized.

***One of the body's prime "emergency services"** enables it to stop an open wound from losing too much blood. For instance, when you cut your finger, breaking blood vessels, a sequence of events comes into play which slows down – and then stops – the potentially dangerous loss of blood.*

The damage to the blood vessel directly stimulates involuntary, or smooth, muscles in the wall, causing them to constrict. This reduces not only the size of the break or wound, but also the amount of blood flowing toward it.

Specialized cells in the blood – thrombocytes or platelets – also respond quickly to any trauma. They release chemicals that further stimulate muscles to contract, aiding the direct constrictive mechanism working in the blood-vessel walls. The muscles may stay contracted for up to 20 minutes, giving immediate protection against excessive blood loss.

Platelets

Under normal circumstances, platelets in the blood slide past eac other and other blood cells – their shape is relatively smooth and the

e non-sticky. But when the walls of a ood vessel are damaged, the atelets come into contact with e protein collagen in the nnective tissue in the essel wall. This stimulates em to become irregular shape and stick to one nother. "Switched on" this way, they can me together and mbine to form a soft ick lump or clot in out one minute. In many rcumstances, this platelet ot can block the flow of blood rough the damaged vessel wall, pecially if the damage is not too vere. After a platelet clot forms, a much ore substantial plug is made over the wound the blood itself coagulates or clots.

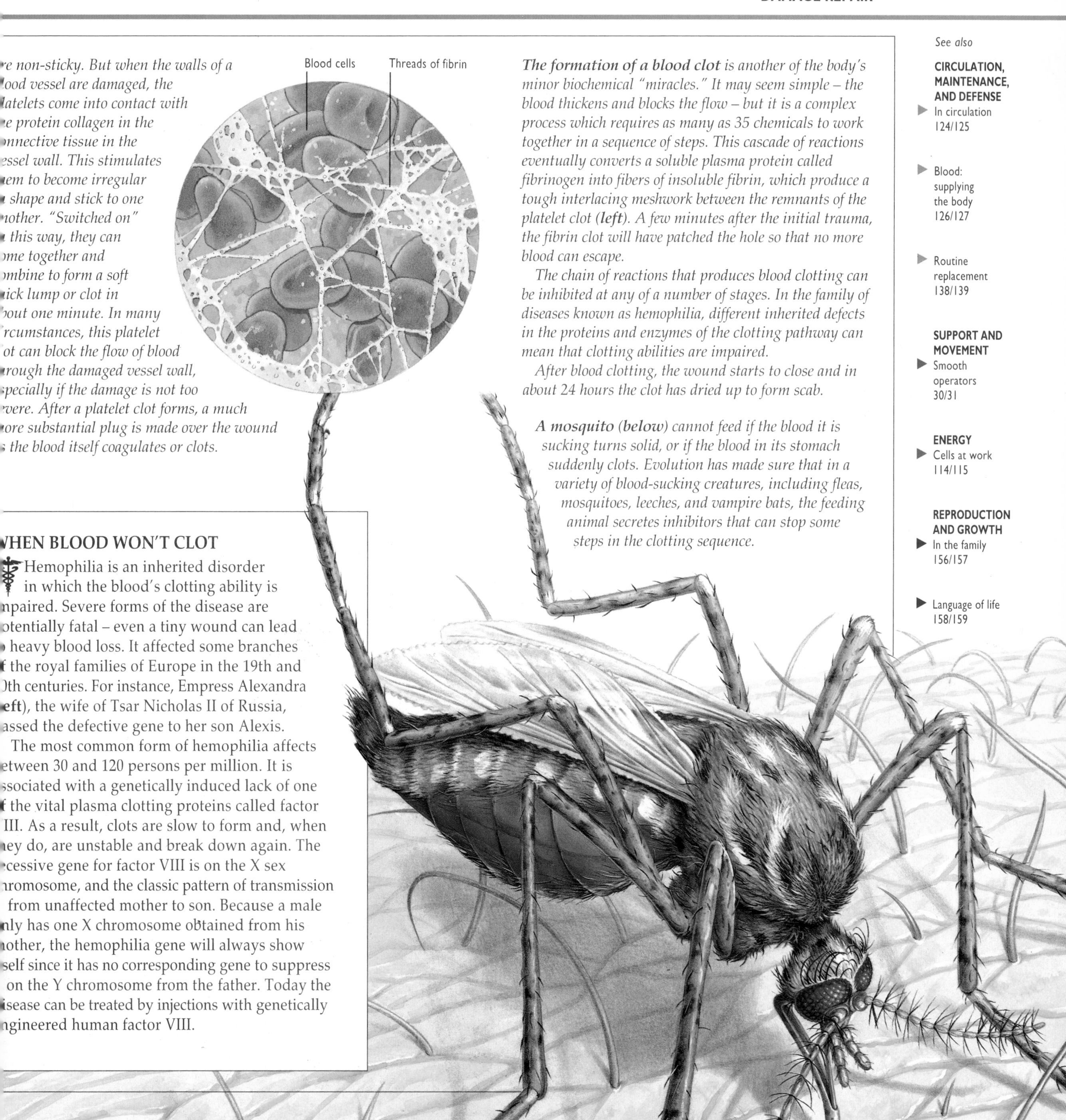

***The formation of a blood clot** is another of the body's minor biochemical "miracles." It may seem simple – the blood thickens and blocks the flow – but it is a complex process which requires as many as 35 chemicals to work together in a sequence of steps. This cascade of reactions eventually converts a soluble plasma protein called fibrinogen into fibers of insoluble fibrin, which produce a tough interlacing meshwork between the remnants of the platelet clot (**left**). A few minutes after the initial trauma, the fibrin clot will have patched the hole so that no more blood can escape.*

The chain of reactions that produces blood clotting can be inhibited at any of a number of stages. In the family of diseases known as hemophilia, different inherited defects in the proteins and enzymes of the clotting pathway can mean that clotting abilities are impaired.

After blood clotting, the wound starts to close and in about 24 hours the clot has dried up to form scab.

***A mosquito (below)** cannot feed if the blood it is sucking turns solid, or if the blood in its stomach suddenly clots. Evolution has made sure that in a variety of blood-sucking creatures, including fleas, mosquitoes, leeches, and vampire bats, the feeding animal secretes inhibitors that can stop some steps in the clotting sequence.*

HEN BLOOD WON'T CLOT

Hemophilia is an inherited disorder in which the blood's clotting ability is npaired. Severe forms of the disease are otentially fatal – even a tiny wound can lead heavy blood loss. It affected some branches the royal families of Europe in the 19th and)th centuries. For instance, Empress Alexandra **eft**), the wife of Tsar Nicholas II of Russia, assed the defective gene to her son Alexis.

The most common form of hemophilia affects etween 30 and 120 persons per million. It is ssociated with a genetically induced lack of one the vital plasma clotting proteins called factor III. As a result, clots are slow to form and, when ey do, are unstable and break down again. The cessive gene for factor VIII is on the X sex aromosome, and the classic pattern of transmission from unaffected mother to son. Because a male nly has one X chromosome obtained from his other, the hemophilia gene will always show self since it has no corresponding gene to suppress on the Y chromosome from the father. Today the isease can be treated by injections with genetically ngineered human factor VIII.

See also

CIRCULATION, MAINTENANCE, AND DEFENSE
- In circulation 124/125
- Blood: supplying the body 126/127
- Routine replacement 138/139

SUPPORT AND MOVEMENT
- Smooth operators 30/31

ENERGY
- Cells at work 114/115

REPRODUCTION AND GROWTH
- In the family 156/157
- Language of life 158/159

Outer defenses

Surrounding our fragile insides is a thin yet strong covering that protects us from the outside world.

It is easy to think of bare skin as being soft and vulnerable. In fact, human skin forms a remarkably effective outer defense for the body. It is tough, resilient, self-replenishing, self-healing when damaged, excludes invading microorganisms, generates hairs and nails, and is the location of the sense of touch. And all these functions are packed into a layer of tissue that is usually only a fraction of an inch thick.

The main protective function of the skin is provided by the layers of dead cells that form the outer part of the epidermis. Each new cell, produced by cell division at the base of the epidermis, moves outward toward the surface. As it does, its cytoplasm, the fluid portion of a cell's interior, is transformed into a tough, sulfur-rich protein known as keratin, and in the process the cell dies. Keratin is robust and is difficult to break down with normal digestive enzymes because of the sulfur–sulfur bonds linking the chains of amino acids that make up the keratin. This strength means that potentially invading bacteria and fungi find it difficult to gain a purchase on the outer layer of the skin.

Outer dead keratinized cells are continuously falling from the skin surface. In fact, a large proportion of household dust is made up of dead human skin. Where whole collections of skin cells fall off together in flakes from the scalp, the flakes are called dandruff. As cells are lost in this way, they are replaced by new ones coming up from below, so maintaining a virtually impenetrable barrier.

SKIN FACT FILE

Among skin's many roles are defense against germs and production of sweat and breast milk in special glands.

Area covered by skin	ave.	21½ sq. feet (2 m²)
Thickness of skin	ave.	1/25–2/25 inch (0.1–0.2 cm)
	thickest	1/5 inch (0.5 cm) on upper back
	thinnest	1/50 inch (0.05 cm) on eyelids
Life span of skin cells		19–34 days
Perspiration	inactive person	1/12 ounce/hour (3 ml/hour)
	active person	1 quart/hour (1 liter/hour)

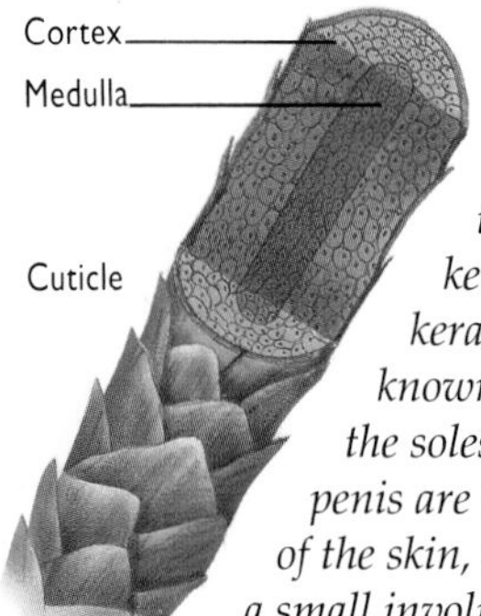

Hairs project from hair follicles *– specialize pockets of actively dividing cells at the base of the dermis. Like the skin surface, as well as finger- and toenails, hairs are made of cells that have turned into keratin and died. Each hair is a solid cylinder of dead, keratinized skin cells which are organized into layers known as the cortex, medulla, and cuticle. Only areas like the soles of the feet, palms, nipples, lips, and the tip of the penis are without hair. Situated deep within the hypodermis of the skin, the hair root is joined to a small involuntary muscle known as the arrector pili. When this contracts, the hair is raised, and a small bump – known as a goose-pimple – forms on the skin. This helps to trap air and so decrease heat loss from the body.*

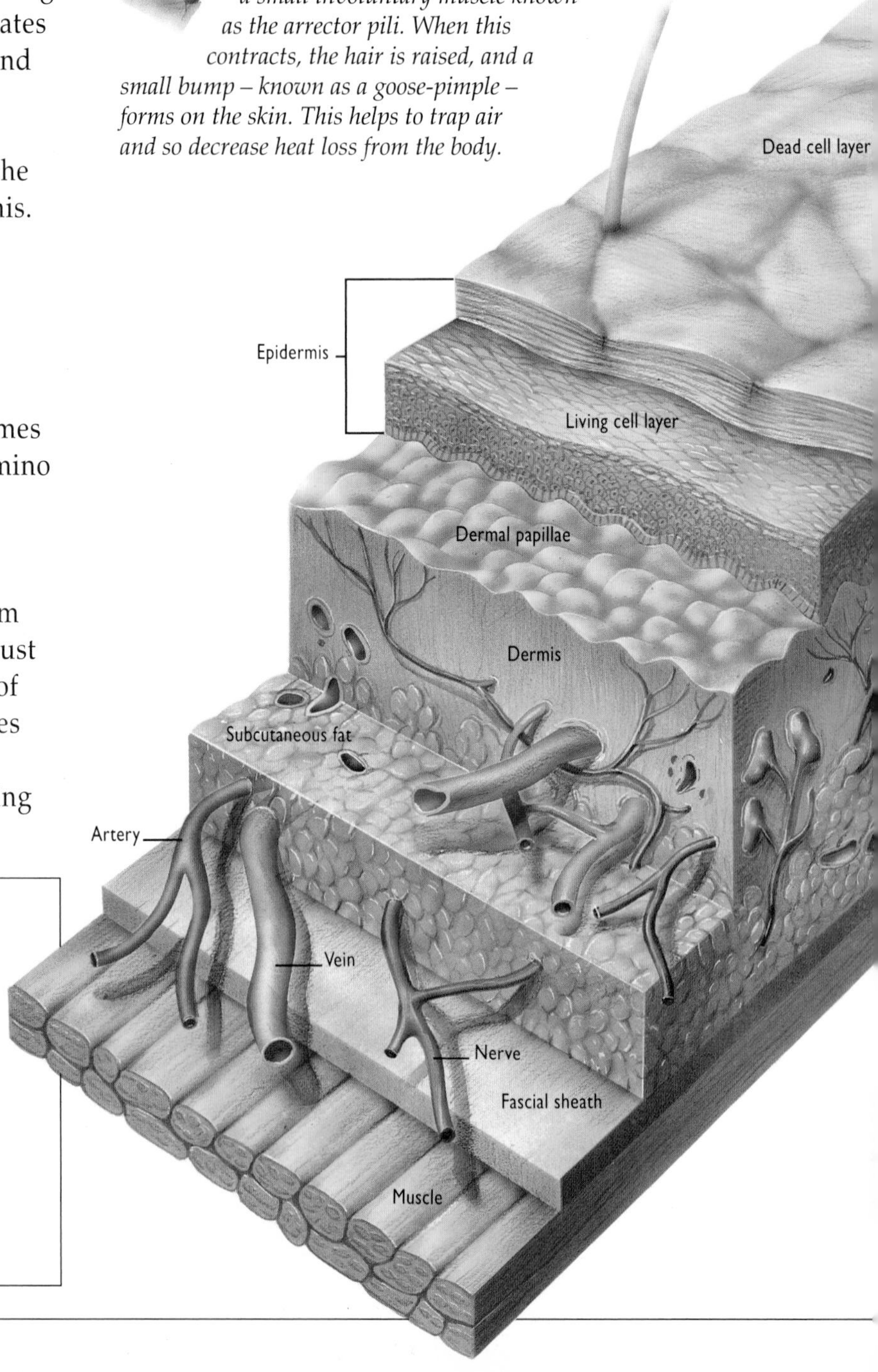

e skin contains three distinct layers even ›ugh it is only as thick as a piece of cardboard. e outermost layer, the epidermis, is a sheet of f-replacing cells; beneath that is the dermis, a cker layer of fibrous connective tissue. Finally, ıyer of looser connective tissue, which often ıtains a great deal of fat, forms the hypodermis. ıderneath lie the superficial muscles.

A person's skin color *depends on how much melanin – a black-brown pigment – is concentrated in the skin. Melanin is found in melanocytes, cells in the lower part of the epidermis. These cells have pigment-filled projections which weave between other skin cells. Melanocytes are present in all areas of the body, with higher concentrations in the external genitals, the nipples, the anal region, and the armpits. Melanin is also responsible for the coloring of hair and eyes.*

Melanin plays a vital role as an ultraviolet (UV) radiation screen, stopping excessive amounts of UV in sunlight from reaching the genetic material in cell nuclei. Too much UV radiation can mutate the genetic material and turn ordinary skin cells into cancerous cells. Prolonged exposure to UV rays increases melanin secretion, producing a tanned skin that is less sensitive to sunlight. But UV light does play a useful role by converting a substance in the skin called 7-dehydrocholesterol into vitamin D, which is vital for laying down calcium salts in bones and teeth.

Sebaceous glands *are found on all surfaces of the body that grow hairs. They are connected to the sides of hair follicles and produce sebum, an oily secretion. Sebum is made when the interior skin cells break down, producing an oily liquid. It coats hairs and the skin's surface with a waterproofing and softening film which also discourages the attachment of harmful bacteria and fungi. If a gland opening becomes plugged with sebum, inflammation occurs and a blackhead results, which could develop into a pimple. Pimples and blackheads are especially common during sexual maturation when sebum secretion levels are high.*

Sweat provides enough nutrition *to support an average of 65,000 bacteria for every square inch of the human body.*

LIVING WITH THE SWEAT EATERS

Perspiration secreted from sweat glands in the skin performs a variety of roles. In addition to cooling the body through evaporation, it provides nutrients for certain bacteria and fungi that live on the skin's surface and produce acidic waste products such as lactic acid, thus reducing the pH of the skin's surface. This surface acidity creates a microenvironment that invading, harmful bacteria find difficult to inhabit.

These beneficial microorganisms are said to be symbiotic. Symbiosis is a state in which both organisms benefit from the relationship. In this case, the bacteria are able to survive off human secretions, while we benefit from the protective qualities of their waste products.

See also

CIRCULATION, MAINTENANCE, AND DEFENSE

▶ Waste disposal 130/131

▶ The water balance 132/133

▶ Maintaining the system 136/137

▶ Damage repair 140/141

▶ Repelling invaders 144/145

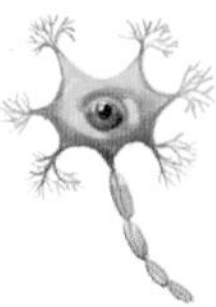

SUPPORT AND MOVEMENT

▶ Staying in shape 12/13

▶ Holding it together 20/21

CONTROL AND SENSATION

▶ Steady as you go 34/35

▶ The contact senses 68/69

Repelling invaders

A range of reflexes and killer chemicals are among the hazards facing foreign bodies that invade the body.

To fight successfully against unwelcome and actively dangerous disease-causing microbes and parasites, the human body has to employ sophisticated defenses. Although the protective mechanisms are diverse, they can be fitted into two broad categories – innate and adaptive, which is generally known as the immune response.

The innate defenses such as the skin and sneezing are fixed defenses – ones which are always available at the same degree of readiness. They are relatively unselective and can be used against many different types of invader. Adaptive defenses, such as antibodies, are more subtle. They are usually selective and specific in their action, operating against only a very narrow band of disease-causing agents. They are adaptive in the sense that, in using them, the body reacts to a particular attack in a responsive rather than a set manner.

The physical barrier of the skin is the first in line of the innate defenses, but beneath it there are a variety of physiological, cellular, and biochemical weapons. These, acting alone or together, are usually capable of removing or destroying the invaders. The physiological weapons include unconscious reflex actions that can damage or remove harmful material o microbes – scratching, sneezing, coughing, vomiting, and th production of tears.

Cellular defenses include cleansing currents in the airways of the lungs which are driven by cilia (tiny waving hairs). Th bronchioles, bronchi, and trachea are all lined with waving ciliated cells that propel sticky mucus from the inner recesse of the lungs up to the throat. This stream of mucus, acting li flypaper, traps fine dust, pollen grains, fungus spores, bacter viruses, and yeast cells from the inhaled air and transports these potentially dangerous particles to the throat. They are then swallowed to be destroyed by hydrochloric acid and digestive enzymes in the stomach, or coughed up. Smoking can destroy this protective flow of mucus, laying the smoke lungs open to easy colonization by bacteria and viruses.

Among the most potent of the many biochemical innate defenses are lysozyme, the antibacterial enzyme found in secretions such as tears, saliva, and milk, and the so-called complement system in the blood. Complement proteins provi a generalized chemical defense against bacteria. When they a activated, either directly by contact with the outer surface of germ or indirectly by antibodies, they shoot holes in bacteri cells. In a series of enzyme activations, non-active compleme proteins are converted to active forms. The clusters of activat digestive enzymes then dissolve holes in the cell membrane the targeted bacteria – killing and repelling the invaders.

LOCAL ACTION

The body responds to local irritation in many protective ways. Scratching, coughing, sneezing, crying, and vomiting can all, in their own ways, help to remove dangerous materials or organisms from some part of the body.

Most of these actions, for example a cough or a sneeze, are involuntary – they are reflexes involving automatically coordinated muscle contractions. Conscious will is normally used only to inhibit these actions.

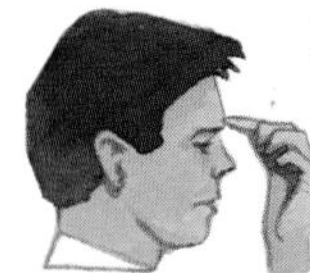

Itching *is caused by low-grade damage to the body's surface layers. Scratching can protect against invasion by scabies or mites, which burrow into the skin's surface. It keeps the parasites under control by excavating their shallow skin burrows, causing them to dry out and die. When itching is due to infection, scratching can cause more soreness and thus more itching – a vicious circle.*

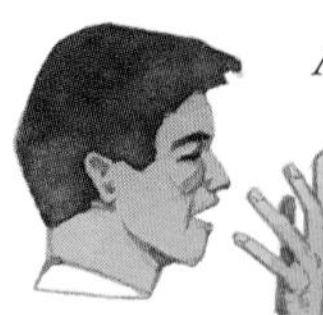

A sneeze *is a sudden exhalation like a cough, but directed through the nose. During a sneeze, air leaves the nose at about 100 mph (160 km/h).*

The cell lining of the passages in the nose is vulnerable to invasion by viruses and bacteria, which can give rise to respiratory infections. Local inflammation of these lining cells, with extra mucus production due to infection, results in irritation in the nasal cavity and the reflex response of a sneeze which can remove infected mucus. Sneezing is also brought on by irritants such as dust or pollen.

The coughing ref *is a rapid and stro exhalation throug the open mouth caused by musc contraction in th diaphragm. The expelled air force particles and microbe-laden muc out of the lungs and upper trache*

A cough is triggered when particles of matter, such as inhal foreign bodies or mucus resulting from increased secretion due to infection, stimulate the upper airways. The cilia-driven stream mucus in the lungs, which delive matter for coughing up, helps to protect the airways from inhaled particles and hostile microbes.

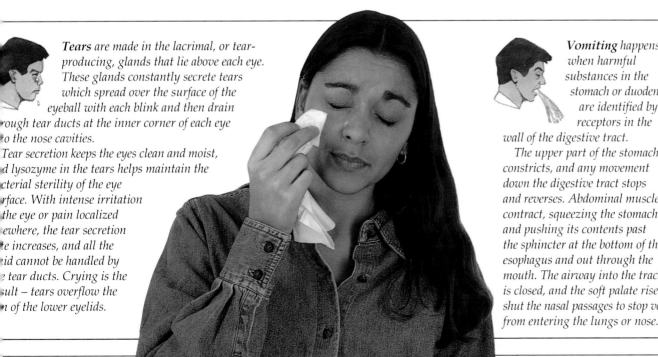

icteria *look like dots even under powerful microscopes. Yet they are 'f-contained unicellular organisms. The bacterium* E. coli *(**right**) en has its own propulsion unit, a rotating flagellum. Bacteria can use humans problems when their metabolic by-products are toxic.*

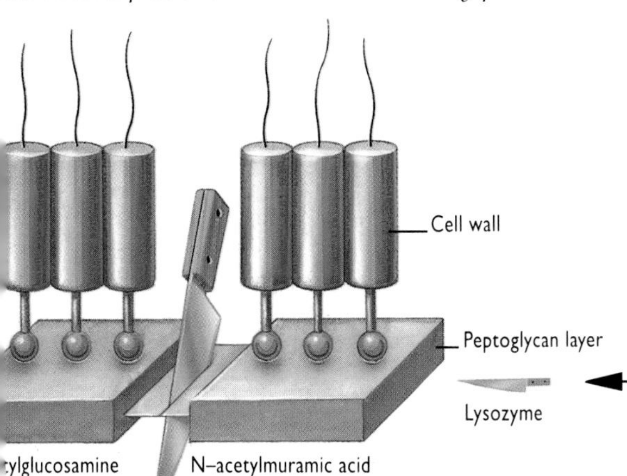

ne of the innate defenses present in the body all the ne is the enzyme (protein catalyst) lysozyme. It is a otective enzyme that can be found in the secretions of vertebrates – creatures with backbones. In people it is esent in mucus secretions, tears, saliva, and breast lk. These secretions are generally found at the open rders of the body where invading bacteria do not have face the highly effective barrier of the skin. For stance, the saliva of the mouth has to deal with a large lux of bacteria every day.

Lysozyme is useful because of its ability to cut through articular chemical bond found in one layer of the protective cell wall that surrounds most bacteria. In the so-called peptoglycan layer of the wall, lysozyme cuts the chemical link between two wall sub-units, N-acetylglucosamine and N-acetylmuramic acid. This breakdown is like removing all the mortar from a brick wall, causing the wall to collapse. With the bonding between the sub-units broken down, the bacterial wall falls to pieces and the bacterium is killed.

Tears *are made in the lacrimal, or tear-producing, glands that lie above each eye. These glands constantly secrete tears which spread over the surface of the eyeball with each blink and then drain rough tear ducts at the inner corner of each eye to the nose cavities.*

Tear secretion keeps the eyes clean and moist, d lysozyme in the tears helps maintain the cterial sterility of the eye rface. With intense irritation the eye or pain localized ewhere, the tear secretion e increases, and all the id cannot be handled by tear ducts. Crying is the ult – tears overflow the n of the lower eyelids.

Vomiting *happens when harmful substances in the stomach or duodenum are identified by receptors in the wall of the digestive tract.*

The upper part of the stomach constricts, and any movement down the digestive tract stops and reverses. Abdominal muscles contract, squeezing the stomach and pushing its contents past the sphincter at the bottom of the esophagus and out through the mouth. The airway into the trachea is closed, and the soft palate rises to shut the nasal passages to stop vomit from entering the lungs or nose.

See also

CIRCULATION, MAINTENANCE, AND DEFENSE
- Maintaining the system 136/137
- Outer defenses 142/143
- Sentinels of immunity 146/147
- Helpers and killers 148/149
- Knowing me, knowing you 150/151

SUPPORT AND MOVEMENT
- Smooth operators 30/31

CONTROL AND SENSATION
- The contact senses 68/69

ENERGY
- Chewing it over 98/99
- The food processor 100/101
- A deep breath 108/109

Sentinels of immunity

Any disease-causing organism invading the body is in for a shock – the body's defenses are ready for it.

In the front line of the immune system are our antibodies – and there are more than a million different types – watching out for any foreign material that enters the body. An antibody is a protein molecule that is able to recognize and stick onto a small, precisely defined part of another protein – normally only a protein fragment that is not part of its own body tissue. Once antibodies have recognized the alien proteins of invaders – the most important of which are disease-causing organisms, such as viruses, fungi, bacteria, and parasites – the defensive machinery of the immune system goes into action. But why are there so many different antibodies, and why do they not recognize the proteins of their own body?

The cells that make antibodies are a class of lymphocyte (white blood cells formed in lymph tissues) known as B-cells. Each B-cell carries, on its surface, molecules of its own individual and specific antibody. During the early development of a baby in the womb, the genes that specify which antibody a B-cell will make undergo a type of

An individual B-cell *makes copies of its own specific antibody and attaches them to the outer surface of its cell membrane. The antibodies extend outward like minute, highly tuned antennas waiting for contact with the specific bit of protein that they can recognize.*

When a disease-causing organism *gets into the body, its surface has proteins (antigens) that are different from those of the body's own tissues. There will thus be some B-cells whose antibodies recognize and stick to those antigens. When they do, the shape of the antibody's Y changes. Like a switch being turned on, this alteration activates the B-cell carrying the antibody and causes it to divide. Huge numbers of cell "offspring" are produced as the cell divides; its progeny then divide, as do their progeny, and so on. The copies have the same antibody-making ability as the parent B-cell. This means that the cell divisions rapidly increase the number of the body's B-cells that can respond to the invading organism.*

An antibody *is made of two light and two heavy chains of amino acids (the sub-units of proteins) held together in a Y shape. The base of the Y is made of heavy chains; its arms are made of both light and heavy chains. The constant regions of chains are the same in many different types of antibodies. But the variable regions – the tips of the arms – each have a uniquely shaped cavity that fits exactly onto the shape of the antibody's "chosen" protein fragment.*

When millions of copies have been made, *most B-cells stop dividing and become plasma cells, type of cell whose interior is packe with the apparatus to make one product – antibodies. These are not stuck to the surface of cells, but are sent as free antibodies intc the blood plasma, where they can come into contact with, recognize, and bind to invading antigens.*

Some of the dividing B-cells continue dividing indefinitely, keeping the population of cells tha respond to the invading antigen "filled up" for a long period, usually many years. These cells a known as memory cells and are th basis of long-term immunity to a disease had in the past.

e shape of the "teeth" on a y fits only a specific lock. If the y is the wrong shape, it will not ork the lock. Like a key, an tibody must be the same shape a fragment of protein on an vading antigen before it can cognize and stick to it. There are er a million different shapes of tibody, so there will always be at st one that will fit the shape of a ce of any antigen's protein.

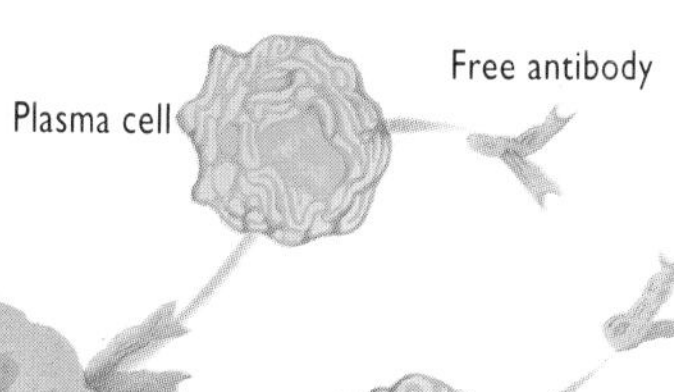

Plasma cell

Free antibody

mory cell

IMMUNIZATION

The process of immunization artificially primes the immune system to repel disease organisms (pathogens) as soon as they come into contact with it. When the body is injected with antigens (foreign proteins) identical or very similar to those of the pathogen, B-cells are activated. They multiply, produce secreted antibodies, and set up memory cells, even though the pathogen is not yet in the body. If the pathogen later invades the immunized person, the pre-existing antibodies and memory cells mount a massive, rapid attack, and the person does not succumb to the disease.

Often the material that is injected is a mild version of the real pathogen, which stimulates but does no harm. Alternatively, a close relative of the pathogen can be injected that induces immunity but does not cause the disease itself.

In immunization's early days, *patients were injected with cowpox (a mild viral disease in humans). This switched on immune protection against the closely related and often lethal smallpox virus. Contemporary satirists lampooned the procedure.*

Free antibodies *manufactured in countless millions by plasma cells roam around in the blood and lymph fluid. When they bump into an invading antigen that they recognize, they bind onto it. Antigens with an antibody attached to them have thus been "branded" by the antibody in a way that means that other parts of the immune system know that the antigen should be attacked.*

When an antibody binds onto its target antigen, it changes shape. It is the shape change of the antibody that makes it "stick" to the outside of macrophages (large white blood cells that engulf and destroy foreign matter). A bacterium with an antibody adhering to it is easily engulfed and killed. This process by which invaders are made more "attractive" to macrophages by antibody attachment is known as opsonization.

genetic shuffling. The result of this is that each of the millions of early B-cells can make a different antibody – but only that one – each able to stick to a different protein fragment. But any B-cell that bumps into a protein that it recognizes binds to it and dies. Since some of the antibodies will recognize proteins that are part of the fetus's body, this huge population of B-cells is reduced. By removing all the responding B-cells when the baby is in the womb – where there are no foreign proteins about – the body makes sures that the remaining set of B-cells can only recognize foreign proteins, or antigens.

Macrophage

See also

CIRCULATION, MAINTENANCE, AND DEFENSE
- Maintaining the system 136/137
- Repelling invaders 144/145
- Helpers and killers 148/149
- Knowing me, knowing you 150/151

CONTROL AND SENSATION
- Key chemicals 42/43
- Moods and emotions 84/85
- Mind over matter 86/87

ENERGY
- The cell and energy 112/113

Helpers and killers

Moving into action against invading organisms is a deadly cocktail of cells that cooperate in the kill.

The immune system has many interlocking components, but its underlying sophistication and precision depend on two major weapon systems: antibodies and the cell-mediated immune system. The key players in cell-mediated immunity are the T-cells (lymphocytes, or white blood cells), but the process depends on a chain of interactions between various cell types.

First in the chain is an antigen-presenting cell (APC) that "eats" any foreign material in the body. Examples of APCs are macrophages – white blood cells found in most tissues that can engulf and destroy bacteria and other disease-causing agents (pathogens). They do this by taking them into a cavity, or vesicle, in their cytoplasm – the part of the cell outside the nucleus – and adding digestive chemicals to them. These chemicals break the bacteria into fragments of the proteins from which they were made, fragments that are now harmless, but which can also be utilized.

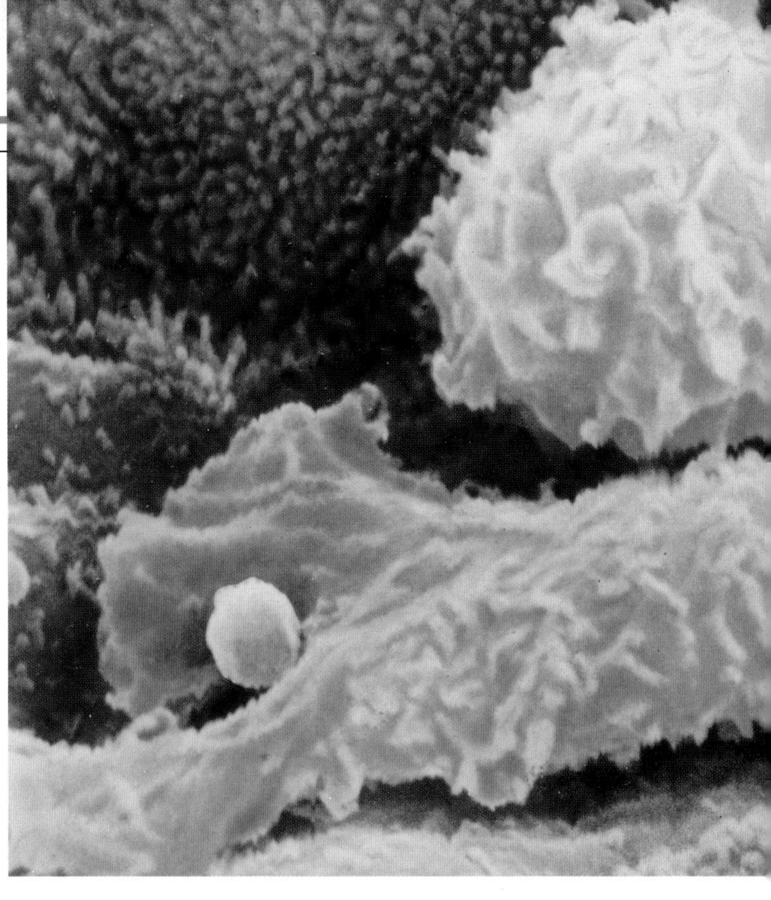

When a defending macrophage, *or engulfing cell, has eaten an invading germ, it sends bits of the germ's foreign (antigenic) proteins to its surface. Here the fragments, or peptides, are presented on the cell membrane stuck to a molecule of major histocompatibility (MHC) protein. Only when presented on an MHC molecule can the antigen fragment be recognized by a helper T-cell.*

Peptide fragment of antigen

Peptide–presenting site

Antigen–presenting cell (macrophage)

TISSUE TYPING

The major histocompatibility complex (MHC) proteins on the surface of human cells such as macrophages are extremely important in determining the "foreignness" of tissues during transplant surgery. Donor cells with identical MHC proteins to those of the recipient are least likely to be rejected after the transplant. This is why transplants between identical (genetically identical) twins are very rarely rejected.

When the transplant comes from a genetically non-identical donor, it is important to match up MHCs as accurately as possible. This is done by "tissue-typing" the recipient and several possible donors. Once typing has been carried out, an organ from the donor with the closest match to the recipient is used for the transplant. If tissues are not very closely matched, it is highly likely that they will be attacked by the recipient's immune system, resulting in rejection.

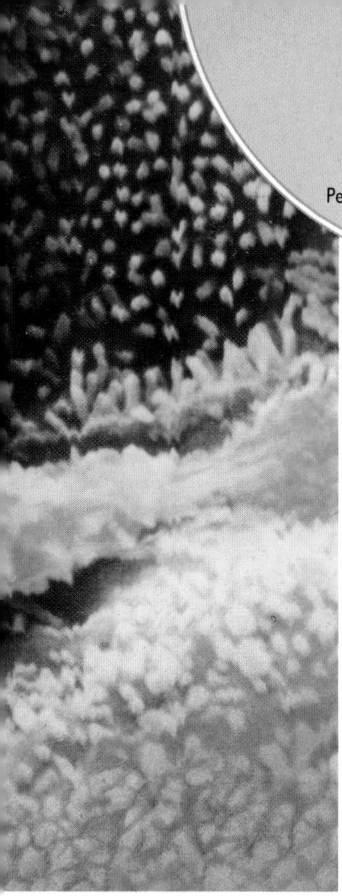

Peptide–presenting site

Peptide

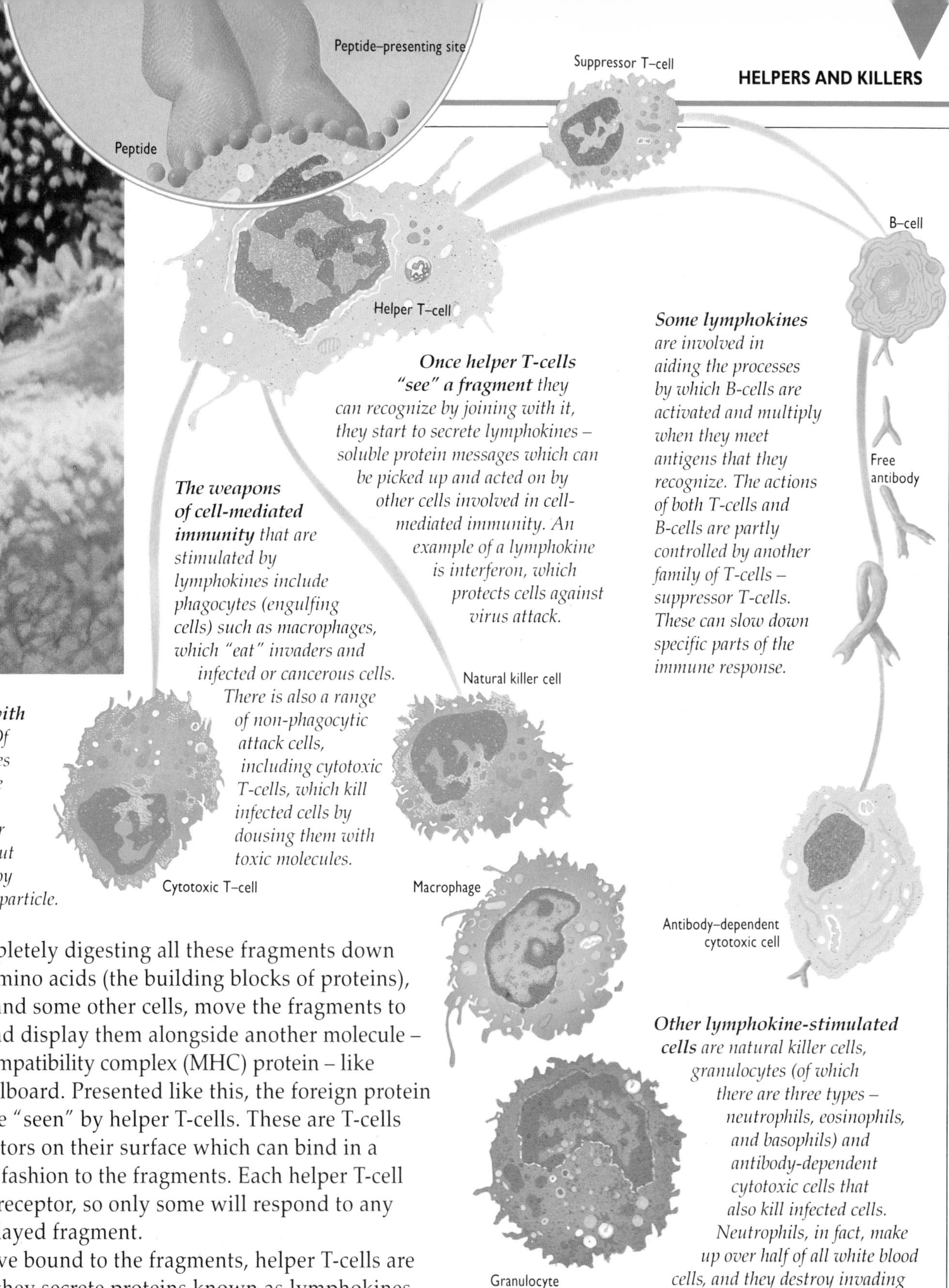

Suppressor T–cell

Helper T–cell

B–cell

Free antibody

Cytotoxic T–cell

Natural killer cell

Macrophage

Antibody–dependent cytotoxic cell

Granulocyte

Once helper T-cells "see" a fragment they can recognize by joining with it, they start to secrete lymphokines – soluble protein messages which can be picked up and acted on by other cells involved in cell-mediated immunity. An example of a lymphokine is interferon, which protects cells against virus attack.

***Some lymphokines** are involved in aiding the processes by which B-cells are activated and multiply when they meet antigens that they recognize. The actions of both T-cells and B-cells are partly controlled by another family of T-cells – suppressor T-cells. These can slow down specific parts of the immune response.*

***The weapons of cell-mediated immunity** that are stimulated by lymphokines include phagocytes (engulfing cells) such as macrophages, which "eat" invaders and infected or cancerous cells. There is also a range of non-phagocytic attack cells, including cytotoxic T-cells, which kill infected cells by dousing them with toxic molecules.*

***…e body teems with …otective cells.** Of …e two macrophages …otographed in the …ng **(above)**, …e elongated darker …nk one is just about … engulf and destroy …e green invading particle.*

…stead of completely digesting all these fragments down … individual amino acids (the building blocks of proteins), …acrophages, and some other cells, move the fragments to …eir surface and display them alongside another molecule – … major histocompatibility complex (MHC) protein – like …osters on a billboard. Presented like this, the foreign protein …agment can be "seen" by helper T-cells. These are T-cells …at have receptors on their surface which can bind in a …ock and key" fashion to the fragments. Each helper T-cell …as a different receptor, so only some will respond to any …articular displayed fragment.

Once they have bound to the fragments, helper T-cells are …tivated, and they secrete proteins known as lymphokines. …hese act as chemical messages that enable other T-cells, …-cells, and phagocytes (engulfing cells) to become active …nd fight against the foreign material in the body.

***Other lymphokine-stimulated cells** are natural killer cells, granulocytes (of which there are three types – neutrophils, eosinophils, and basophils) and antibody-dependent cytotoxic cells that also kill infected cells. Neutrophils, in fact, make up over half of all white blood cells, and they destroy invading cells by engulfing them. Cytotoxic cells secrete chemicals that make the contents of invading cells leak out, thus destroying them.*

See also

CIRCULATION, MAINTENANCE, AND DEFENSE

▶ Waste disposal 130/131

▶ Maintaining the system 136/137

▶ Repelling invaders 144/145

▶ Sentinels of immunity 146/147

▶ Knowing me, knowing you 150/151

CONTROL AND SENSATION

▶ Moods and emotions 84/85

ENERGY

▶ Key chemicals 42/43

▶ The cell and energy 112/113

Knowing me, knowing you

The immune system's ability to tell self from non-self, in terms of tissues, is not always a good thing.

Exquisitely precise, the defensive mechanisms of the human immune system are vital for our wellbeing. Antibodies, protective T-cells, and phagocytes (engulfing cells) together provide defense against all types of invading pathogens and many types of cancer cells. But there are circumstances in which the system can cause problems: both in autoimmune diseases and following tissue and organ transplant, for example. Autoimmune diseases are ones in which our own tissues become damaged because our immune defenses are inappropriately directed against them in a type of immunological "friendly fire." Examples of such diseases are multiple sclerosis and rheumatoid arthritis.

Tissue and organ transplants can be a potential problem for the immune system. Unless donor and recipient are one and the same person – in the case of a skin graft, for example – or the donor is an identical twin of the recipient, there is a strong possibility that some of the tissues of the transplant from the donor will be regarded as foreign by the immune system of the recipient. In such cases, with antibodies and T-cells targeting the transplanted tissue, their combined attack may cause the transplant to be rejected.

A number of approaches are used to overcome this problem. Some transplants, like the corneas of the eyes, contain no blood vessels or lymphatic vessels. This means that lymph- and blood-borne T-cells cannot get to the transplant, so it cannot be rejected. Because of this, cornea transplants between genetically non-identical people are usually successful.

When transplanted organs – a kidney, liver, lung, or heart, for example – do have blood vessels, something has to be done to prevent rejection. Tissue typing is a checking procedure prior to the transplant to make sure that the donor's tissues are as similar as possible to the recipient's. This minimizes the "foreignness" of the transplant. Immunosuppressive drugs can also dampen down the rejection reaction. These drugs reduce the immune attack on the transplant in a number of ways. Some, like 6-mercaptopurine, inhibit the multiplication of white blood cells; others, such as cortisone, reduce inflammation; while cyclosporins reduce the activity of helper T-cells.

The body's immune defenses *attack matter made of proteins that are foreign to the body. A substance like Dacron (**right**) has no proteins in it so can be used to make artificial body parts.*

A person donating blood *is giving tissue to be transplanted. Blood is as much a tissue as the more obvious transplanted organs such as the heart, lun and kidneys. It is also a tissue that can be donated easily. Most people can do it on a regular basis without feeling any ill-effects. Some people should not give blood, however, including those who have suffered from certain diseases such as hepatitis and those who are HIV positive.*

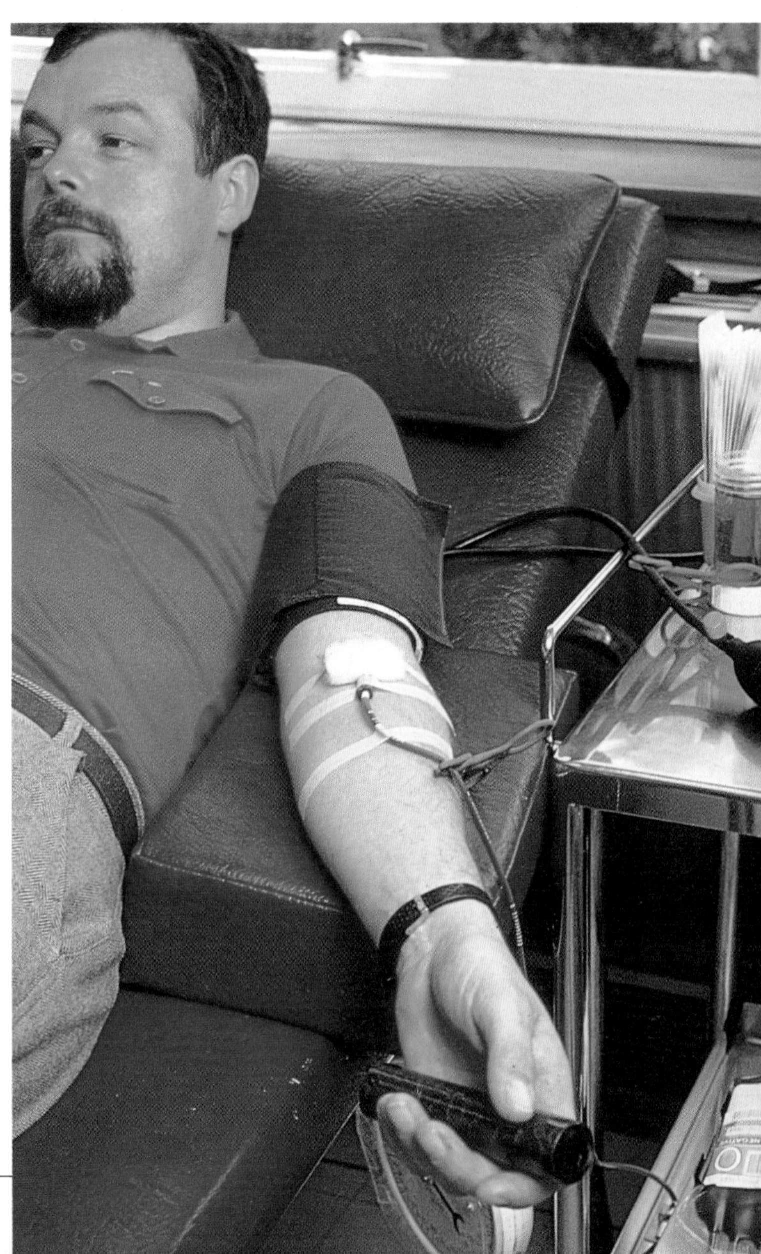

Non-living material *can sometimes be successfully placed inside the body without the immune system attacking it. For this to occur, the material introduced must be chemically inert, so that it does not react with substances in the body to corrode or break down. It should also contain no organic substances that might be recognized as foreign by the immune system. Most of the materials used have surfaces that are ultra-smooth and chemically "uninteresting" to the immune system. Examples include stainless steel and titanium for bone prostheses, and the plastics Teflon and Dacron for heart valves and blood vessel walls. The inert properties of these materials are also exploited in more everyday situations, like the non-stick Teflon coating of a frying pan and stainless steel in kitchen knives, the Dacron in clothing and the titanium of the outer case of a wrist watch.*

Using such materials, scientists have been able to develop a number of artificial parts that can be used to replace human components. For instance, knee, hip, ankle, toe, finger, and shoulder joints as well as inter-vertebral disks can be inserted where the body's own joints have become non-functional. Parts of the circulatory system, including major and minor arteries and the heart, can also be replaced. Testicular implants are available for cosmetic replacement of a missing testis. There is even a hydraulic implant available with a pump to inflate the penis where this is no longer possible by natural means.

THE CORRECT TYPE OF BLOOD

Often a life-saving procedure, the transfusion of donated blood into the circulatory system of a patient is the most common type of tissue transplant. As with other types of tissue transplant, careful matching of the donor's and the recipient's blood types is necessary to make sure that transfused blood is not rejected.

In the early years of blood transfusion, it was not understood why transfused blood was sometimes incompatible with that of the recipient. And by 1900, through the investigations of the Viennese pathologist Karl Landsteiner (1868–1943), it was realized that it was not possible to use blood safely from just any donor when giving blood to a particular patient. The reason for this is that there are many different blood types, corresponding with the different types of proteins (antigens) on the surface of blood cells.

A person generally has one specific type of blood-cell antigen (the two main ones are termed A and B) and antibodies to the antigen or antigens they do not possess. This means, for instance, that someone with blood antigen type A has antibodies to B, those with blood type B have antibodies to A. People with both antigens, A and B, have antibodies to neither. Conversely, those with neither A nor B (the so-called O group) have antibodies to both. In transfusion, it is the antigens contained in the transfused blood that cause potential problems. If the recipient has antibodies to the transfused antigens, a dangerous immune reaction occurs, and the red cells clump together.

See also

CIRCULATION, MAINTENANCE, AND DEFENSE
- Blood: supplying the body 126/127
- Maintaining the system 136/137
- Sentinels of immunity 146/147
- Helpers and killers 148/149

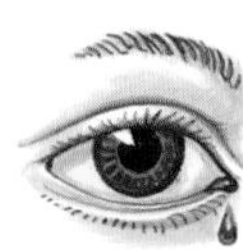

SUPPORT AND MOVEMENT
- Bone junction 18/19

CONTROL AND SENSATION
- Down the wire 46/47
- The biological camera 56/57

ENERGY
- Cells at work 114/115

REPRODUCTION AND GROWTH
- In the family 156/157

Reproduction and Growth

The cycle of life is repeated endlessly down the generations: people are born, mature, reproduce, and in the fullness of time, die. The story of human life from the fertilization of an egg by a sperm right up to adulthood is one of the most triumphant demonstrations of the complexity and precision of nature at work. And, as with most things to do with the human body, the crucial workings of the process take place at a cellular and molecular level. Despite its minute size, scientists have worked out how the information of heredity is stored, duplicated, passed on, and then translated into the many types of tissue that make up a human. And this knowledge of genetics now places all of the stages and processes of reproduction and growth in a much clearer light.

__Left (clockwise from top)__: genetic material – the stuff of life; the learning process; starting out; the shuffling of genes; growing to a plan.
__This page (above)__: nearly ready for the world; __(right)__ DNA – instructions for life.

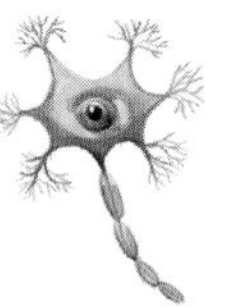

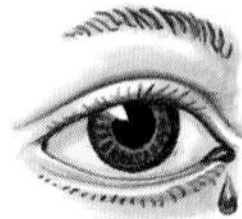

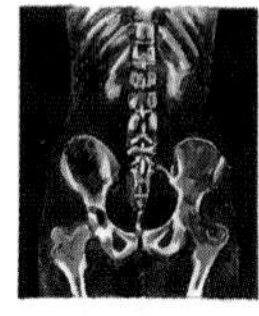

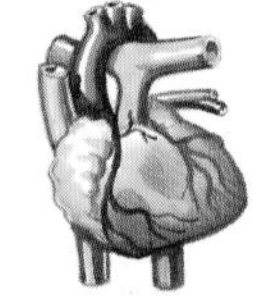

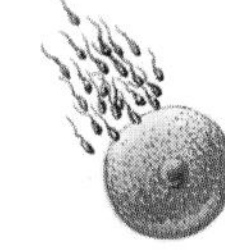

Cradle to grave

The average human life span depends on a number of factors.

Extraordinary – and unpredictable – the story of a human life is altered and molded by a variety of far-reaching influences. For instance, the exact physical and intellectual development of individuals is influenced both by their genetic make-up and by the conditions – physical and cultural – in which they grow up. And while genes from parents certainly determine a great deal – from eye color to the ability to roll the tongue into a tubelike shape – to a large extent they govern potentials rather than actual outcomes.

The effect of genetic traits is strongly modified by conditions in the environment that a person inhabits. Diseases, nutritional state, upbringing, social opportunities, and a multitude of other factors all help to shape the final pattern of a person's life. For instance, the maximum height that people can attain is determined by their genes, but if they do not receive proper nutrition at the appropriate times in their life, they will not reach it.

Maximum life span, like maximum height, would appear to be genetically programmed – the longest being about 110 years, although few people actually live to that age. But this too is highly influenced by environmental and social factors, as seen in the remarkable increase in life expectancy over the past 100 years, particularly in the developed, industrialized countries of the world. In the United States 100 years ago, average life expectancy at birth was not more than middle age. In the last decade of the 20th century this has risen to almost 79 years for women and 71–72 years for men. Gradually, actual life spans are creeping toward the genetic potential.

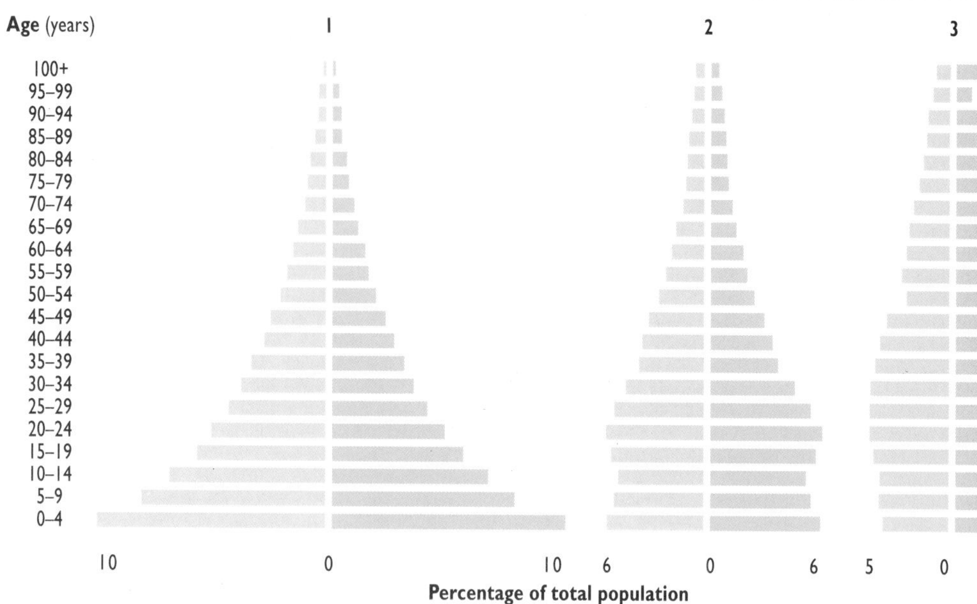

This increasing longevity is the result of a number of interlinked factors, including improved hygiene and clean water supplies. The introduction of sanitation systems and the use of immunization and drugs to control infectious diseases have also played a part. And better nutrition and an increased understanding of the effects of diet and the dangers of habits such as smoking have all contributed to the trend toward longer lives.

***A bar chart** showing percentages of people alive in five-year age ranges reveals the population structure of a country. A wide-based steeply tapering pyramid* (**1**) *is typical of an undeveloped country with high birth rates, high death rates generally, and low life expectancy. The lower birth rate, lower death rates in childhood, and large numbers of a considerable age in a developed country give a more even shape to the chart* (**3**). *A country in a mid-developmental stage* (**2**) *has an intermediate shape.*

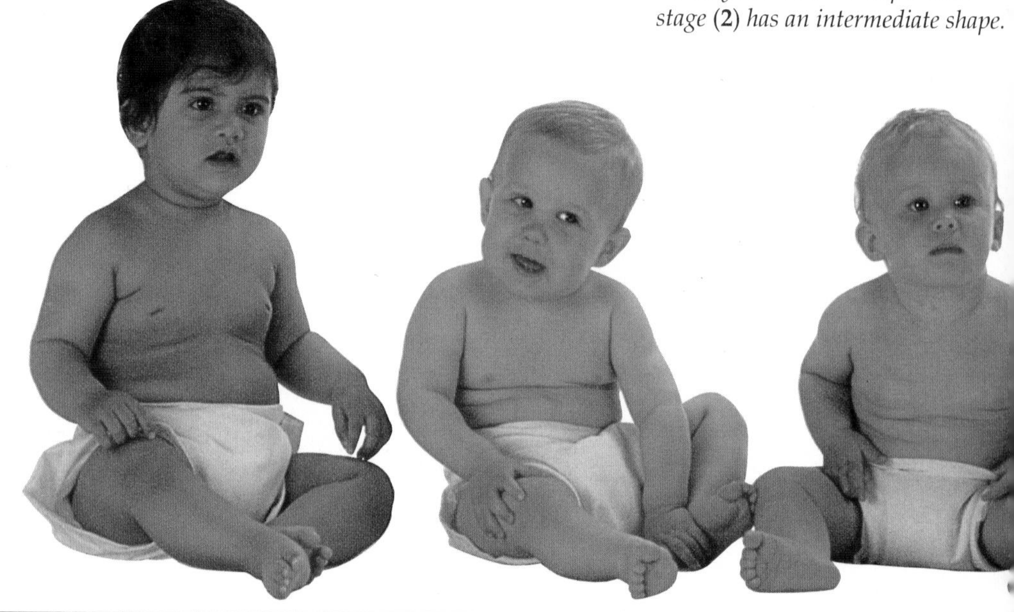

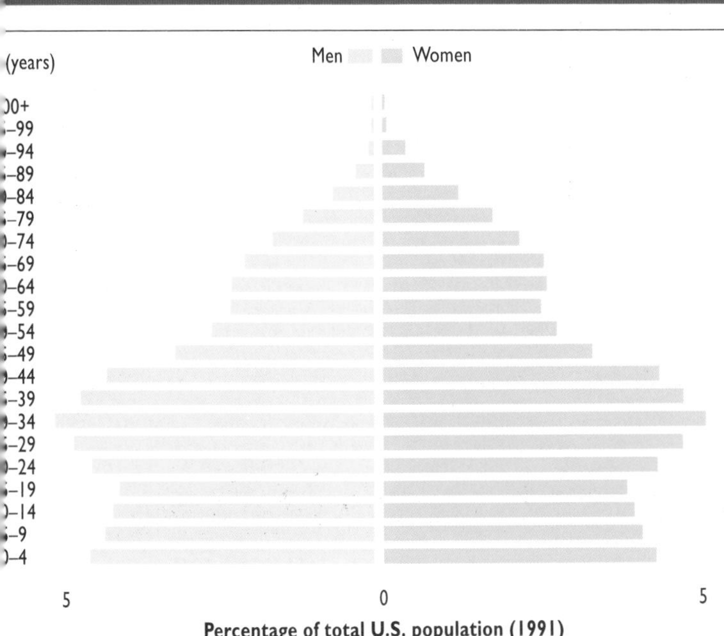

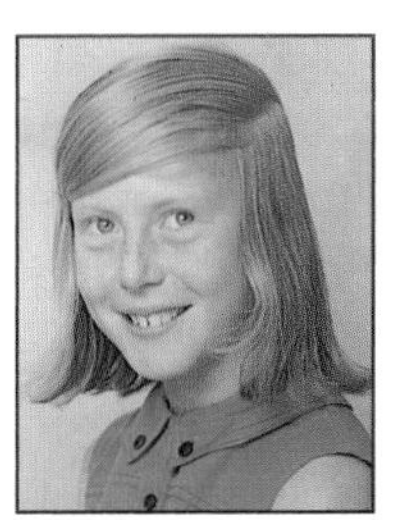

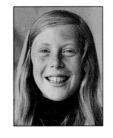

The population chart *for the United States of America in 1991 is typical of a developed country. It has some unusual features, however, including the bulge in the pyramid for people aged between about 20 and 45. Perhaps the most easily attributable cause of the bulge is the so-called baby boom that occurred after World War II. This accounts for the older people in the bulge, born shortly after their fathers returned from the war, and some of the people aged 20 or so, the children of the "baby boomers."*

The faces of four people (right) *photographed at three stages in their lives – childhood, adolescence, and maturity – show how time brings about gradual changes in physical appearance. Despite obvious signs of aging, each person is clearly identifiable at each stage, and his or her basic look – evidenced in facial structure and overall expression – remains remarkably and characteristically constant.*

These six babies *have their lives ahead of them. Already, though, their genetic inheritance, which gives them their potential abilities, is interacting with their environment to shape them into what they will become.*

The babies, between five and nine months old, are embarked on the learning processes of childhood – already, for instance, they have learned how to sit up unaided. As the learning continues, growth and the accumulation of physical, communication, and social skills will transform these totally dependent babies into self-reliant adults who might themselves have babies. How long they live will depend on what was programmed into them by their genes and on how their lives evolve. Some will reach close to their maximum life span, and others will fall far short because of disease, perhaps caused by lifestyle or by factors imposed by the environment.

See also

REPRODUCTION AND GROWTH
- In the family 156/157
- Language of life 158/159
- The newborn baby 176/177
- The growing child 178/179
- Sexual maturity 180/181
- Getting on 182/183

CONTROL AND SENSATION
- Learning skills 80/81
- Problem solving 82/83

In the family

Similarities can be seen between parent and child, but everyone is unique.

Genes, made of DNA, determine height, skin, eye and hair color, facial features, and so on. We get these genes from our parents, and this simple fact explains the hereditary links between generations – children look like their parents because they share genes with them. A person's genes come together at the moment of fertilization and remain the same throughout their life. Half of the genes come from the father's sperm and half from the mother's egg.

In each body cell's nucleus, genes are arrayed along 46 chromosomes, which are arranged in 23 pairs. One chromosome in each pair is a replica of a maternal chromosome, the other a copy of a paternal one. Genes for specific functions are found at precise locations on particular chromosomes, and there are always two copies of any given gene, one on each chromosome of a pair. The two genes in a pair can be identical or different, and this arrangement influences a person's characteristics. A dominant gene is one that affects the body form even when it is paired with a different, or recessive, gene type. Only when a person has the same recessive gene on both chromosomes of a pair is the characteristic of the recessive gene apparent in their body.

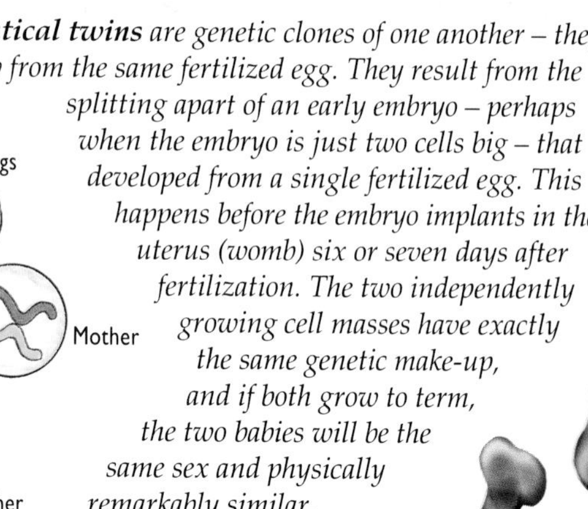

Identical twins *are genetic clones of one another – they grew from the same fertilized egg. They result from the splitting apart of an early embryo – perhaps when the embryo is just two cells big – that developed from a single fertilized egg. This happens before the embryo implants in the uterus (womb) six or seven days after fertilization. The two independently growing cell masses have exactly the same genetic make-up, and if both grow to term, the two babies will be the same sex and physically remarkably similar.*

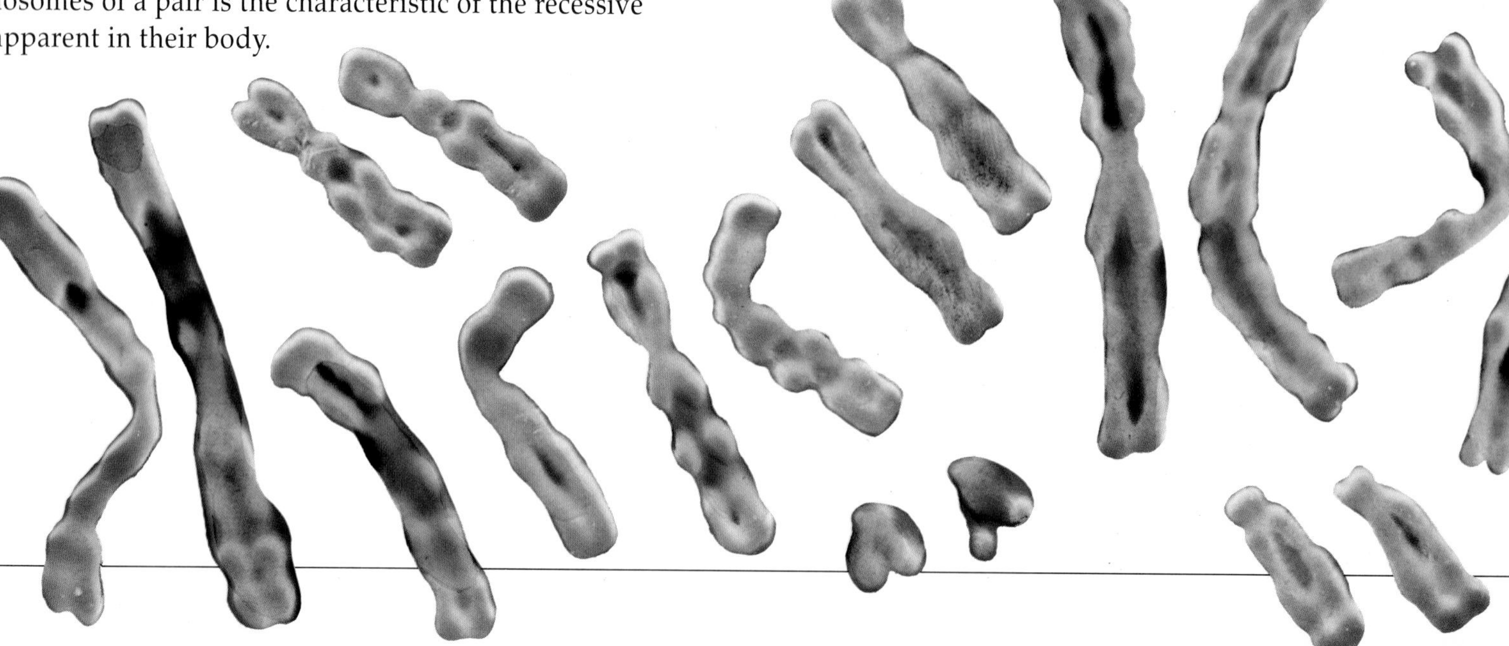

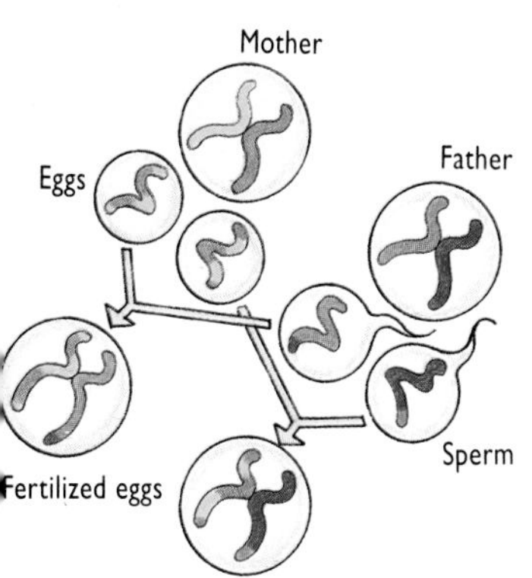

A woman's chromosomes are distinct from a man's, since there are two X sex chromosomes (*shown right*), instead of an X and a Y.

Sex chromosomes

on-identical twins are born when two eggs, ther than one, are released at ovulation and both are rtilized. Since the two children that develop are the sult of two different eggs being fertilized by two fferent sperm, they are not genetically identical. deed, they are no more alike than any two blings. Non-identical twins may be of the me or different sex.

Each person's genetic information is carried by the DNA molecules found in every cell that has a nucleus. DNA molecules, together with their protein support molecules, form the chromosomes; humans have 46 in each cell. At certain stages in a cell's life cycle, the chromosomes – normally diffuse and tangled with each other – literally pull themselves together to form the 46 distinct shapes shown across these pages. By staining the set of 46 chromosomes from a person's cells with special dyes, photographing them with a microscope, and then arranging the chromosomes into a "league table" of 23 pairs, an ordered image called a karyotype is produced.

Each chromosome in each pair has a doubled shape because it has already copied itself in preparation for cell division (mitosis). There are normally 22 pairs of autosomes – chromosomes unconnected with the determination of gender – and one pair of sex chromosomes. Karyotypes are sometimes used in the diagnosis of certain genetic diseases.

WHICH SEX?

A person's sex is determined when an egg from the mother is fertilized by a sperm from the father. Females have two X sex chromosomes (**see above**), while males have one X and one smaller Y chromosome. This means that a child's sex is specified solely by the nature of the sperm that fertilizes the egg from which the child develops.

The mother possesses no Y chromosomes, so all her eggs will have only an X chromosome. By contrast, half of the man's sperm will possess an X chromosome, half a Y. If an X chromosome sperm fertilizes the egg, an XX, or female, embryo results. If a Y chromosome sperm is responsible for fertilization, this produces an XY, or male, embryo. The equal numbers of X and Y sperm in any population of sperm explains why, statistically, males and females are produced in almost equal numbers.

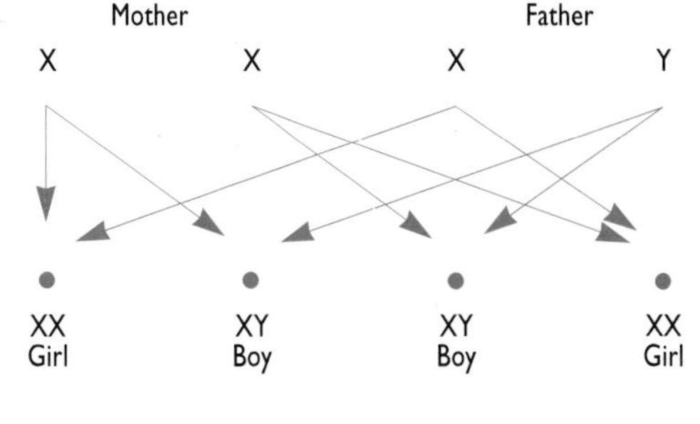

See also

REPRODUCTION AND GROWTH
▶ Language of life 158/159
▶ Building bodies 160/161
▶ The sexual advantage 162/163
▶ Cycle of life 164/165
▶ Fertilization 166/167

ENERGY
▶ Cells at work 114/115

CIRCULATION, MAINTENANCE, AND DEFENSE
▶ Routine replacement 138/139

Language of life

The data of heredity is carried by a set of complex molecules that tells the body how to grow and live.

To build and run such an extraordinarily complex organic machine as a human being takes a vast set of instructions. And yet all the necessary knowledge is crammed into the nucleus of each and every human cell. The information is written in segments, or genes, of a set of huge deoxyribonucleic acid (DNA) molecules. In fact, the DNA molecules carry tens of thousands of genes, which are the instruction units of heredity. But DNA is not found alone in a cell nucleus; it is linked to supporting protein molecules; the DNA molecules and their support proteins together make up chromosomes. The nucleus of each human cell has 23 pairs of chromosomes.

Each gene in a DNA molecule contains the coding for the manufacture of a specific protein – the molecules out of which our bodies are constructed. Each protein is made of a long, linear sequence of sub-units called amino acids; there are 20 different amino acids. The sequence of amino acids is defined by each gene in a list of instructions that spells out the structure of a protein.

The instructions are given in codes made up of four sub-units, or nucleotide bases, on the DNA molecule. There are four bases – adenine, thymine, cytosine, and guanine – usually represented by the letters A, T, C, and G. The codes consist of three-letter words. Every three-letter word that can be made using the four letters (such as AAT, GAT) either stands for an amino acid or is a signal to stop or start a protein chain.

All the information *of human heredity – the instructions for how the body operates and the data for the construction, maintenance, and everyday functioning of individual cells – is packaged into the nucleus of each individual cell, which itself is usually only a minute fraction of an inch across. This information, called the total human gene set, or genome, is carried on chromosomes – DNA molecules and protein support molecules. Each DNA molecule is made of an immensely long sequence of chemical sub-units, known as nucleotides. The genome contains around 6 billion nucleotides, divided between 46 DNA molecules.*

Chromosomes, *which can be seen when they are stained with dye and observed using high-power microscopes, spend most of their time as a diffuse network of extended filaments, known as chromatin. When a cell is dividing chromosomes shrink down and form clearly recognizable paired units (**below**).*

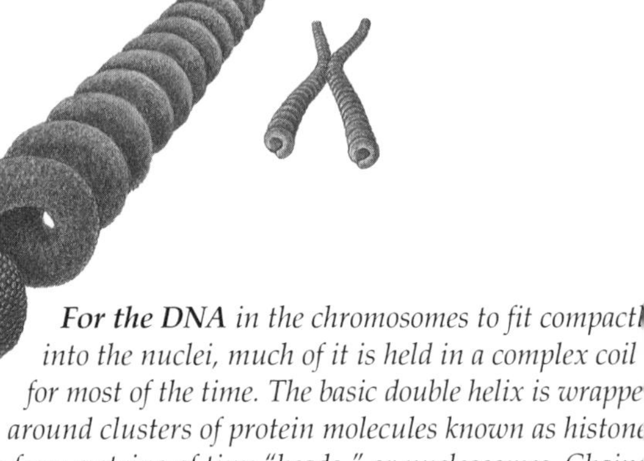

For the DNA *in the chromosomes to fit compactly into the nuclei, much of it is held in a complex coil for most of the time. The basic double helix is wrapped around clusters of protein molecules known as histones to form a string of tiny "beads," or nucleosomes. Chains of nucleosomes are then coiled up.*

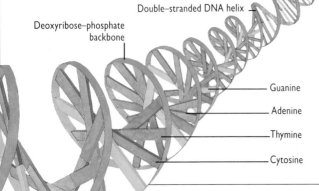

DNA is a double spiral *or helix shape, like a twisted ladder. It is made from the sub-units of DNA, called nucleotides. Each nucleotide is made up of a phosphate group, a sugar unit, and one of four bases (adenine, guanine, thymine, and cytosine). Each side of the ladder is made of a chain of alternating sugar and phosphate units. Two bases link between the sides to make the rungs.*

MOLECULES OF HEREDITY FACT FILE

Written down, the genetic information in a cell's nucleus would be 3 billion words long, enough to fill 3,000 books of 1,000 pages with 1,000 words a page.

No. of chromosome pairs	23
nucleotides in DNA molecule	130 million (ave.)
base pairs in DNA molecule	65 million (ave.)
base pairs in total	3 billion (ave.)
Possible 3-base permutations	64
Width of DNA helix	80 billionths of an inch (2 nm)
Gap between base pairs	13 billionths of an inch (0.34 nm)
Length of 1 twist of DNA spiral	135 billionths of an inch (3.4 nm)

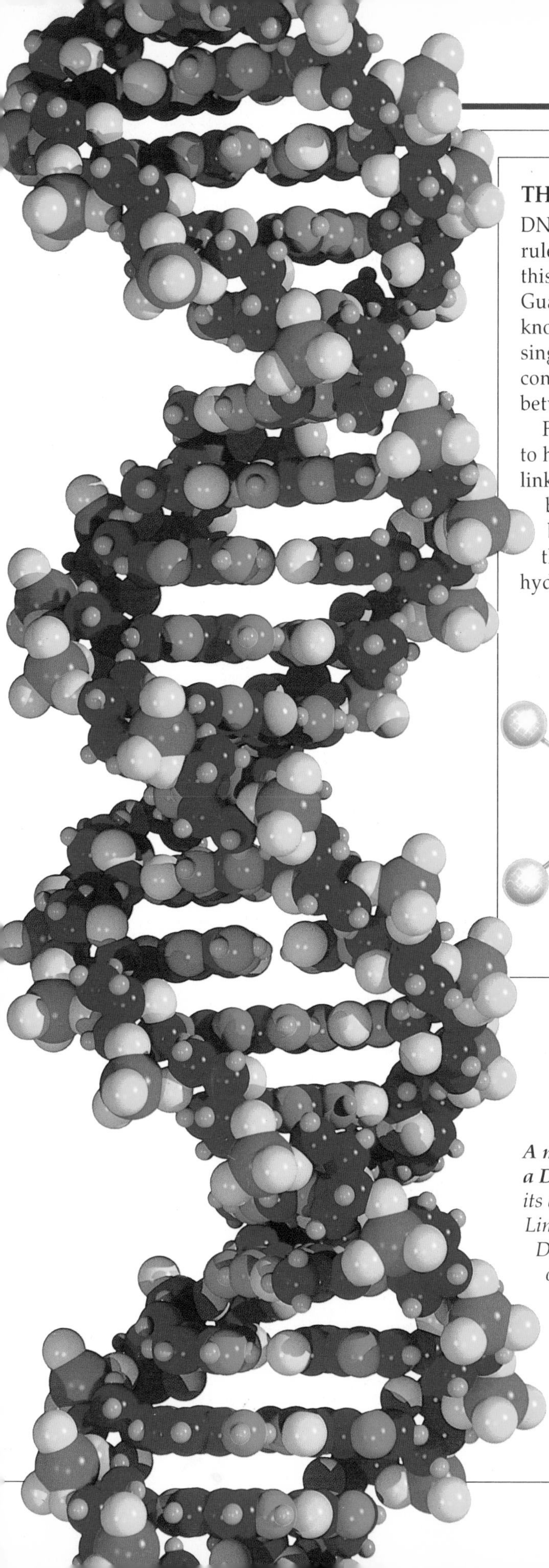

THE INEVITABLE LINKS

DNA can copy itself exactly because of the so-called base pairing rule, which states that A fits only with T, and C only with G; this is explained by the chemical structures of the DNA bases. Guanine and adenine are wider double-ring molecular sub-units known as purines, while cytosine and thymine are narrower single-ring units called pyrimidines. One rung of two bases has to contain one purine and one pyrimidine to be the right length to fit between the two strands of the DNA ladder.

For two bases to be able to "dock" with one another, they have to have the right arrangement of atoms to match the number of links, or hydrogen bonds, offered by the opposite base. A and T both have two bonds, so they can link; C and G both have three bonds and they, too, can link. But A cannot link with C, despite their being a purine–pyrimidine couple, because A forms two hydrogen bonds and C three – they are incompatible.

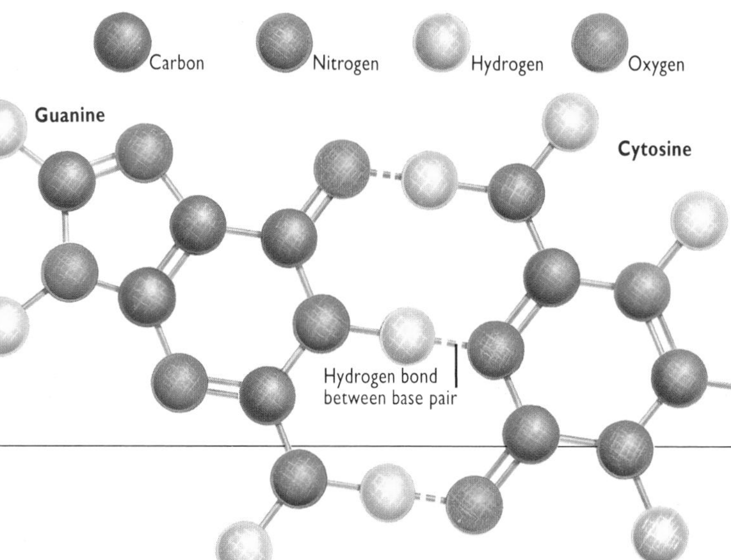

A model of a tiny section of a DNA molecule (left) *shows its double helix construction. Linking the two sides of the DNA helix are the bases. The only base pairings which are possible are adenine (A) with thymine (T) and cytosine (C) with guanine (G). This enables DNA to copy itself when the molecule unzips lengthwise down the middle, since each base on each half can only link to one other base. Every A needs a replacement T and every G needs a C. Each strand of unzipped DNA thus picks up the missing bases (which are swirling around in the chemical "soup" of the nucleus) in the exact same order, creating two identical new DNA molecules. This capacity for molecular copying of genetic material underlies the reproduction of all life on Earth.*

See also

REPRODUCTION AND GROWTH
▶ In the family 156/157
▶ Building bodies 160/161
▶ The sexual advantage 162/163
▶ The growing plan 170/171
▶ The future body 184/185

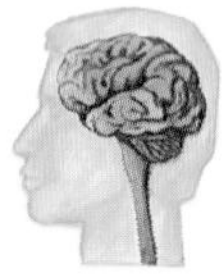

SUPPORT AND MOVEMENT
▶ Staying in shape 12/13

CONTROL AND SENSATION
▶ Making sense 72/73

ENERGY
▶ The cell and energy 112/113

CIRCULATION, MAINTENANCE, AND DEFENSE
▶ Routine replacement 138/139

Building bodies

How does the human genetic blueprint – written on molecules in the cell nucleus – become living tissue?

The structure and physical functioning of the different parts of the body are based upon the structure and functioning of tens of thousands of different proteins. Some of these, such as the digestive enzymes which break up the food we eat, perform specific metabolic jobs; others, like the proteins of muscle tissue, make up structural components of the body.

Proteins are made in the body's cells, and the instructions for their manufacture are stored in genes – segments of deoxyribonucleic acid (DNA) molecules which are found inside the nucleus of almost all body cells. Some genes carry plans for proteins with neither metabolic nor structural jobs. These are control proteins – they determine when other proteins should be made and when they should not. In effect, their function is to switch on and off the genes that build specific proteins or groups of proteins. Some genes are switched on in all cells, but others only operate in specialized cells. For example, the genes that make digestive enzyme proteins are switched on in digestive-system cells, and muscle-protein-making genes are switched on in muscle cells.

DNA thus contains the plans and instructions for the body. Under normal circumstances, these plans not only do not change – if they

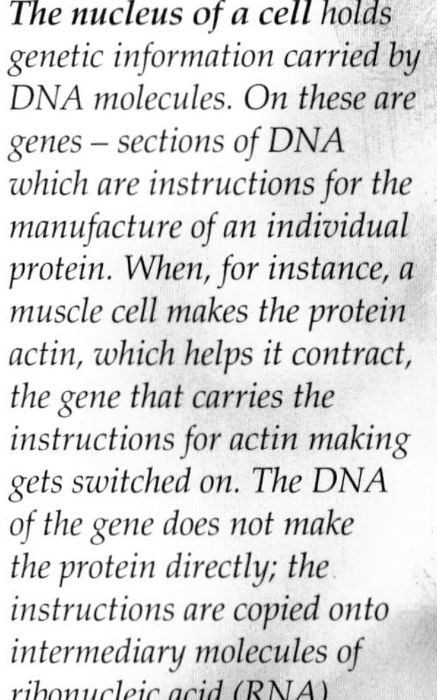

***The nucleus of a cell** holds genetic information carried by DNA molecules. On these are genes – sections of DNA which are instructions for the manufacture of an individual protein. When, for instance, a muscle cell makes the protein actin, which helps it contract, the gene that carries the instructions for actin making gets switched on. The DNA of the gene does not make the protein directly; the instructions are copied onto intermediary molecules of ribonucleic acid (RNA).*

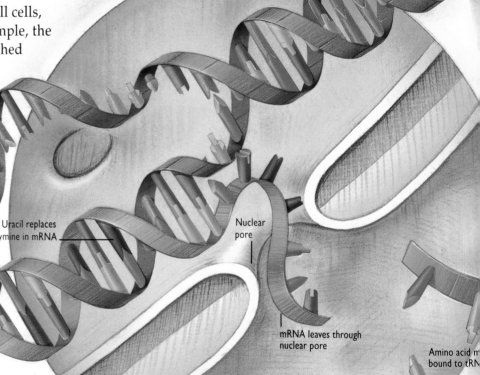

***To copy a gene,** a process known as transcription, the section of the ladderlike DNA molecule carrying the gene splits. Each half rung, made of a chemical sub-unit, or base, can join up with just one other base. Thus a sequence of bases is copied in the exact order as they appear on the gene, forming a single-stranded nucleic acid – RNA.*

***The RNA** carrying the copied gene is known as messenger RNA (mRNA). The base uracil replaces the DNA's thymine.*

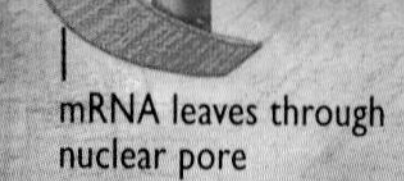

Granular endoplasmic reticulum

Endoplasmic reticulum is an organelle, or structure, found in cells outside the nucleus. It is made of flattened sacs of membrane, and it both circulates materials around the cell and acts as a store for enzymes (molecular catalysts) and proteins. In some areas the outer surface of the reticulum is covered with thousands of tiny ribosomes, giving it a rough appearance. This is called granular endoplasmic reticulum. Ribosomes are the actual sites of protein making. They are hamburger-shaped structures, each consisting of a large and a small sub-unit, which contain roughly equal amounts of protein and a type of RNA known as ribosomal RNA (rRNA).

Ribosome

did the result would be mutation and chaos – but they also never leave the nucleus. So how does an activated (switched-on) protein-building gene bring about the manufacture of the protein for which it carries plans?

When a gene is activated in the nucleus, the first thing that happens is a process called transcription – a copy of that gene's instructions is made. The copy is created in the form of a molecule of messenger ribonucleic acid (mRNA). The mRNA moves out of the nucleus, through a nuclear pore, to a ribosome, a site where the information the mRNA carries is used to make the protein. In this process, known as translation, the chemical sub-units of the protein – amino acids – are brought together in the correct order and number with the help of molecules known as transfer ribonucleic acid (tRNA). After assembly, the protein leaves the ribosome and goes to where it is needed.

Protein chain forms actin fiber in muscle cell

Large ribosomal sub-unit

Assembled protein chain

tRNA links with mRNA

Small ribosomal sub-unit

After leaving the nucleus via nuclear pores, mRNA moves to a ribosome. Here the genetic code sequence in the mRNA is converted into the sequence of amino acids that make a protein molecule. Conversion of the mRNA code into a protein is known as translation. The process relies on the fact that every sequence of three bases along the DNA, and thus along the mRNA copy, is a code for an individual amino acid. For translation to occur, another type of RNA, transfer RNA (tRNA), is needed. At one end of a tRNA molecule is a triplet of bases that corresponds to a code for an amino acid; at the other end of the tRNA is the amino acid itself.

On the ribosome, triplets of the mRNA base code are activated in sequence and link with the appropriate tRNA molecules which bring the amino acids. The many hundreds of triplet code words are rapidly changed into a sequence of amino acids to make a protein molecule, which then leaves the ribosome. In the case of actin, it forms part of the fiber in a muscle cell. Thus the DNA, which remains inside the nucleus, has been copied to make protein that makes part of a tissue.

See also

REPRODUCTION AND GROWTH
- In the family 156/157
- Language of life 158/159
- The sexual advantage 162/163
- Fertilization 166/167
- The growing plan 170/171

SUPPORT AND MOVEMENT
- Muscles at work 26/27

ENERGY
- Absorbing stuff 104/105
- The cell and energy 112/113
- Cells at work 14/115

CIRCULATION, MAINTENANCE, AND DEFENSE
- Routine replacement 138/139

The sexual advantage

All organisms of all species live by the same, inescapable, law of nature – reproduce or die out.

Every individual organism ultimately dies, and the only way in which this can be partly overcome is to leave offspring that carry characteristics of the organism into the future. Many "simple" species reproduce asexually: individuals produce copies of themselves containing identical genetic information. But many species – humans included – reproduce sexually.

In humans the result of sexual intercourse is that an egg cell meets a sperm cell and a new, unique cell is made, a fertilized egg. Many cell divisions later, the egg becomes a baby. We take it for granted that a child shares some characteristics with its parents, but has others that are unique to itself. In fact, apart from pairs of identical twins, each of the 5 to 6 billion people now alive is genetically unique. This is because of the two-part gene shuffling of sexual reproduction.

First, a human's genes are the result of the coming together of the gene sets of the egg and sperm, mixing together in one individual genes that were previously in separate individuals: the mother and father. Second, the production of the half gene set in each egg and sperm involves a type of shuffling, called meiosis, which precedes sperm and egg formation. In a normal cell there are 46 chromosomes (structures carrying the DNA molecules that spell out the genes) arranged in 23 pairs. Meiosis halves the chromosome number from 46 to 23, but each of the new 23 contains genes selected at random from one of the pairs. The assembling on one chromosome of genes that were previously on two separate parental ones is a process known as crossing over. The double shuffling of crossing over and egg–sperm union generates the almost infinite variety of humans.

But does sexual reproduction have any particular advantages over asexual reproduction? The answer, generally, is yes. While it may seem a less complicated way to replicate, asexual reproduction does have drawbacks. For instance, if conditions are favorable for identical organisms, all is well, but if a disease strikes to which they have no genetic resistance, they will all perish. By contrast, the chances are that at least some of the genetically varied offspring of sexual reproducers will be able to survive a disease that kills others, so the species as a whole will survive.

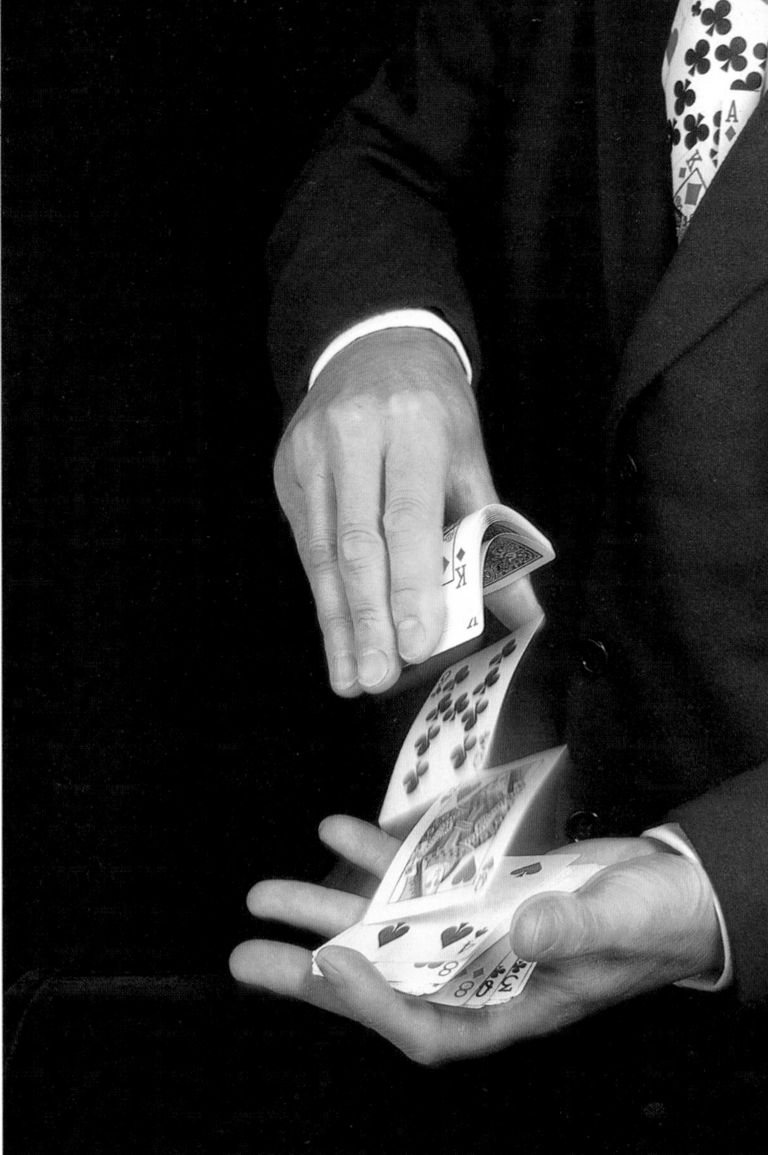

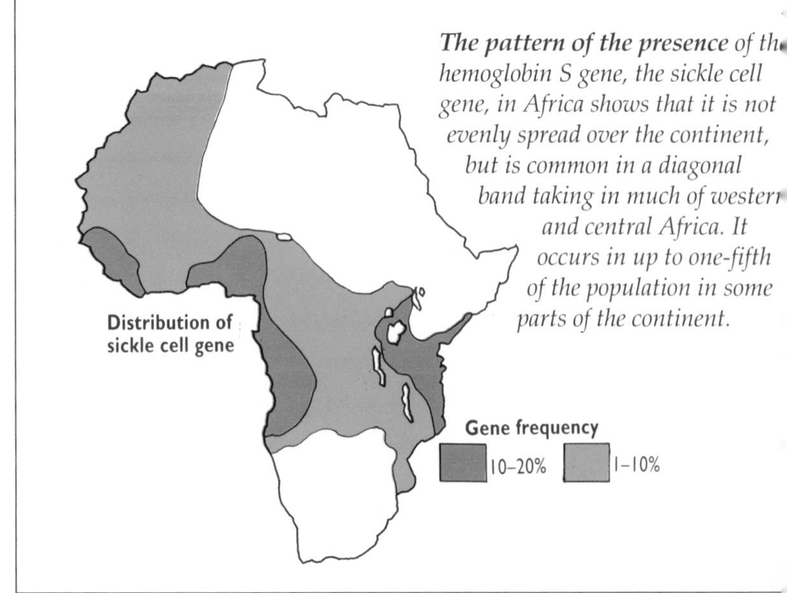

The pattern of the presence *of the hemoglobin S gene, the sickle cell gene, in Africa shows that it is not evenly spread over the continent, but is common in a diagonal band taking in much of western and central Africa. It occurs in up to one-fifth of the population in some parts of the continent.*

Like the shuffling of genes in sexual reproduction, *with each shuffle of a pack of 52 cards, a magician can deal out a different arrangement. We obtain half our genes from our mother and half from our father, and each half set has been shuffled just as well as the cards.*

Sexual reproduction is not the only way to produce offspring. There are simpler, non-sexual methods by which new plants or animals bud off from a parent. Since only one parent is involved in this process, it normally gives rise to offspring that are genetically identical with one another and with the parent.

*If a shoot, or cutting, is taken from a plant and placed in soil or water, a new plant will grow that is genetically the same as the parent plant. This method is vital for keeping the consistency of varieties of cultivated crop plants. The genetic make-up of particular types of apples (**right**) is originally formed by accidental or selective sexual breeding. Once a desirable form has been achieved, it can be maintained indefinitely by non-sexual propagation techniques.*

Malaria is a potentially lethal parasitic disease *caused by rapidly multiplying protozoan parasites that live inside human red blood cells and destroy them. People become infected when they are bitten by an infected mosquito. But if the parasite enters a cell with sickle cell hemoglobin, the altered red blood cell cannot sustain the parasite, which dies instead of multiplying and causing malaria.*

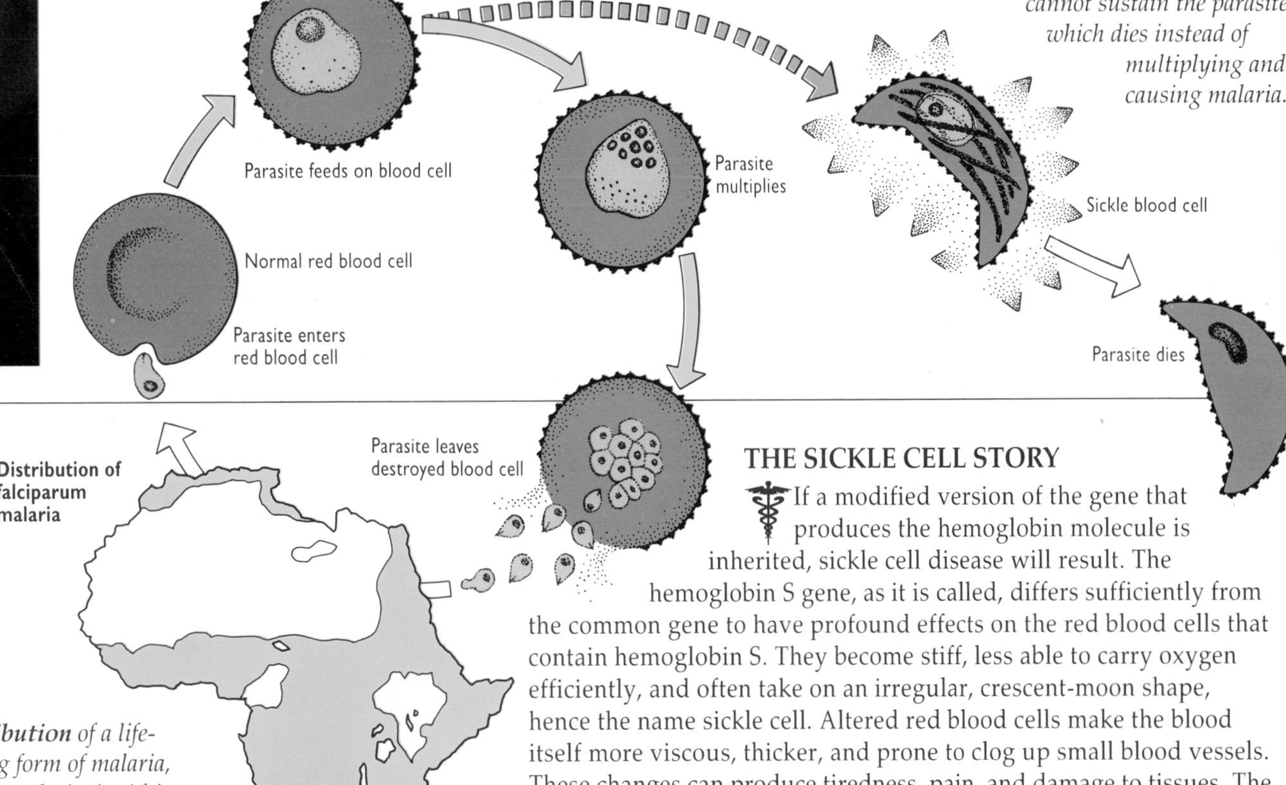

The distribution *of a life-threatening form of malaria, falciparum malaria, in Africa corresponds closely to that of high numbers of carriers of the sickle cell gene.*

THE SICKLE CELL STORY

If a modified version of the gene that produces the hemoglobin molecule is inherited, sickle cell disease will result. The hemoglobin S gene, as it is called, differs sufficiently from the common gene to have profound effects on the red blood cells that contain hemoglobin S. They become stiff, less able to carry oxygen efficiently, and often take on an irregular, crescent-moon shape, hence the name sickle cell. Altered red blood cells make the blood itself more viscous, thicker, and prone to clog up small blood vessels. These changes can produce tiredness, pain, and damage to tissues. The persistence of this apparently harmful gene in human populations is thought to be due to the protection it gives to people infected with the dangerous parasitic disease malaria. It seems that in areas of high malaria risk, the partial protection the gene provides for people infected with the disease offsets the disadvantages of the sickle cell disease.

See also

REPRODUCTION AND GROWTH
▶ In the family 156/157
▶ Language of life 158/159
▶ Building bodies 160/161
▶ Cycle of life 164/165

SUPPORT AND MOVEMENT
▶ Staying in shape 12/13

CONTROL AND SENSATION
▶ Hormones of change 40/41

ENERGY
▶ Cells at work 114/115

CIRCULATION, MAINTENANCE, AND DEFENSE
▶ Blood: supplying the body 126/127

Cycle of life

The sex organs – especially a woman's – are controlled and coordinated by subtle hormonal signaling.

Whether or not a woman of reproductive age becomes pregnant depends on two things above all. She cannot conceive if she has not produced an egg; neither can she conceive if there are no sperm around in her reproductive tract. Put simply, the male role in pregnancy is to produce and deliver sperm so one can meet the woman's egg and fertilize it. But the making of an egg is only part of a rather more intricate cycle of events that take place in the woman's body. For instance, the woman also has to provide a place for the fertilized egg to develop in her womb, or uterus.

The production of an egg or sperm takes place in a specialized form of cell division known as meiosis. Unlike mitosis, the ordinary type of division where an exact replica of a cell is made, in meiosis the genes on a cell's 23 pairs of chromosomes are first shuffled to produce genetic novelty. Then the number of chromosomes is halved as each half of a newly shuffled chromosome pair splits when the cell itself divides. The end result is cells with 23 instead of the usual 46 chromosomes. This happens so that when egg and sperm – each with a half set – combine at fertilization, the new cell has the full 46 chromosomes.

Men normally ejaculate between 300 and 500 million sperm at a time and can ejaculate many times each month. The level of the hormone responsible for sperm production – testosterone – remains relatively constant over time. Women, however, produce on average just one egg per month, and the levels of the hormones responsible for egg production and the provision of a lining in the womb suitable for the nurturing of a fertilized egg change during the month.

Sperm (male sex cells) are made in the two testes, *or testicles, which hang in a protective bag, the scrotum. Between each testicle and the penis is a system which first takes sperm up through the fine ducts of the epididymis into the vas deferens. There are two vasa deferentia – one from each testicle – which take sperm into the lower pelvic region. A number of accessory glands make nutritive and protective fluids with which the sperm are mixed to form semen. By far the bulk of semen is made of these fluids.*

At the moment of ejaculation, muscles around the urethra contract involuntarily, squeezing semen out of the end of the penis in three or four rapid bursts, followed by a few less strong, irregular contractions.

There is more to semen *– the fluid ejaculated on male orgasm – than sperm alone. The seminal vesicles and the prostate gland produce fluids that contain the sugar fructose, which provides energy for the swimming sperm, and alkaline substances that neutralize the acidic conditions in the vagina. The bulbourethral gland secretes a slippery fluid that leaks from the penis during sexual arousal to lubricate intercourse.*

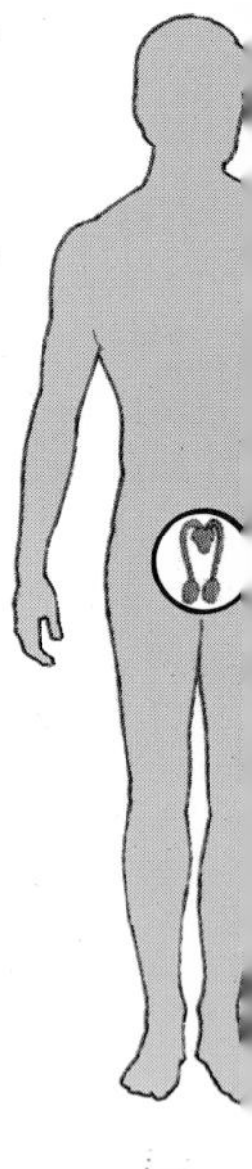

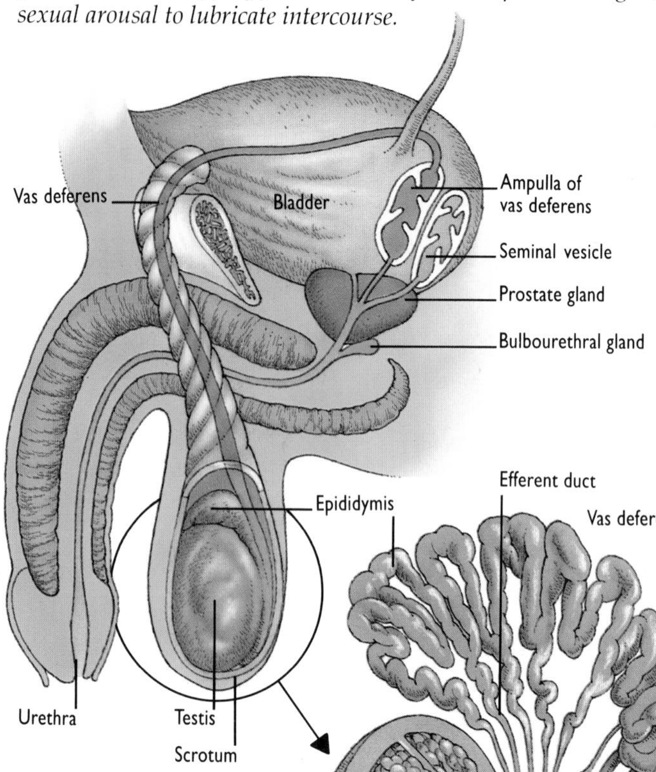

Each testis *– divided into lobules – holds nearly 1,000 tiny seminiferous tubules in which sperm are formed. A sperm takes two months to form, but the production rate is vast – up to 500 million are released per ejaculation.*

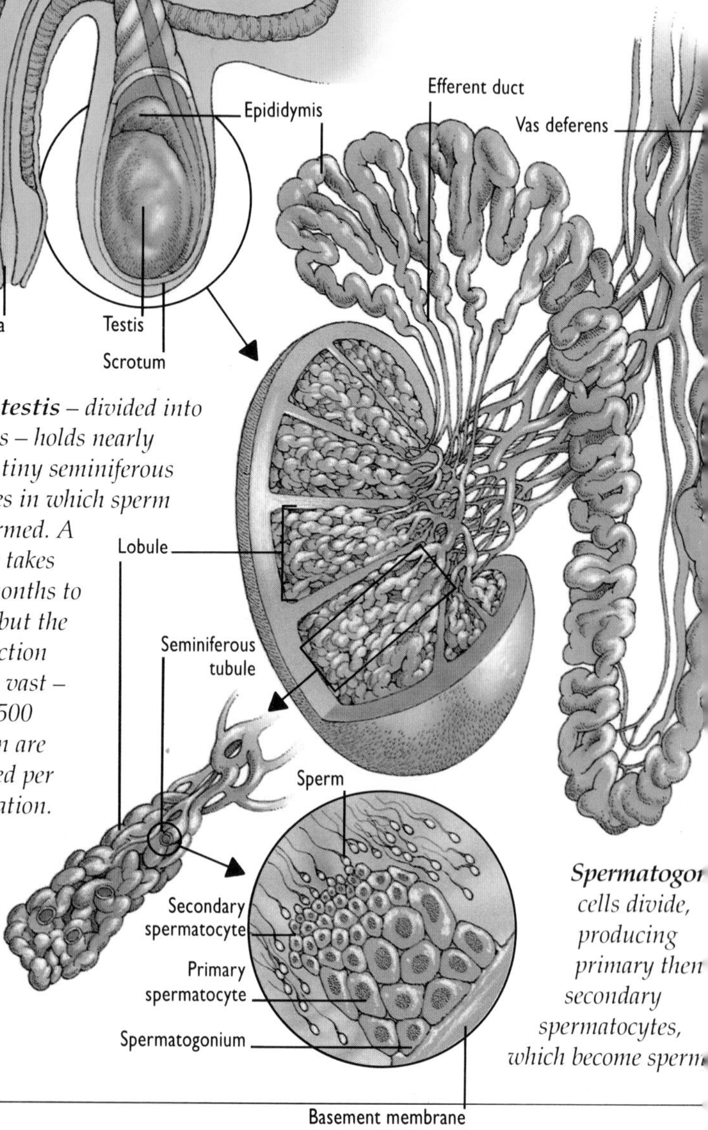

Spermatogo[n] *cells divide, producing primary then secondary spermatocytes, which become sperm.*

THE MENSTRUAL CYCLE

Female sex organs are more intricate than those of the male, and they also fulfill a more complicated, multistage series of tasks. The two ovaries produce eggs – female sex cells, or secondary oocytes. These are normally released singly from alternate ovaries – in a process known as ovulation – in a monthly cycle, the menstrual cycle. At puberty a woman has about 200,000 oocytes, but she only releases about 500 of them during her reproductive life span up to menopause. The number of oocytes is determined during fetal development: a woman cannot "make" more eggs.

Egg development takes place in a follicle, a glandular sac, which has an important role. Not only does it contain and then release a matured egg, but it also releases hormones that regulate several other processes. For instance, prior to egg release, the follicle makes large amounts of the hormone estrogen. This stimulates the lining of the uterus to make itself receptive to an egg, should one be fertilized. And after egg release, the follicle, now called the corpus luteum, continues to make estrogen and starts to make progesterone, which maintains the uterus lining. If an egg is not fertilized, the prepared lining breaks down and is shed in a monthly process known as menstruation, or bleeding, which lasts a few days.

In an ovary, *the egg in a follicle changes from a primary to a secondary oocyte (the sex cell) as the follicle matures. The oocyte is released at mid-cycle.*

Secondary oocyte
Mature follicle
Follicle approaching maturity
Released oocyte
Ruptured follicle
Coagulated blood
Developing corpus luteum
Mature corpus luteum
Primary oocyte
Primordial follicle
Corpus albicans

After egg release, *the follicle changes into the corpus luteum, which makes estrogen and progesterone. It later turns into the corpus albicans, a sort of scar tissue.*

Hormones play the key role *in the timing of events during the 28-day menstrual cycle. The start of bleeding is usually taken as the start of the cycle. In the first few days, raised levels of luteinizing hormone (LH) and follicle-stimulating hormone (FSH) – both made in the pituitary gland – cause follicles to develop in the ovaries. By about day six, one follicle – the most developed – is "selected." It starts making estrogen, which rebuilds the lining of the womb in the pre-ovulatory phase.*

A rise in estrogen levels on about day 12 raises the LH and FSH levels, which prepares the follicle and egg for release. On about day 14, ovulation happens. The corpus luteum formed from the follicle now makes high levels of estrogen and progesterone to maintain the uterine lining in the post-ovulatory phase. But after 28 days, the corpus luteum degenerates (unless the egg is fertilized), estrogen and progesterone levels drop, the uterus lining breaks down, and menstrual bleeding takes place.

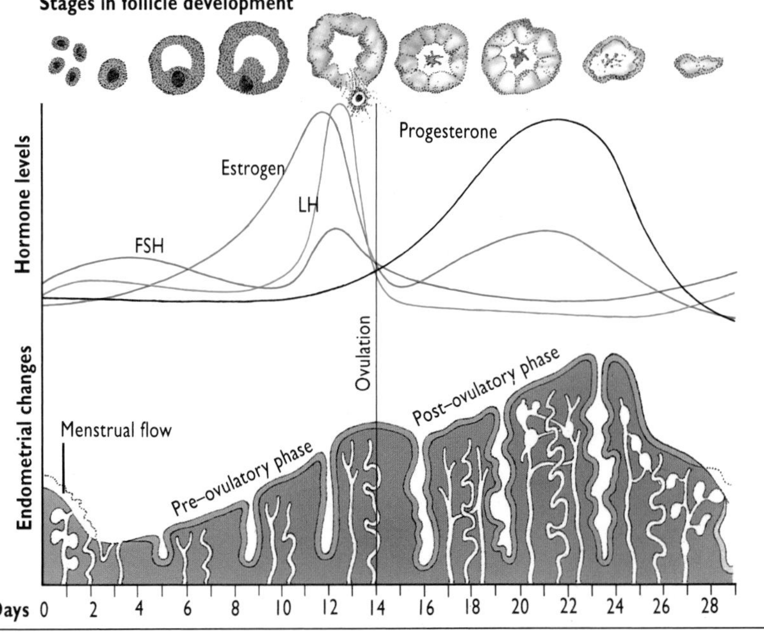

See also

REPRODUCTION AND GROWTH
▶ Fertilization 166/167
▶ The first month 168/169
▶ The growing plan 170/171
▶ Baby in waiting 172/173
▶ Being born 174/175
▶ The newborn baby 176/177
▶ The growing child 178/179
▶ Sexual maturity 180/181

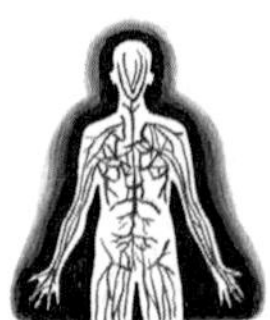

CONTROL AND SENSATION
▶ Hormones of change 40/41

ENERGY
▶ Cells at work 114/115

CIRCULATION, MAINTENANCE, AND DEFENSE
▶ Transportation systems 118/119
▶ Routine replacement 138/139

Fertilization

Each of us starts as a single microscopic cell – an egg fertilized by the winner of an amazing competition.

When fertilization takes place, an egg from a woman and a sperm from a man meet and fuse, marking the beginning of the life of a human being. The fusion of the two cells normally takes place in one of the two Fallopian, or uterine, tubes that link a woman's ovaries to her uterus, or womb. This meeting in the right place at the right time is the result of two quite different journeys.

The journey of the egg – a relatively large spherical cell surrounded by a jellylike coat, or zona pellucida – starts when it is released from an ovary. The egg cannot move by itself. Instead, it is carried along by waving cilia – hairlike extensions of cells on the walls of the female reproductive tract. Cilia on the fingerlike fimbriae at the open end of each Fallopian tube waft the egg from the ovary into the tube. Additional cilia on the inner wall of the tube move the egg along the tube like a pearl on a conveyor belt.

Meanwhile, coming in the opposite direction are the sperm, delivered into the vagina when a man ejaculates during sexual intercourse. Not only do they have a longer journey than the egg, but they also make it under their own steam. Hundreds of millions of sperm are released together during one ejaculation to become competitors racing to gain the prize of fertilizing the egg in the Fallopian tube some 6–8 inches (15–20 cm) away from the upper vagina. Only one sperm wins.

Each sperm is potentially capable of swimming to the top of the vaginal canal, through the mucus in the cervical canal that leads from the vagina to the uterus, up through the uterus itself and, finally, into the Fallopian tubes. This 8-inch (20-cm) journey is some 4,000 times the length of the 1/500-inch (0.05-mm) long sperm cell and is equivalent to a 3-foot (1-m) eel swimming 2½ miles (4 km). The trip probably takes between one and five hours for the fastest, most successful sperm. They swim through the fluid contents of the female tract by beating their propulsive tails (flagella). The long distance, the acid conditions in the tract, and the cilia of the uterine wall beating in the opposite direction to their own movement mean that the hundreds of millions of sperm in the vagina are reduced to a few hundred or even less in the region of the Fallopian tubes where fertilization occurs. At fertilization, the genetic information in sperm and egg, from father and mother respectively, joins together; and as the newly fertilized egg makes its first cell division, embryonic development begins. Approximately six or so days later, the egg implants itself in the uterus and pregnancy begins.

On average, about 400 million sperm *are ejaculated into the vagina, of which some 350 million are actively swimming and viable. Only 10,000 or so make it through the cervical canal to the uterus. Of these, perhaps 1,000 to 3,000 get to the upper parts of the uterus and then turn left or right into the Fallopian tubes. In the area of most likely fertilization, the upper region of a Fallopian tube, there may be only about 100 sperm in the vicinity of the egg.*

400 million ejaculated

10,000 reach uterus

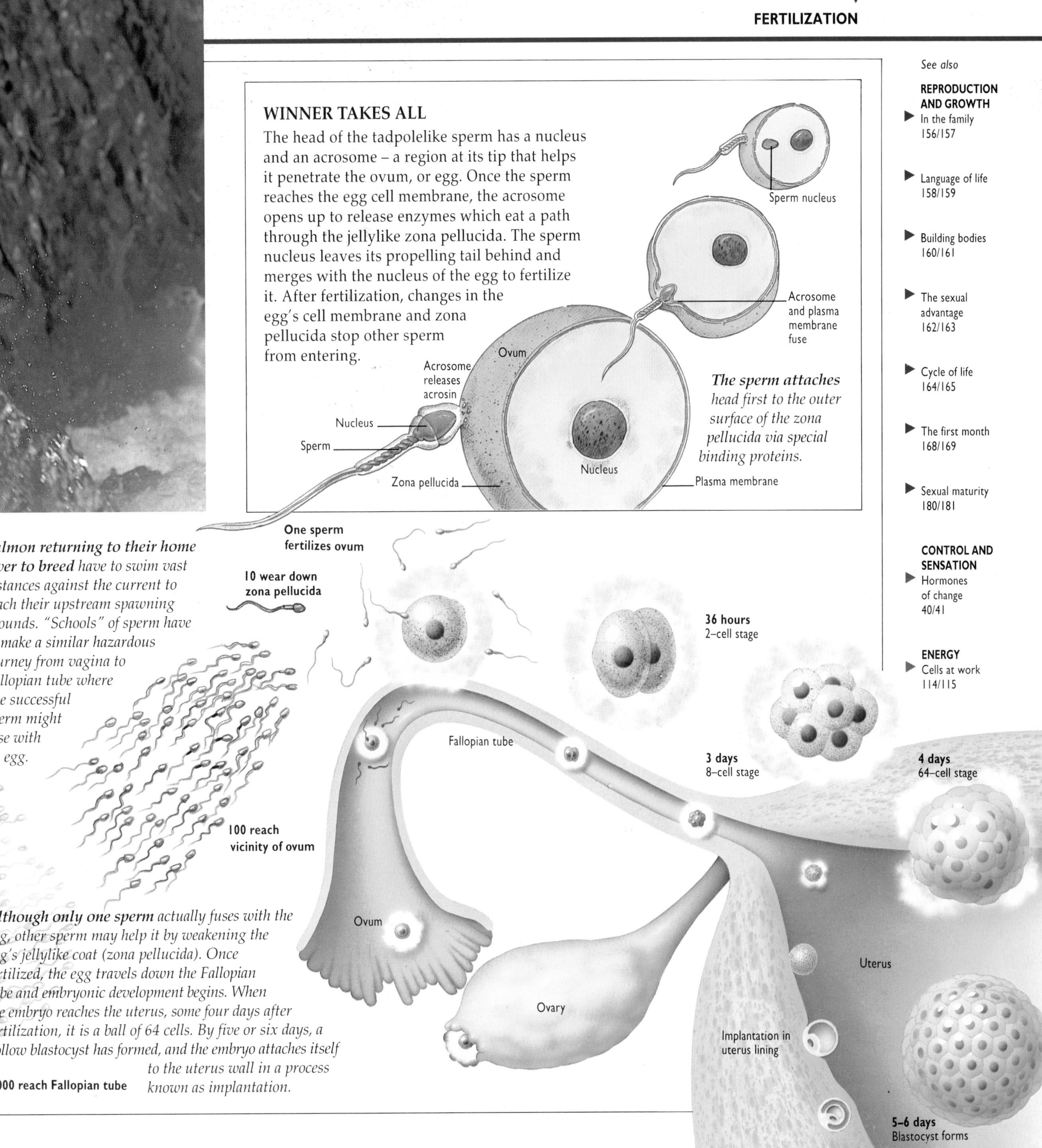

WINNER TAKES ALL

The head of the tadpolelike sperm has a nucleus and an acrosome – a region at its tip that helps it penetrate the ovum, or egg. Once the sperm reaches the egg cell membrane, the acrosome opens up to release enzymes which eat a path through the jellylike zona pellucida. The sperm nucleus leaves its propelling tail behind and merges with the nucleus of the egg to fertilize it. After fertilization, changes in the egg's cell membrane and zona pellucida stop other sperm from entering.

The sperm attaches *head first to the outer surface of the zona pellucida via special binding proteins.*

ılmon returning to their home ver to breed *have to swim vast stances against the current to ach their upstream spawning ounds. "Schools" of sperm have make a similar hazardous urney from vagina to allopian tube where e successful erm might se with 1 egg.*

lthough only one sperm *actually fuses with the g, other sperm may help it by weakening the g's jellylike coat (zona pellucida). Once rtilized, the egg travels down the Fallopian be and embryonic development begins. When e embryo reaches the uterus, some four days after rtilization, it is a ball of 64 cells. By five or six days, a ollow blastocyst has formed, and the embryo attaches itself to the uterus wall in a process known as implantation.*

See also

REPRODUCTION AND GROWTH
- In the family 156/157
- Language of life 158/159
- Building bodies 160/161
- The sexual advantage 162/163
- Cycle of life 164/165
- The first month 168/169
- Sexual maturity 180/181

CONTROL AND SENSATION
- Hormones of change 40/41

ENERGY
- Cells at work 114/115

The first month

Within just four weeks, the entire fetal support system has formed in the womb.

For about six days after a human egg has been fertilized by a sperm, the resulting embryo develops as a free clump of cells. At this stage the minute embryo is not in direct contact with its mother's tissues, but is floating in the fluid inside one of the Fallopian tubes (the tubes leading from the ovaries) or in the uterus (the womb).

At six days the early embryo is known as a blastocyst. It is a tiny hollow ball and has an inner cell mass, the embryoblast, concentrated at one end. This inner cell mass will become the new baby; the outer cells of the ball – the trophectoderm or trophoblast – will later help to form the placenta, which will nourish the growing fetus. Both the trophoblast cells and the embryoblast cells have come from the original fertilized egg, which has by now divided many times.

This early stage of development in the womb already shows the subtle complexities of human embryology. The fertilized egg not only generates a baby, which will ultimately be an independent, unique human being, but is also responsible for the formation of the placenta, which is not part of the baby itself. In fact, the situation is even more complex. The trophoblast cells form four membrane systems – the fetal or extraembryonic membranes – none of which is part of the fetus itself. These membranes – the chorion, amnion, allantois, and yolk sac – protect and support the developing baby. The chorion, together with uterine tissue, forms the placenta itself. The amnion provides the fetus with a protective, fluid-filled space for growth. And the allantois – part of the body stalk – and yolk sac eventually form the umbilical cord.

By day seven after fertilization, the blastocyst has become attached to the inner lining of the uterus – the endometrium – in a process known as implantation. Eventually, the blastocyst is entirely surrounded by uterine tissue, which brings the developing fetus into contact with sources of nutrients and oxygen and provides a means of removing waste products. Initially, this happens through contact between the outer cells of the blastocyst – the trophoblast cells – and blood vessels in the endometrium; later, it is through the placenta.

If fertilization and implantation do not occur, the womb lining is shed in a monthly process known as menstruation. A fertilized egg, however, is equipped to produce a special hormone – human chorionic gonadotrophin – which prevents this from happening. As a result, menstruation ceases from the moment of fertilization until a few weeks after the baby is born – the endometrium is not sloughed off and the embryo continues to develop.

Amniotic cavity
Chorion
Uterine cavity
Chorionic cavity
Amnion
Placenta
Mucous plug
Cervix

4 weeks
3/16 inch (0.5 cm)

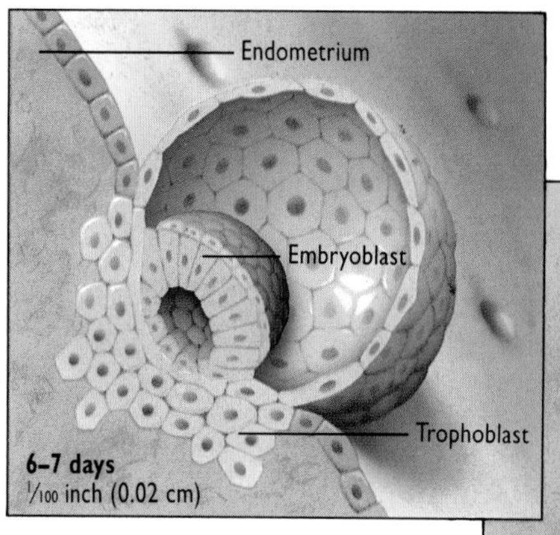

***At 6–7 days** the trophoblast cells become attached to the endometrium – the lining of the uterus. A week later, at 12–13 days, the trophoblast is completely enclosed by uterine tissues and implantation is complete. The protective amniotic cavity starts to form.*

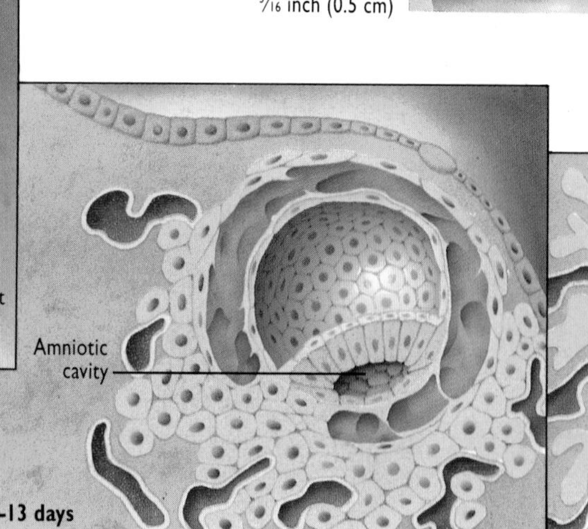

***At 14–15 days** chorionic villi reach uterine blood vessels. The body stalk (which will form part of the umbilical cord) and the primitive streak (which will form the baby) appear.*

he human embryo *in position in the uterus out four weeks after fertilization is usually ss than $^3/_{16}$ inch (0.5 cm) long, although it already has developing eyes and a functioning heart. The great thickness of the muscular wall of the uterus together with multiple protective membranes and fluid in both the amniotic and chorionic cavities cushion the embryo from knocks and shocks. The outer of these membranes, the chorion, has penetrated the surrounding uterine tissue to make contact with maternal blood vessels. In one area, the placenta, this contact becomes highly organized, so the embryo can gather oxygen and nutrients om its mother's bloodstream and discharge s waste products into her blood so that she n discard them. The cervix – the entrance to e uterus – is sealed by a plug of mucus.*

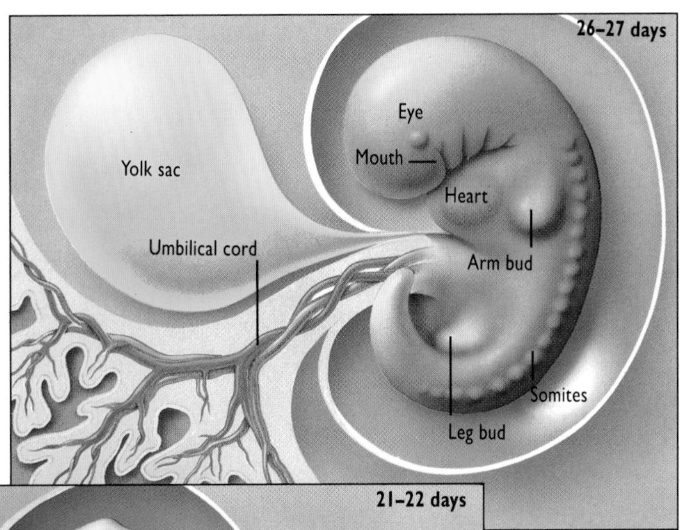

At 26–27 days *the embryo has in miniature the beginnings of most organ systems – limb buds, a heart, and a head with eyes and a mouth. The umbilical cord, linked to the placenta, provides a supply of blood and nutrients. This is just one of the life-support systems for the embryo established during the first month after fertilization.*

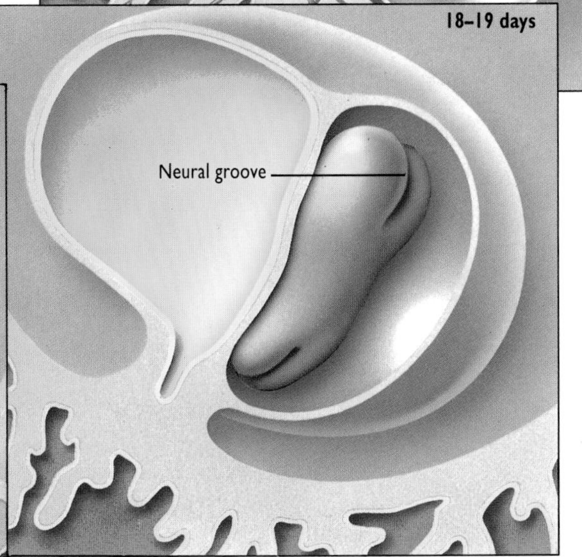

At 21–22 days, *just three weeks after fertilization, paired tissue blocks known as somites begin to form. These lay down a pattern of nerves and muscles in the body of the developing embryo. Proper blood vessels are now present in the umbilical cord, which connects the embryo to the placenta.*

At 18–19 days *a neural groove forms on the dorsal (back) surface of the embryo. This is the beginning of the development of the baby's nervous system. At this stage, growth is rapid, and cells in the embryo – nourished by blood from the maternal blood vessels – divide quickly to produce new tissue.*

See also

REPRODUCTION AND GROWTH
- Building bodies 160/161
- Cycle of life 164/165
- Fertilization 166/167
- The growing plan 170/171
- Baby in waiting 172/173
- Being born 174/175

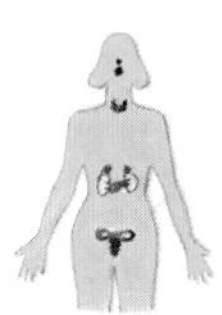

CONTROL AND SENSATION
- The chemicals of control 38/39

ENERGY
- The cell and energy 112/113
- Cells at work 114/115

CIRCULATION, MAINTENANCE, AND DEFENSE
- Growing heart 122/123

The growing plan

How do cells, which hold the genetic information to perform any role, specialize in structure and function?

In humans a single fertilized egg divides, then divides again and again, multiplying over and over until there are billions of cells. In a coordinated sequence, controlled by genes, these cells become organized into tissues, organs, and organ systems. Most of this sorting out process, or cellular differentiation, happens when the human embryo is minute, buried among protective membranes and fluids deep in its mother's womb.

Genes are passed in identical form from the original fertilized egg to every subsequent cell. They carry a complete set of the genetic instructions for making the many tens of thousands of different human proteins. These are the building blocks, machine tools, and control chemicals used to construct the body. What makes cells different is which of the many proteins they actually make. Muscle cells in the biceps, for instance, have to make contractile proteins, whereas cells in the adrenal glands must be able to secrete epinephrine.

Which genes are switched on – and thus which proteins are made – in a cell is a process that is itself controlled by genes. The control genes make proteins that activate sets of genes that in turn make the proteins that are appropriate for the type of cell.

The differentiation process starts a week or so after fertilization, when the early embryo – no more than a ball of cells – folds up into layers. Cells are programmed to move to a region of the early embryo that determines their range of differentiation possibilities. These regions are given the names ectoderm, mesoderm, and endoderm because they correlate with cell layers (the names refer to outer, middle, and inner cell layers).

Once committed to a layer, and thus to a particular range of possibilities, a developing cell cannot normally change its fate by moving into a different set. An embryonic ectoderm cell, for instance, may eventually form a skin cell or a nerve cell, but it cannot change into a muscle cell or a liver cell, which come, respectively, from the mesoderm and endoderm regions.

Scientists are beginning to identify the genes that first determine the range of possibilities of cells in the embryo and then further differentiate them into specific tissue types and organs. To the surprise of many, the controlling genes for major structural characteristics, such as head to tail differentiation in animals, are shared by almost all animals: very similar genes control the pattern of development in fruit flies, mice, and humans. One family of these genes are the so-called homeobox genes which code for control proteins that help specify the position of organs in an animal. These protein seem to act as signals to cells, making them switch on differen subsets of the total gene set depending on their own position in the body.

Studies of flowers *have thrown light on how cells differentiate. The sepals, petals, stamens, and carpels of a typical flower – such as the passion flower (**below** and **right**) – are made of cells that are the descendants of cells that were once identical. The early cells of the flower bud switch on one of three control genes, depending on the distance from the bud's tips, with concentric rings of activation A, B, and C. From then on, the fates of all the cells in the bud are set, because the protein products of the control genes switc on genes in bud cells that form different flower structures. Those cells that receive only an A signal make sepals at the base of the flowe those that get A and B together make petals; B and C signals mixe induce stamen formation; while C alone at the center of the flower causes carpels to form.*

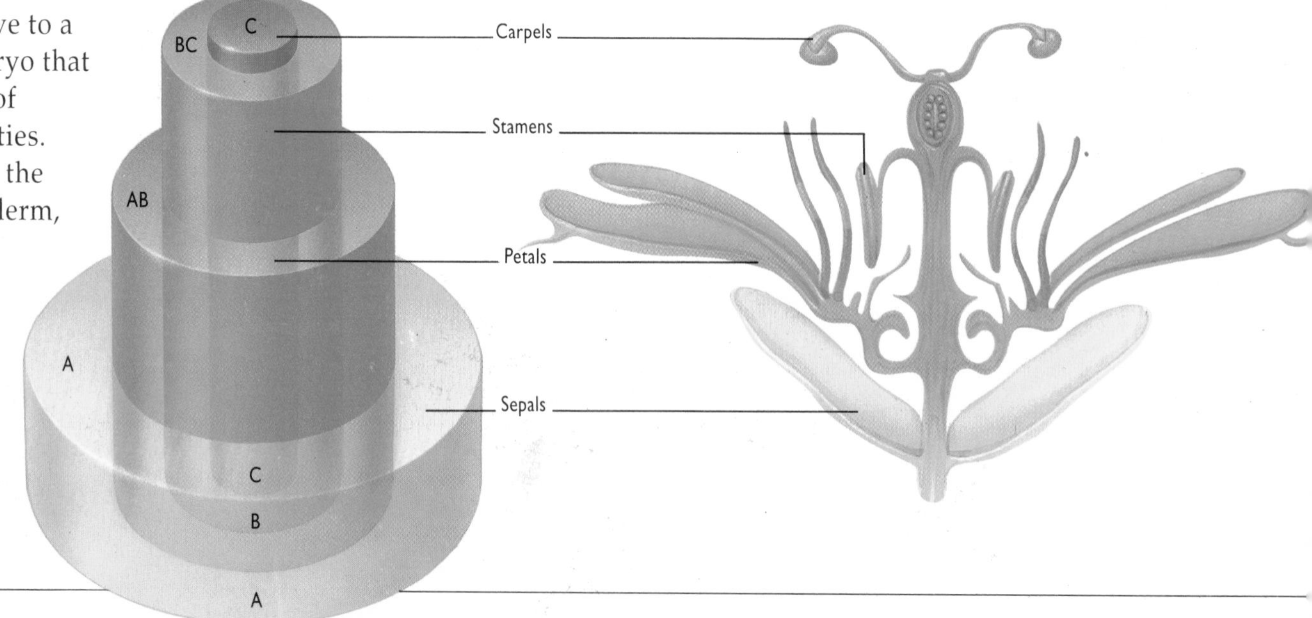

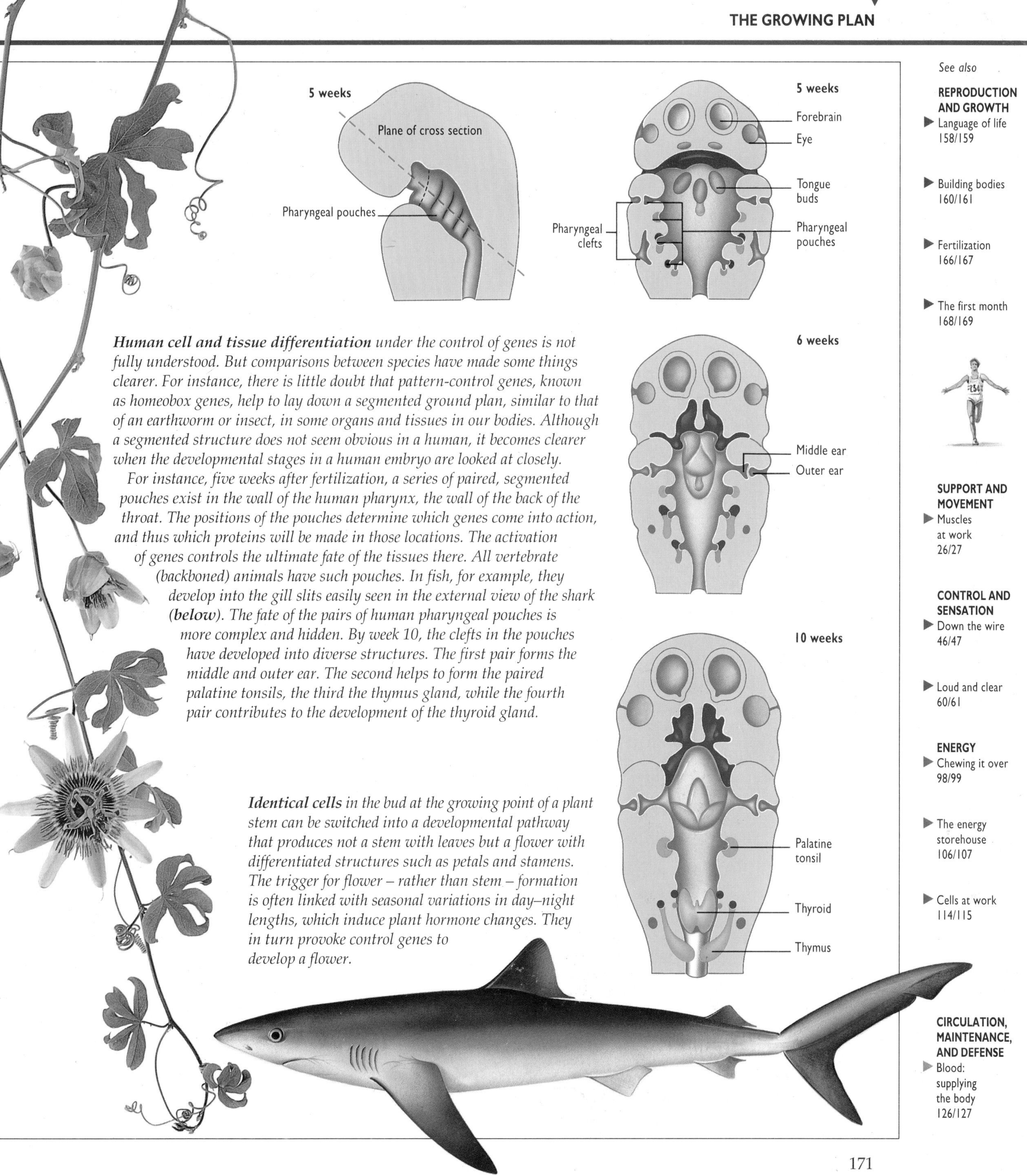

Human cell and tissue differentiation *under the control of genes is not fully understood. But comparisons between species have made some things clearer. For instance, there is little doubt that pattern-control genes, known as homeobox genes, help to lay down a segmented ground plan, similar to that of an earthworm or insect, in some organs and tissues in our bodies. Although a segmented structure does not seem obvious in a human, it becomes clearer when the developmental stages in a human embryo are looked at closely.*

For instance, five weeks after fertilization, a series of paired, segmented pouches exist in the wall of the human pharynx, the wall of the back of the throat. The positions of the pouches determine which genes come into action, and thus which proteins will be made in those locations. The activation of genes controls the ultimate fate of the tissues there. All vertebrate (backboned) animals have such pouches. In fish, for example, they develop into the gill slits easily seen in the external view of the shark ***(below)****. The fate of the pairs of human pharyngeal pouches is more complex and hidden. By week 10, the clefts in the pouches have developed into diverse structures. The first pair forms the middle and outer ear. The second helps to form the paired palatine tonsils, the third the thymus gland, while the fourth pair contributes to the development of the thyroid gland.*

Identical cells *in the bud at the growing point of a plant stem can be switched into a developmental pathway that produces not a stem with leaves but a flower with differentiated structures such as petals and stamens. The trigger for flower – rather than stem – formation is often linked with seasonal variations in day–night lengths, which induce plant hormone changes. They in turn provoke control genes to develop a flower.*

See also

REPRODUCTION AND GROWTH
▶ Language of life 158/159
▶ Building bodies 160/161
▶ Fertilization 166/167
▶ The first month 168/169

SUPPORT AND MOVEMENT
▶ Muscles at work 26/27

CONTROL AND SENSATION
▶ Down the wire 46/47
▶ Loud and clear 60/61

ENERGY
▶ Chewing it over 98/99
▶ The energy storehouse 106/107
▶ Cells at work 114/115

CIRCULATION, MAINTENANCE, AND DEFENSE
▶ Blood: supplying the body 126/127

Baby in waiting

In just two-thirds of a year, a fully formed human being grows from something the size of a fingertip.

About eight weeks after fertilization, the embryo has the external characteristics of a recognizably human infant. It is, however, only about 1 inch (2.3 cm) long and is totally dependent on its umbilical link with the placenta and the protective fluid and membranes around it. At about this time, it becomes known as a fetus rather than an embryo.

By now, most of the different types of tissues that will make up the baby have already formed, so during the rest of the mother's pregnancy, the fetus grows – remarkably fast. Fueled by nutrients from the uterus, the fetus increases in weight from about 1 ounce (25 g) at eight weeks to more than 7 pounds (3 kg) at term – an increase of about 120 times in bulk in just seven months.

Item by item, the detailed modeling of the bodily features of a human occurs. Hair, eyebrows, and eyelashes are in place in the 20th week. After that, fingernails and toenails form. Downy hair, or lanugo, peculiar to the fetal period, soon covers the baby's torso and limbs. From 20 weeks onward, the fetus's heartbeat can be picked up using a sensitive stethoscope. Miniature lungs have now formed, but they are filled with fluid and cannot yet act as respiratory organs. The hands start to show the gripping reflex, and a mother feels tiny kicks. By about 24 weeks, the fetus is sufficiently developed to be able to stand a chance of survival outside the womb if it is intensively supported in a hospital incubator.

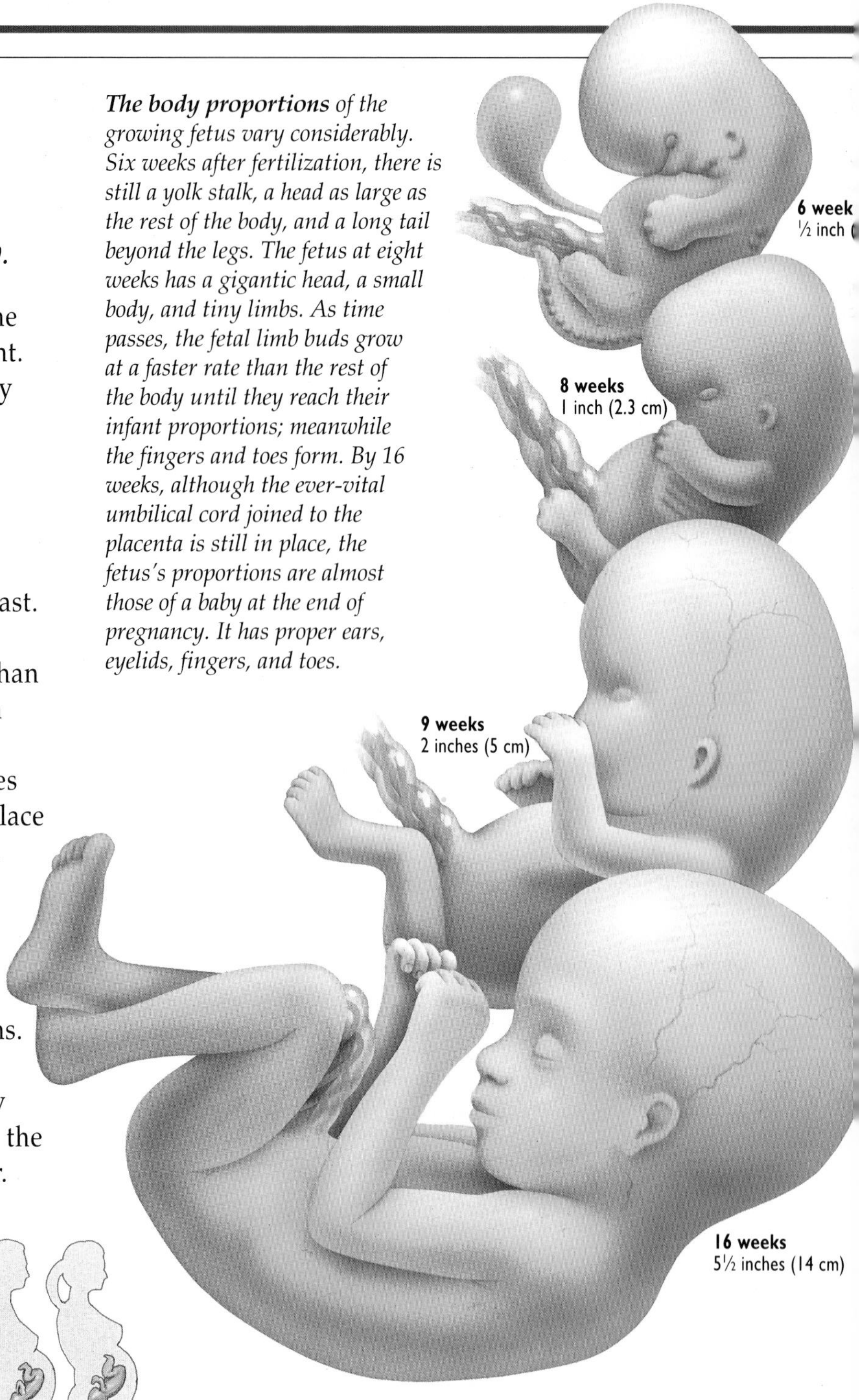

***The body proportions** of the growing fetus vary considerably. Six weeks after fertilization, there is still a yolk stalk, a head as large as the rest of the body, and a long tail beyond the legs. The fetus at eight weeks has a gigantic head, a small body, and tiny limbs. As time passes, the fetal limb buds grow at a faster rate than the rest of the body until they reach their infant proportions; meanwhile the fingers and toes form. By 16 weeks, although the ever-vital umbilical cord joined to the placenta is still in place, the fetus's proportions are almost those of a baby at the end of pregnancy. It has proper ears, eyelids, fingers, and toes.*

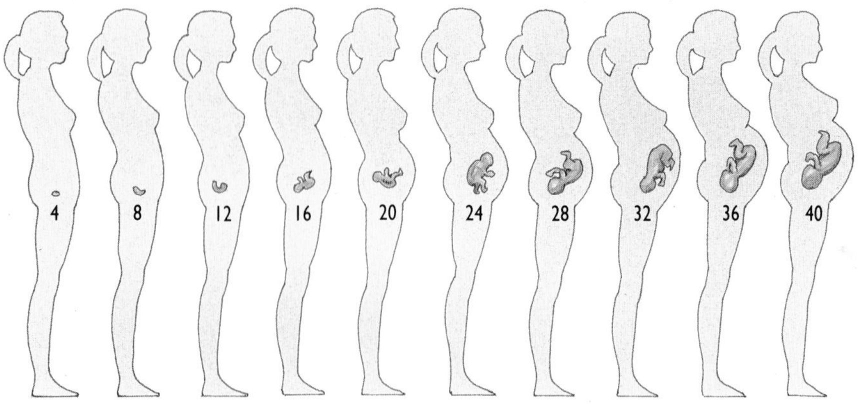

6 week
½ inch (

8 weeks
1 inch (2.3 cm)

9 weeks
2 inches (5 cm)

16 weeks
5½ inches (14 cm)

4 8 12 16 20 24 28 32 36 40

Weeks after fertilization

***During the 40 weeks of pregnancy,** there is a radical change in the size o the womb and of the baby inside. After about 16 weeks, the "bump" becomes obvious and the woman looks pregnant. At 36 weeks, the uterus has risen up to be level with the woman's ribs, and she has to lean backward to stay balanced when standing up. After birth, the uterus shrinks back to its normal non-pregnant dimensions in about six weeks.*

Most babies are born head first, and by 28 weeks after fertilization many are already positioned head down in the womb ready for birth. The baby is cushioned by amniotic fluid, and the neck of the womb, or cervix, is sealed by a plug of mucus.

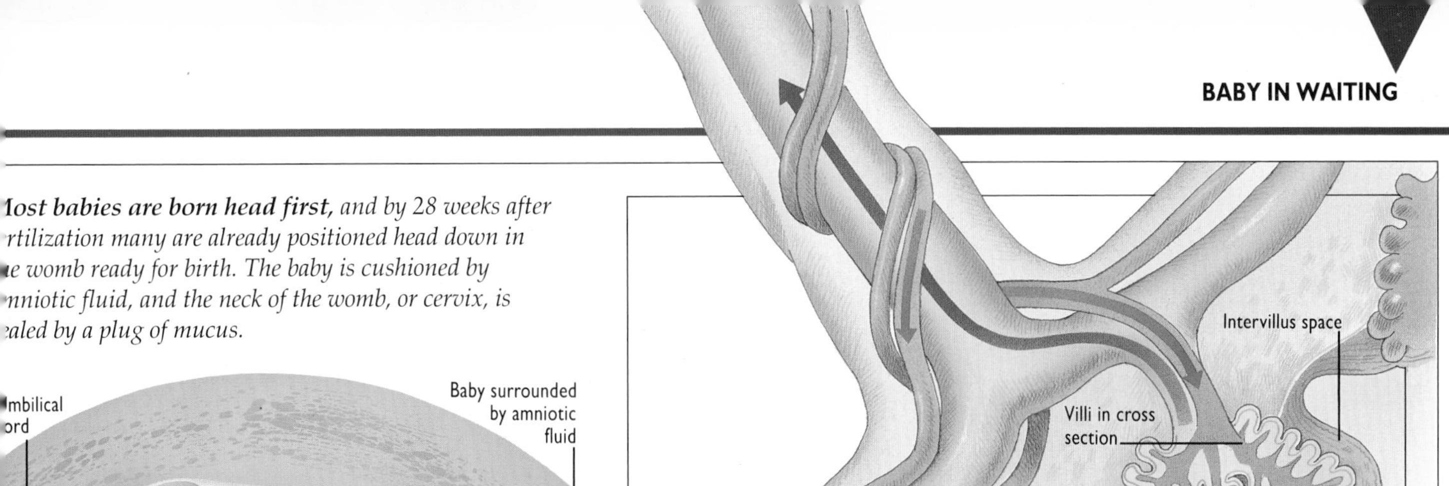

The mother's blood *seeps into the placenta's intervillus spaces. Nutrients, oxygen, and waste products are exchanged between mother and baby via villi.*

THE PLACENTA – LIFE-SUPPORT FOR A BABY

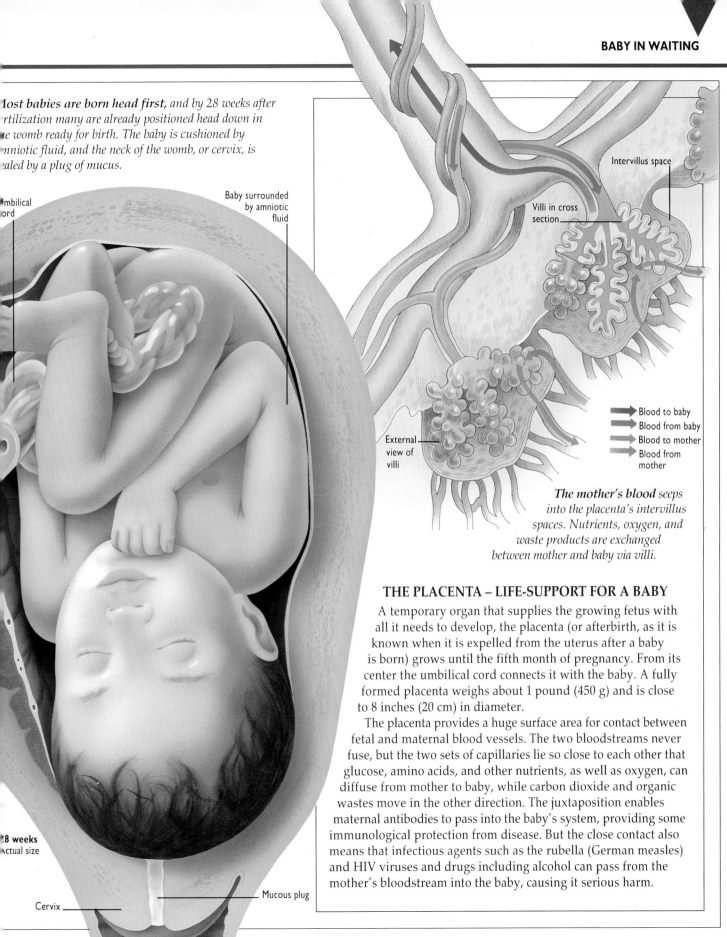

A temporary organ that supplies the growing fetus with all it needs to develop, the placenta (or afterbirth, as it is known when it is expelled from the uterus after a baby is born) grows until the fifth month of pregnancy. From its center the umbilical cord connects it with the baby. A fully formed placenta weighs about 1 pound (450 g) and is close to 8 inches (20 cm) in diameter.

The placenta provides a huge surface area for contact between fetal and maternal blood vessels. The two bloodstreams never fuse, but the two sets of capillaries lie so close to each other that glucose, amino acids, and other nutrients, as well as oxygen, can diffuse from mother to baby, while carbon dioxide and organic wastes move in the other direction. The juxtaposition enables maternal antibodies to pass into the baby's system, providing some immunological protection from disease. But the close contact also means that infectious agents such as the rubella (German measles) and HIV viruses and drugs including alcohol can pass from the mother's bloodstream into the baby, causing it serious harm.

See also

REPRODUCTION AND GROWTH
- Cradle to grave 154/155
- Cycle of life 164/165
- Fertilization 166/167
- The first month 168/169
- The growing plan 170/171
- Being born 174/175
- The newborn baby 176/177

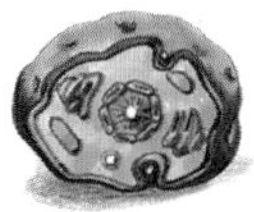

CONTROL AND SENSATION
- Key chemicals 42/43

ENERGY
- The gas exchange 110/111
- The cell and energy 112/113

CIRCULATION, MAINTENANCE, AND DEFENSE
- Growing heart 122/123
- In circulation 124/125

Being born

One of the shortest journeys in the world is also one of the most important – a baby's trip down the birth canal.

In the final month of a pregnancy, the growing child in the uterus puts ever greater demands on its mother's body. Put simply, it is getting too big. In the last four weeks, the mother's ribs have to spread out to provide space for her lungs since they are pushed upward by the increasingly large uterus (womb). By week 36, the top of the uterus is up at the level of the lowest ribs. Most women at this stage have to stand and walk leaning backward to counterbalance the bulge at the front. The time for the baby to move into the outside world is approaching fast – if it grew much more, its head might become too large to pass through the pelvic birth canal.

Contractions of the uterus – albeit weak ones – can be felt by the mother or a doctor as early as about week nine of pregnancy. But in the final month, as the time of birth draws nearer, uterine contractions become more powerful and frequent. Approximately 266 days after a microscopic egg was fertilized, a baby – the result of that fertilization and now weighing some 7 pounds (3.2 kg) – is ready to be born.

Powerful contractions of the uterus stimulate the mother's pituitary gland to secrete the hormone oxytocin, which induces even stronger contractions. The positive feedback between contractions and a hormone that increases contractions seems to be an important part of the complex trigger that initiates labor and birth.

Sometimes, however, babies are born before the nine month are up. Babies born prematurely are small and are unable to control their body temperature effectively. They also have particular problems with efficient breathing, partly because the muscles in the chest wall and diaphragm are not yet strong enough to support powerful breathing movements. Poor breathing is also due to the fact that the spaces inside the lungs are still underdeveloped. In the last stages of fetal development, the baby's lung lining secretes large volumes of a natural detergent solution, or surfactant. It eventually reduces the liquid surface tension in the inner spaces of the lungs which holds back the movement of air into those spaces. A premature baby has not produced this surfactant, so its breathing is much more labored until the lung lining has matured.

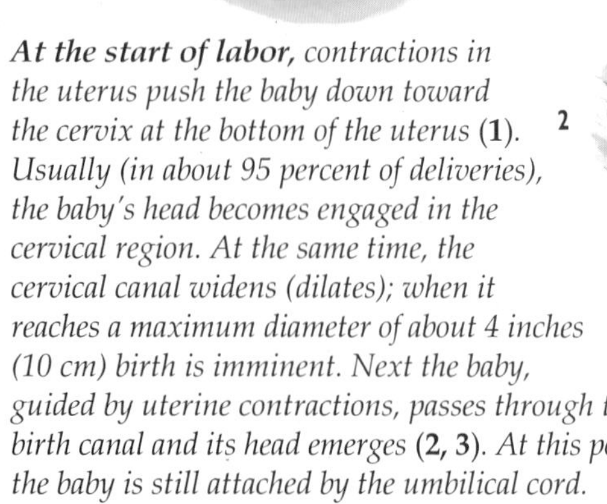

LEAVING THE WOMB, ENTERING THE WORLD

There are three stages of labor, which takes anything from a few hours to three days in total. In the first, contractions of the uterus prepare the birth canal, and toward the end muscular contractions break the amniotic sac around the baby, and the liquid escapes through the vagina – the water breaks. The second stage is the birth itself: the baby passes head-first down the birth canal. In the third stage, the afterbirth is expelled.

At the start of labor, *contractions in the uterus push the baby down toward the cervix at the bottom of the uterus* (**1**). *Usually (in about 95 percent of deliveries), the baby's head becomes engaged in the cervical region. At the same time, the cervical canal widens (dilates); when it reaches a maximum diameter of about 4 inches (10 cm) birth is imminent. Next the baby, guided by uterine contractions, passes through the birth canal and its head emerges* (**2, 3**). *At this point, the baby is still attached by the umbilical cord.*

While the baby is still attached to the placenta (**right**), its lungs are filled with amniotic fluid. Two shunts, one between the heart's atria (the oval window, or foramen) and one between the pulmonary artery and aortic arch (the ductus arteriosus), allow oxygenated blood from the placenta to pass directly to the heart's left side for pumping to the body.

At birth, the oval window closes, and the baby's first breath forces blood from the right side of the heart into the lungs via the pulmonary artery, and the baby receives oxygenated blood from its own lungs (**far right**). Within minutes, the ductus arteriosus begins to close and is sealed 10 days after birth. Links to the umbilical cord soon shrivel.

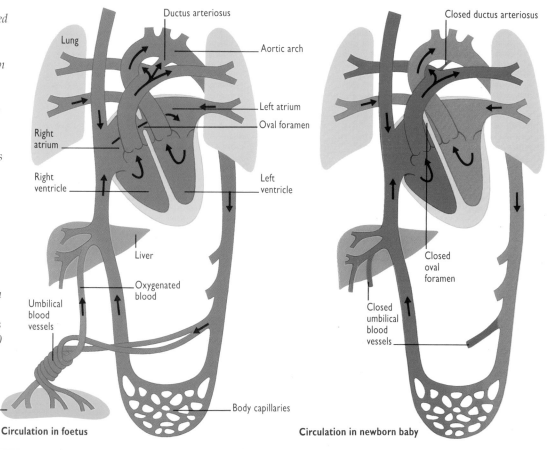

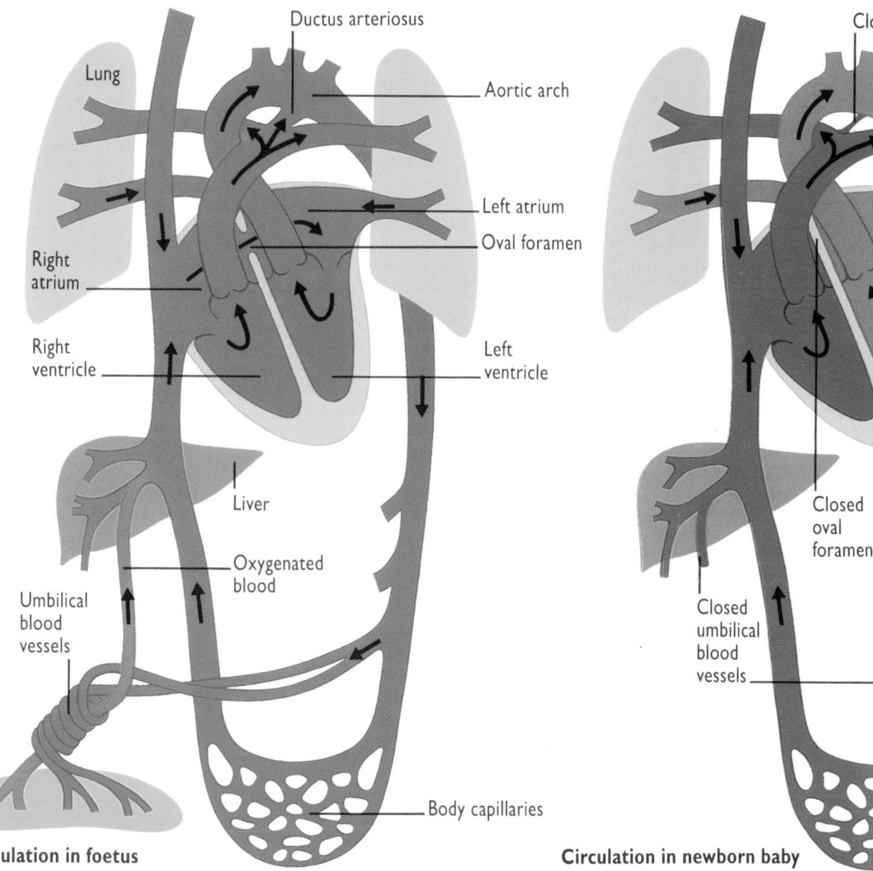

Circulation in foetus

Circulation in newborn baby

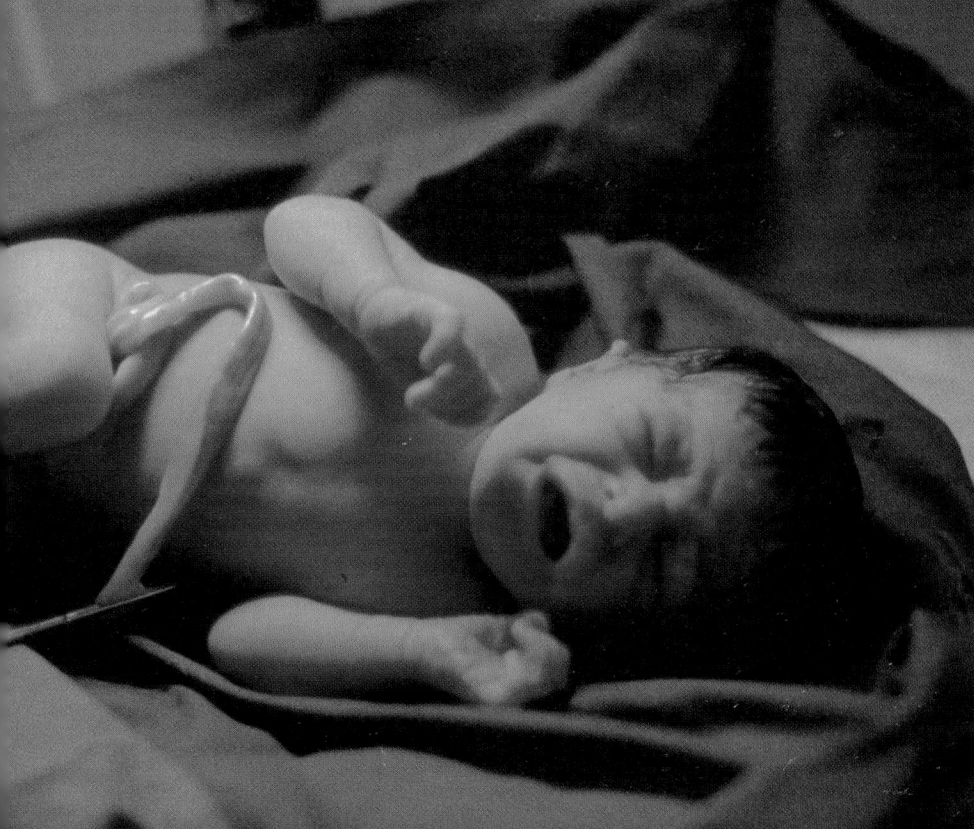

It is normal delivery practice to clamp and then cut the umbilical cord as the baby emerges. Soon after it enters the world, the baby takes its first breath and utters its first cry. The baby must change from an "underwater" human (with lungs full of amniotic fluid and provided with oxygen via a life support blood system linked with the placenta) to a normal air-breathing human.

After the baby has been delivered, the third stage of labor begins. This usually takes only about 20 to 30 minutes and involves the painless continuing contractions of the uterus to expel the afterbirth – the remains of the amniotic fluid, cord, and placenta.

See also

REPRODUCTION AND GROWTH
- ▶ Cradle to grave 154/155
- ▶ Building bodies 160/161
- ▶ Fertilization 166/167
- ▶ The first month 168/169
- ▶ The growing plan 170/171
- ▶ Baby in waiting 172/173
- ▶ The newborn baby 176/177

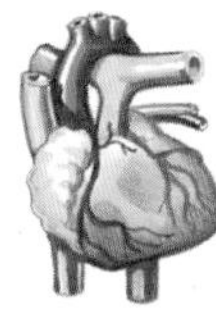

CONTROL AND SENSATION
- ▶ Hormones of change 40/41
- ▶ Learning skills 80/81

CIRCULATION, MAINTENANCE, AND DEFENSE
- ▶ The living pump 120/121
- ▶ Growing heart 122/123
- ▶ In circulation 124/125

The newborn baby

Despite being dependent on its parents, a new baby has some in-built abilities that help it along as it develops.

When a new baby enters the world, it is for the first time a separate human being, although it is still entirely reliant on its parents for warmth, protection, and food. From the time of birth, the developing baby makes rapid advances toward becoming truly independent. By the age of 18 months or so, most children can walk, talk, feed themselves, and handle a whole range of practical everyday problems.

The newborn baby can see and hear the world around it and has an efficient sense of touch. But muscular coordination and strength are not well developed to begin with. It does, however, have some reflex movements that are valuable for survival. For instance, a baby cries when tired, uncomfortable, or hungry, and parents find crying almost impossible to ignore. The suckling reflex is also vital – when a baby's cheek or lips are touched, it turns at once toward the touch, grasps the object in its lips, and begins sucking. This specific behavior pattern allows efficient feeding from the nipple of the mother or a bottle.

More sophisticated movement (motor) skills, culminating in self-feeding and unaided walking, develop as the baby's muscles and nervous system mature. At birth, for instance, the nerves that control muscle contraction are not insulated by the usual fatty sheath of myelin. But by 12 months, almost all nerves are encased in myelin, aiding walking and other feats of motor control. Repetition and practice help solve coordination problems.

***Milk to feed a baby** is made in gland cells in the breasts under the influence of a hormone – prolactin – from the pituitary gland. When a baby sucks on a nipple, this provokes the pituitary to release a surge of another hormone, oxytocin. This has the effect of making the smooth muscles in the breast contract, moving milk toward the nipple and releasing it as the baby needs it. The release of milk in response to the baby's sucking is the "milk letdown" reflex.*

***There is no fixed timetable** or even strict order in which a developing baby is able to perform particular activities. The exception is walking, in which stages are attained in a relatively set pattern and at fairly predictable times. After birth, a baby is unable to lift its head, but four to six weeks later, increasing muscle strength means it can raise its head from the horizontal.*

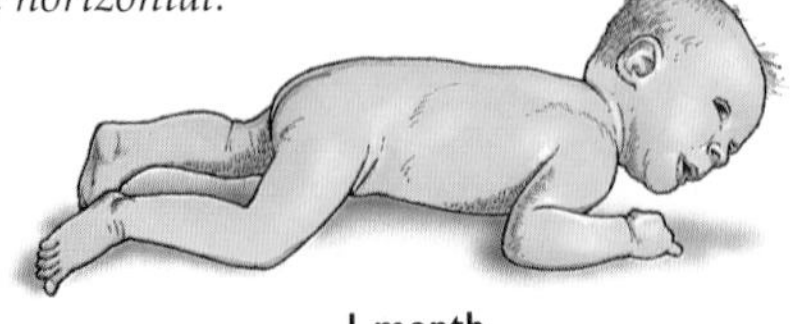

1 month

***At four to five months,** a baby can lie on its front and lift its bottom up by pushing its knees forward or lift its shoulders from the ground by pushing up with the arms.*

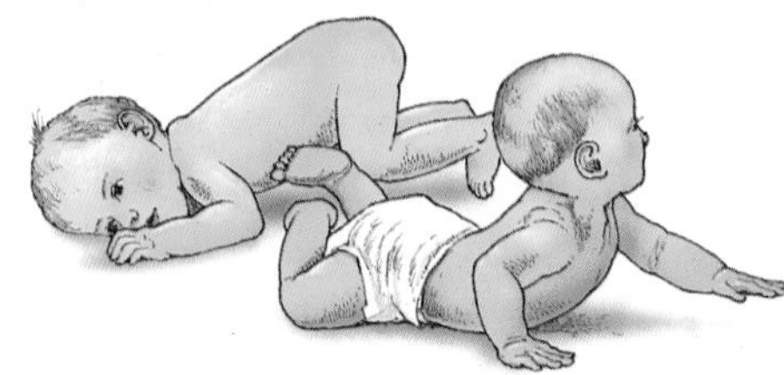

4–5 months

***By nine or ten months,** many babies can crawl on knees and hands. Others use "bottom shuffling" derived from sitting skills.*

9–10 months

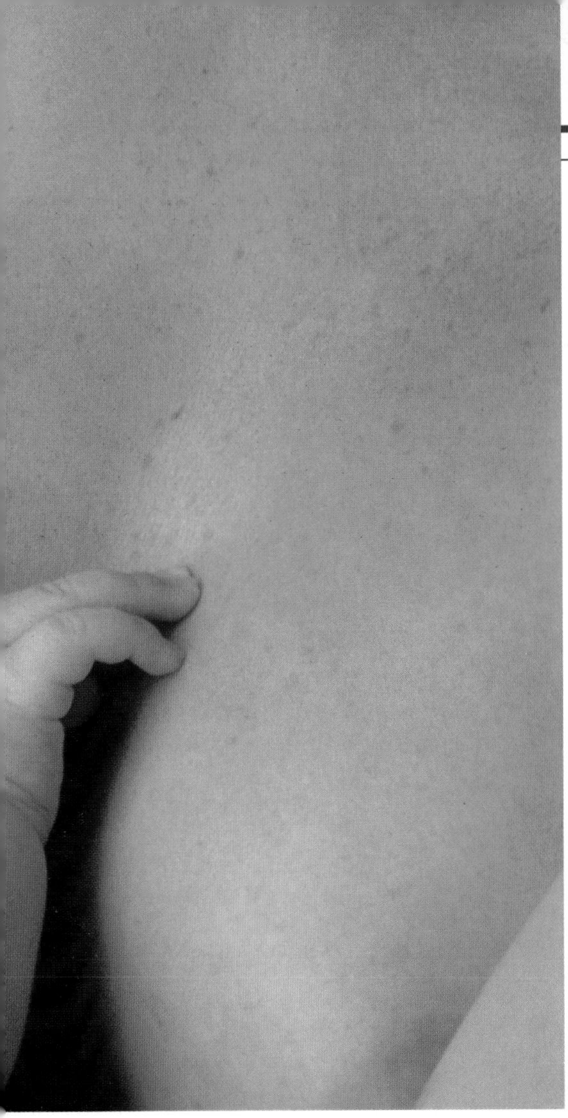

NOURISHMENT FOR THE NEWBORN

A new mother's breasts produce about 1 quart (1 liter) of milk a day, but for up to three days after birth, the breasts give a protein-rich liquid – colostrum. This contains antibodies, to protect the baby from disease, and hormones that stimulate the lining of its digestive tract to digest milk.

Milk's composition varies from mammal to mammal. Human milk has the same fat content but much less protein and more sugar (lactose) than cow's milk. Seal milk has more than 10 times the amount of fat and protein found in human milk. This thick, rich milk enables a young seal to grow fast on its birth beach and lay down a thick layer of blubber before going to sea.

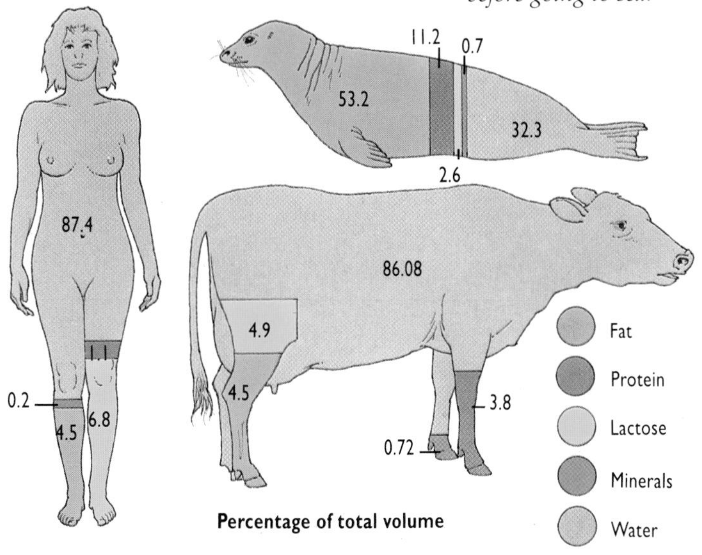

Percentage of total volume

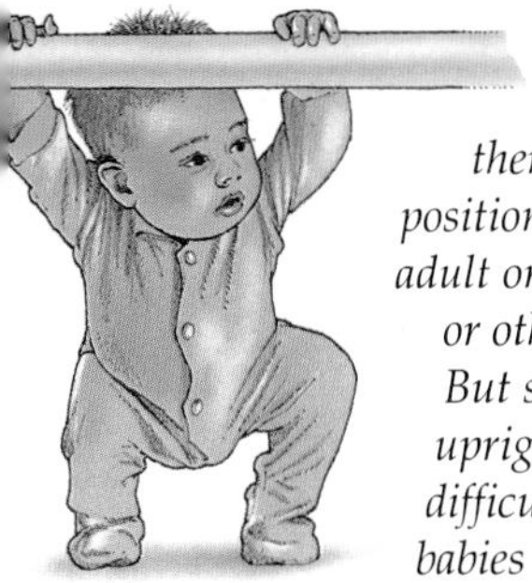

11 months

At around 11 months, *many children can pull themselves into a standing position, either with help from an adult or by hanging onto furniture or other supports.*

But staying upright is difficult, and babies at this age tend to wobble and all over if they attempt to move. Once a child is ccustomed to the idea of taking all its weight on its eet, its confidence soon grows, and standing, the reliminary to walking, becomes second nature.

However, at this stage a child might have trouble itting down again. A child set down on the floor or n a crib will pull up to the standing position and then et stuck and cry for help. This stage only lasts for a ew weeks, and with help – a gentle lowering onto the loor – the child soon learns how to sit down again ithout a hard bump.

12 months

By about a year, *many babies are able to walk in a coordinated fashion by placing one leg in front of the other in a patterned way. Normally, though, they will need a mobile support, such as an adult's hand or a walker, to do this without falling. The first unaided step soon develops into a sequence of steps, and the child becomes mobile.*

Some children seem to skip stages when learning to walk; it is not uncommon for a child to go from the pre-crawling stage straight to the standing stage.

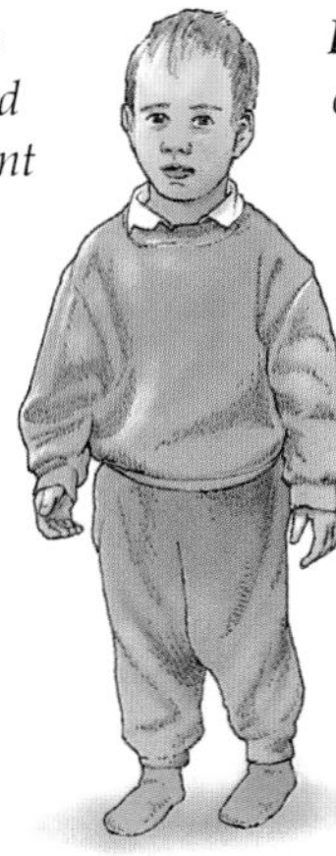

14–16 months

Between 14 and 16 months, *on average, a baby becomes a toddler, able to walk upright and unassisted. There is, however, considerable variation in timing from child to child. Some are up and running by 12 months, while others take their time and do not learn to walk until up to two years after birth. Whatever the case, in less than two years, the child has come a long way – from a helpless, almost immobile newborn baby to an active miniature adult. The attainment of the walking stage is matched by the development of the muscles and the nervous system that coordinates them. A child can now explore under its own steam.*

See also

REPRODUCTION AND GROWTH
- The first month 168/169
- Baby in waiting 172/173
- Being born 174/175
- The growing child 178/179

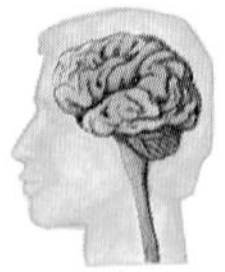

CONTROL AND SENSATION
- Down the wire 46/47
- Into action 50/51
- In the balance 70/71
- Learning skills 80/81

ENERGY
- You are what you eat 96/97

CIRCULATION, MAINTENANCE, AND DEFENSE
- Growing heart 122/123

The growing child

From learning to walk to reaching adolescence, a child undergoes an amazing series of changes.

By the end of its first dozen or so years, a child is on the verge of sexual maturity and is able to walk, speak, write, and communicate socially with sophistication and fluency. Reaching this point from the helplessness of infancy seems to happen with little apparent effort. The child seems programmed to learn new things easily, and the early years are a time when fresh knowledge and skills are absorbed. Thereafter learning new skills becomes increasingly difficult.

It is during the childhood years that humans develop the skills for understanding and manipulating the world around them – a world of objects, activities, social interactions, and the consequences of action. But in addition to the intellectual changes, there are physical changes. The first 18 months or so of life see dramatic rates of growth; but these then slow down and body shapes alter more gradually. A stocky toddler becomes a taller five-year-old with longer limbs, making the child seem thinner. This thin look lasts until muscle growth catches up with the height gain. In the middle phase of childhood, between five and seven years, the first set of teeth, or milk teeth, start to be replaced by the permanent adult set. As the teenage years approach, there is another growth spurt. This marks the beginning of adolescence – the coming of sexual maturity.

Between the ages of one and four, *children develop increasing motor skills and expand their intellectual accomplishments and their ability to see visual patterns in the world. This is demonstrated in the way a child draws.*

At the age of one, a child can hold a pen and make marks on paper. The lines produced are repeated curves generated by swinging movements of the whole hand and arm.

By the time a child starts primary school, *at the age of six or seven, he has already mastered much, including how to communicate verbally and, probably, how to read and write. Between the ages of 6 and 11 or 12, when a child goes to secondary school, formal learning feeds the growing mind.*

A two-year-old's marks *are more deliberate. The lines have clear beginnings and ends, and the dots show more controlled, finer movements.*

The stages in growth *of a girl from birth to the age of 15 reveal a change in proportions. For each age the total body height has been drawn to the same scale so that relative body part sizes can be compared. The baby has a relatively huge head and short limbs, but by 15 the girl has a proportionately small head and limbs that are relatively much longer than the baby's.*

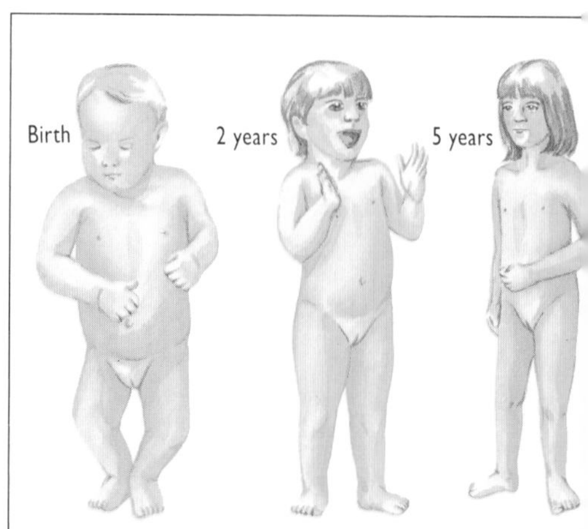

See also

REPRODUCTION AND GROWTH

▶ Cradle to grave 154/155

▶ The newborn baby 176/177

▶ Sexual maturity 180/181

CONTROL AND SENSATION

▶ Hormones of change 40/41

▶ Making memories 78/79

▶ Learning skills 80/81

By the time a child is three, *a picture (**left**) is an attempt to make an image of something. Each of the dense patches of scribbled lines represents a unit of the body – a hand, a nose, or a pair of eyes.*

The image drawn *by a three-and-a-half-year-old (**below**) shows well-controlled lines. The body shape is plain, and there are feet and a head with hair. Together with the dotted shading, this gives the image the beginnings of real representational qualities.*

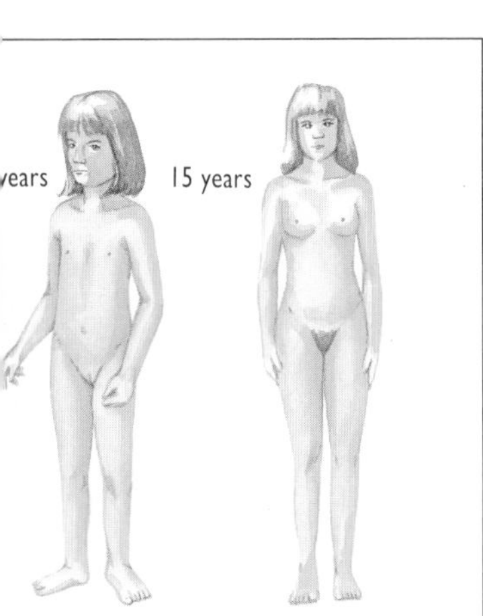

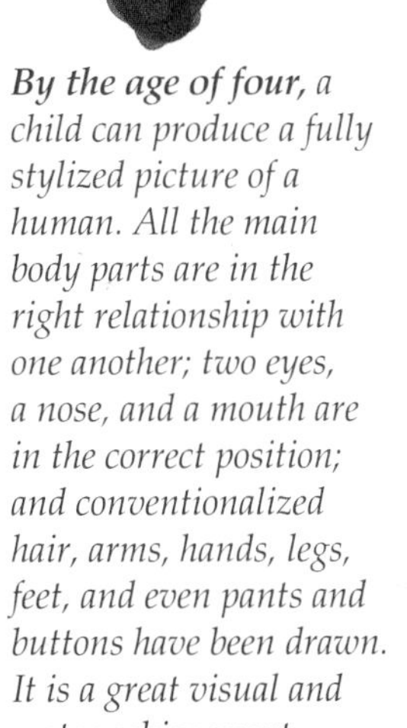

By the age of four, *a child can produce a fully stylized picture of a human. All the main body parts are in the right relationship with one another; two eyes, a nose, and a mouth are in the correct position; and conventionalized hair, arms, hands, legs, feet, and even pants and buttons have been drawn. It is a great visual and motor achievement.*

Sexual maturity

In the space of a few key years, a person changes from a child into a sexually mature adult able to reproduce.

The changes of adolescence and puberty are controlled by a complex mix of interacting hormones. They bring about the enlargement and maturation of sex organs and the parallel development of a number of secondary sexual characteristics that are the external hallmarks of an adult. "Adolescence" refers to the period of general growth in boys and girls and the maturation of the reproductive organs, while, strictly speaking, puberty is the time during adolescence when a maturing child becomes capable of reproduction. The onset of puberty is about 12 years for girls and about 14 for boys, although the age can vary by a few years each way.

At the start of sexual maturation in both sexes, the hypothalamus region of the brain begins to secrete gonadotropin-releasing hormone (GnRH), which stimulates the secretion of follicle-stimulating hormone (FSH) and luteinizing hormone (LH) from the pituitary gland. In a boy these hormones prompt the testes (testicles) to grow further and stimulate them to produce the sex hormone testosterone. In a girl FSH and LH cause her immature ovaries to synthesize and release progesterone and estrogen.

__In childhood__ there is not much difference in athletic performance between boys and girls. But at about the age of 11 or so, boys become better at activities that require speed and strength such as sprinting (__right__).

Yards per second | Meters per second

7 | 6 | 6 | 5 | 5 | 4

Age (years) 5 7 9 11 13 15 17

Male

Female

__From the start of sexual maturation,__ testosterone brings about the development of typical male characteristics. Facial and body hair begins to grow, which can produce a thick beard or a hairy chest in young men. Hair also appears in the armpits and groin. Body reshaping reduces fat and increases muscle size. The Adam's apple (part of the larynx) enlarges and the voice drops to a lower pitch. The testes start to produce sperm, and all the accessory glands of the male reproductive system, for example the prostate, begin to mature.

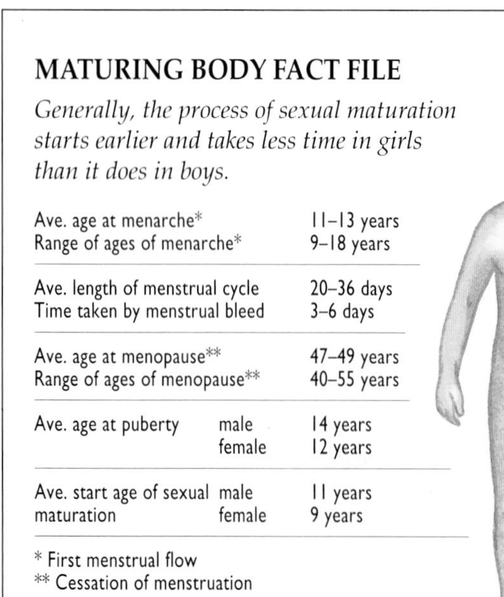

MATURING BODY FACT FILE

Generally, the process of sexual maturation starts earlier and takes less time in girls than it does in boys.

Ave. age at menarche*		11–13 years
Range of ages of menarche*		9–18 years
Ave. length of menstrual cycle		20–36 days
Time taken by menstrual bleed		3–6 days
Ave. age at menopause**		47–49 years
Range of ages of menopause**		40–55 years
Ave. age at puberty	male	14 years
	female	12 years
Ave. start age of sexual maturation	male	11 years
	female	9 years

* First menstrual flow
** Cessation of menstruation

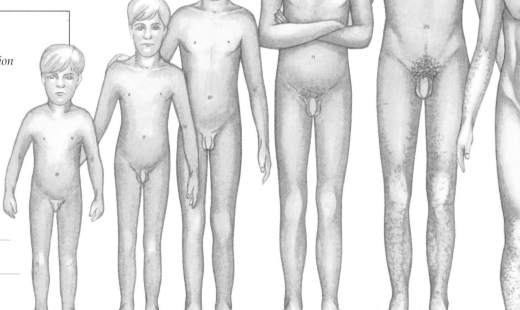

2 years 6 years 10 years 14 years 18 years 22 years

BREAST DEVELOPMENT

Humans are mammals, and one of the things that tells mammals apart from other creatures is that female mammals make milk to feed their young. Prior to the reproductive years, milk production is not needed, so a girl's breasts are similar to those of a boy. But along with other changes of puberty, brought about by hormones which are preparing the human body for reproduction, the breast develops the capability to make milk.

Changes in the nipple *are the first external signs of sexual maturation. They can begin as early as 8 and are usually obvious by 11. The nipple grows bigger and more prominent, and the tissues under the nipple form a small, cone-shaped enlargement. Next the breast itself enlarges and becomes smoothly curved. The area around the nipple, the areola, gets bigger and darker than the surrounding skin and the breast reaches its maturity in a young adult. During lactation, milk is made in the 15–20 lobes of milk-secreting (mammary) glands which increase in size to give the breast a more rounded shape. After lactation the breast returns to much the same shape as before. After the reproductive years, the glands reduce in size, as does the breast.*

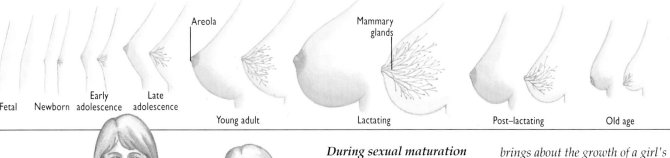

Areola
Mammary glands
Fetal
Newborn
Early adolescence
Late adolescence
Young adult
Lactating
Post–lactating
Old age

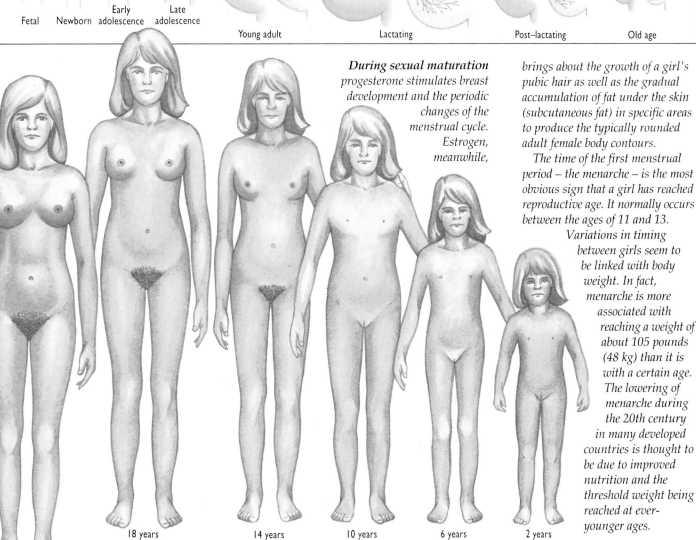

During sexual maturation *progesterone stimulates breast development and the periodic changes of the menstrual cycle. Estrogen, meanwhile, brings about the growth of a girl's pubic hair as well as the gradual accumulation of fat under the skin (subcutaneous fat) in specific areas to produce the typically rounded adult female body contours.*

The time of the first menstrual period – the menarche – is the most obvious sign that a girl has reached reproductive age. It normally occurs between the ages of 11 and 13. Variations in timing between girls seem to be linked with body weight. In fact, menarche is more associated with reaching a weight of about 105 pounds (48 kg) than it is with a certain age. The lowering of menarche during the 20th century in many developed countries is thought to be due to improved nutrition and the threshold weight being reached at ever-younger ages.

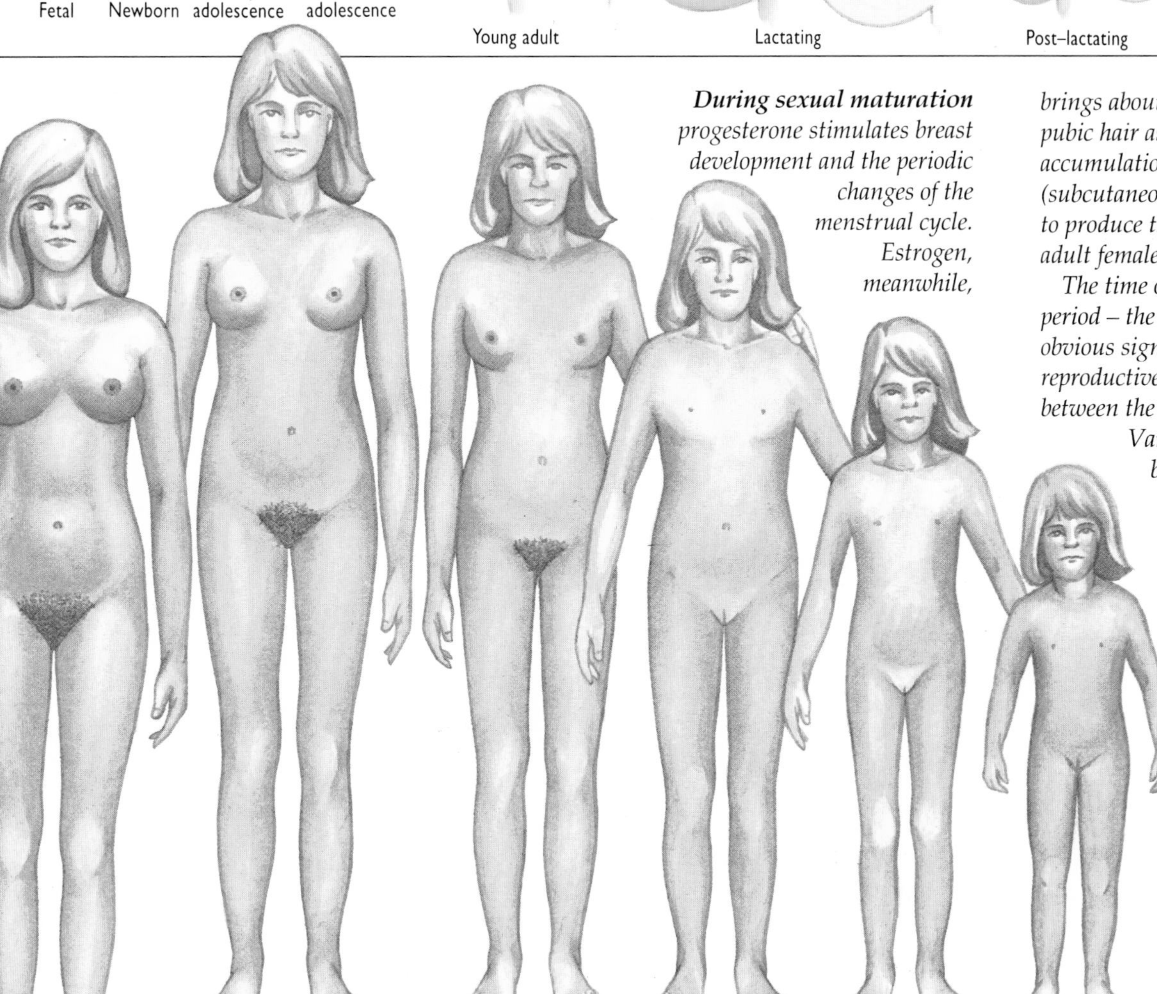

See also

REPRODUCTION AND GROWTH
- Cradle to grave 154/155
- The sexual advantage 162/163
- Cycle of life 164/165
- Fertilization 166/167
- The newborn baby 176/177
- The growing child 178/179

CONTROL AND SENSATION
- Control systems 36/37
- Hormones of change 40/41

ENERGY
- Cells at work 114/115

CIRCULATION, MAINTENANCE, AND DEFENSE
- Transportation systems 118/119

Getting on

As time passes, the human body matures and reaches its prime. Inevitably, though, it changes with age.

In most individuals it is possible to discern some gradual yet obvious signs of the aging process between the ages of 45 and 55. The skin slowly loses its youthful elasticity, the amounts of fat under the skin are reduced, and the skin – already showing wrinkles – begins to wrinkle more and sag. Muscles grow weaker, while sensory systems become less acute and the memory less reliable.

At some point between the ages of 45 and 55, for instance, women stop producing and releasing eggs from their ovaries, and the monthly menstrual cycle becomes irregular and then stops. This occurrence, which marks the end of their reproductive lives, is known as menopause. The hormonal changes associated with the alterations in the way the ovary and uterus (womb) work sometimes have indirect effects, the most common of which are "hot flashes." The changed hormone balance may also bring about increased loss of structural mineral from bones, making them brittle and more liable to break – a condition known as osteoporosis. Some women use hormone replacement therapy to offset some of these postmenopausal symptoms and changes. Men normally experience reduced secretion of the sex hormone testosterone in later years, but the consequences are not as extensive as those of the menopause.

But what exactly causes aging? The body is a community of cells, most of which replace themselves on a regular basis by dividing. So why – if the cells are producing exact replicas of themselves – should the body not stay the same once adulthood has been reached? The signs of aging show that cells are not functioning as well as they once did.

It seems there are a number of factors working together. For instance, as part of its normal activities, a cell produces chemicals known as free radicals. These are highly toxic and can damage cells as time goes by. Other damage can be caused by mutation, or change, in the cell's DNA molecules, which control its genetic information. With changed DNA, the cell might not perform well. Another cause of aging could be a process known as programmed cell death. It is thought that part of the genetic information held in cells is a signal to stop dividing after a certain time.

Whatever the real causes of aging, there is general agreement that stimulation and ongoing activity may slow some of the psychological and physical aspects. But although aging can be influenced by lifestyle, the genetic make-up you inherit from your parents also affects your chances of living a long and healthy life. Like everything else about the body, the way in which it ages is an amalgam of the influences of genes and environment.

Keeping active *keeps you young. In the past there was a feeling that after years of an active working life, the pattern of existence should be one of quiet and inactivity. But there is now evidence that many of the debilitating physical and emotional aspects of aging can be slowed or reduced if an active and stimulating life is enjoyed.*

An exercise such as swimming enhances cardiovascular health and lung function, strengthens weakening muscles, and carries little risk of the sports injuries associated with more strenuous exercise. An active mental life is just as important as keeping on the go physically. If a person does not exercise the mind then it, too, becomes flabby. People who stay involved in life retain their faculties better and have more fun.

Marie Curie *(1867–1934) was a Polish-born French physicist famed for her work on radioactivity. With Henri Becquerel and her husband, Pierre, she received the 1903 Nobel prize for physics. At the age of 43 she received a second Nobel prize, for chemistry.*

Charles Darwin *(1809–1882) was the British naturalist who transformed ideas about the living world by setting out the idea of evolution caused by natural selection. He was 50 when* On the Origin of Species *was published.*

Mao Zedong *(1893–1976) was a key figure in China's communist revolution and in the first decades thereafter. He became leader of China in 1949, at age 55.*

***Pablo Picasso** (1881–1973), born in Spain, was arguably the most influential visual artist of the 20th century. He was a progressive and innovative force in all types of art, including painting, etching, sculpture, and ceramics. He was still painting in the final year of his life, at age 91.*

***Benjamin Franklin** (1706–90), born in Boston, Massachusetts, was a writer, a scientific researcher in the field of electricity, a politician, and an inventor – he is credited, for instance, with the invention of the lightning rod and bifocal glasses. In 1787, at age 81, he was a key member of the Constitutional Convention which framed the U.S Constitution.*

***An incredible human resource** exists among the older people in society as the ages and activities of the men and women here show. Talented people continue to succeed and be creative after the start of their fifth decade.*

The wisdom and sophistication that develops in talented individuals through a lifetime of experiences and diversity can result in great achievement in any decade.

***icolaus Copernicus** (1473–1543), Pole, was 70 when his* evolutions of the Celestial pheres *was published. It shook the world of astronomy by proposing that the Earth rotated and orbited the Sun.*

***Queen Victoria** (1819–1901) reigned as the queen of Great Britain from 1837 until 1901. During this period she oversaw the pinnacle of the country's imperial power – at one time a quarter of the world was in the British Empire – and celebrated the Diamond Jubilee of her reign at age 78.*

CHANGES THROUGH THE AGES

The weight, speed, or function of various bodily attributes alters, on average, through a life span. The values for an attribute – for example, metabolic rate, at ages 25 (red), 50 (mid-orange), and 70 (light orange) – are compared with the maximum value (100 percent). Lung capacity shows the most decline between 25 and 70 years. Other attributes, including brain and liver weight, and nerve-impulse transmission speed, hold up well. In fact, liver weight is at its maximum at age 50, not at 25 as with all the others.

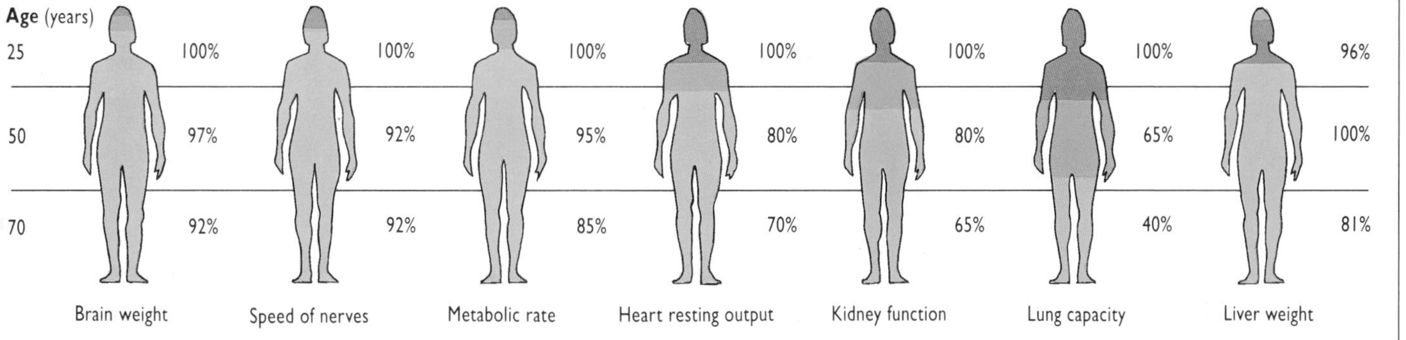

See also

REPRODUCTION AND GROWTH

▶ Cradle to grave 154/155

▶ Sexual maturity 180/181

▶ The future body 184/185

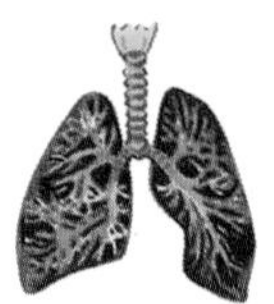

CONTROL AND SENSATION

▶ Hormones of change 40/41

▶ Down the wire 46/47

▶ Making sense 72/73

ENERGY

▶ The energy storehouse 106/107

▶ A deep breath 108/109

CIRCULATION, MAINTENANCE, AND DEFENSE

▶ The living pump 120/121

▶ Routine replacement 138/139

The future body

Scientists now have the power to make changes to the very stuff of life – the genes of human heredity.

Since World War II, the pace at which new therapies for human diseases have emerged has accelerated, and there have been several distinct waves of innovation. The first to have a profound impact was the discovery of antibiotics, natural microbe-killing substances derived from fungi and bacteria; new ones are still being found. Next came organ transplant techniques, and there is now an expanding range of body systems that can be replaced if they become diseased. The most recent boost to the explosion of new therapeutic ideas has come from the burgeoning field of molecular biology.

The structure and importance of the DNA molecule – the substance of heredity and genes – was discovered in 1954. But it was not until the 1970s and 1980s that methods began to be developed to enable DNA to be altered at will in the laboratory. Scientists can now, in fact, directly manipulate the molecules of genes. They can remove a gene from a cell, describe it in minute detail, chop pieces out of it, insert new pieces, and transfer working genes from one organism to another.

Thus it is possible to conceive of the direct treatment of genetic diseases by gene therapies. The life-threatening inherited disease cystic fibrosis, for instance, is caused by a defective gene that controls transportation of certain materials across cell membranes. Attempts have recently been made to introduce the normal, functioning gene into the cells of patients with this disease by using harmless viruses carrying the "good" human gene as "gene infecting" agents.

Chemical plants are now able to produce, in reasonable quantities, therapeutic human proteins that once had to be laboriously extracted from dead tissues and purified. One such is human growth hormone, produced in the pituitary gland. It controls the rate of growth, especially in children and adolescents. The condition known as dwarfism, in which a person fails to grow tall enough, can occur if insufficient growth hormone is secreted. At one time growth hormone to correct dwarfism was obtained from the pituitary glands of dead bodies – it took many thousands to yield enough to treat a child. Now, however, the gene for growth hormone can be extracted from human cells and inserted into bacteria, which are grown in culture vessels. The growth hormone that they make can then be isolated and used to treat dwarfism. Other genetically engineered bacteria make interferon and tumor necrosis factor for treatment of cancers and factor VIII for treatment of some forms of hemophilia.

These new powers over the workings of the human body have spawned a novel series of complex and troubling ethical and social issues. Is it wise to tinker with a person's genetic make-up? Should people be allowed to "choose" aspects of their offspring? What are the consequences of altering the genes that a person might pass on to future generations? The clinical benefits of many of the new therapies are obvious. The undoubted benefits, however, have to be carefully considered against the possible harm that human "genetic engineering" might cause.

***In the past,** insulin was extracted from pig pancreases to treat human diabetes, a disease in which sufferers have to inject insulin to regulate their blood glucose levels, since they do not produce enough of their own. With modern genetic engineering techniques, however, it is now possible – and, indeed, standard practice – to make human insulin in a chemical plant. Bacteria implanted with the human insulin gene are grown in culture vats where they manufacture human insulin.*

Human chromosome

DNA fragments

DNA fragments

DNA fragment with interferon gene

Bacterial cell

Bacterial plasmid

Transformed plasmid

Bacterial cell

Bacterial chromosome

Transformed bacterium

Bacterial ribosome

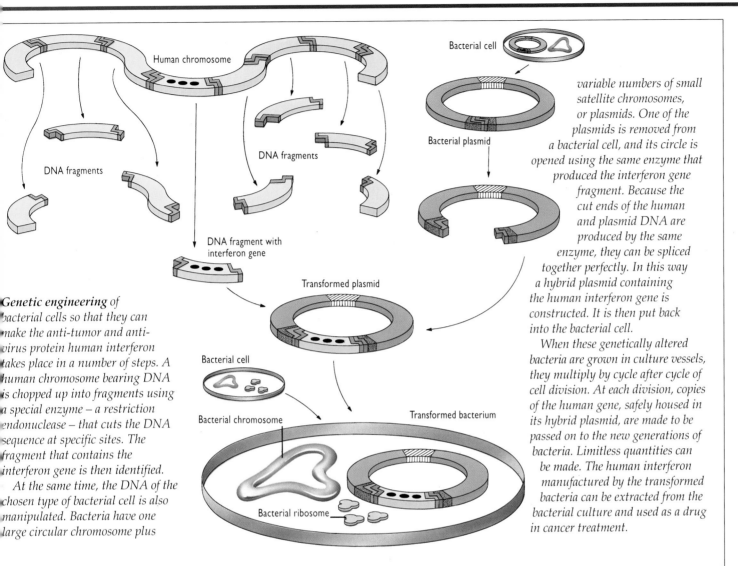

***Genetic engineering** of bacterial cells so that they can make the anti-tumor and anti-virus protein human interferon takes place in a number of steps. A human chromosome bearing DNA is chopped up into fragments using a special enzyme – a restriction endonuclease – that cuts the DNA sequence at specific sites. The fragment that contains the interferon gene is then identified.*

At the same time, the DNA of the chosen type of bacterial cell is also manipulated. Bacteria have one large circular chromosome plus variable numbers of small satellite chromosomes, or plasmids. One of the plasmids is removed from a bacterial cell, and its circle is opened using the same enzyme that produced the interferon gene fragment. Because the cut ends of the human and plasmid DNA are produced by the same enzyme, they can be spliced together perfectly. In this way a hybrid plasmid containing the human interferon gene is constructed. It is then put back into the bacterial cell.

When these genetically altered bacteria are grown in culture vessels, they multiply by cycle after cycle of cell division. At each division, copies of the human gene, safely housed in its hybrid plasmid, are made to be passed on to the new generations of bacteria. Limitless quantities can be made. The human interferon manufactured by the transformed bacteria can be extracted from the bacterial culture and used as a drug in cancer treatment.

See also

REPRODUCTION AND GROWTH
▶ Cradle to grave 154/155

▶ Language of life 158/159

▶ Building bodies 160/161

CONTROL AND SENSATION
▶ Key chemicals 42/43

ENERGY
▶ Cells at work 114/115

CIRCULATION, MAINTENANCE, AND DEFENSE
▶ Routine replacement 138/139

▶ Damage repair 140/141

▶ Repelling invaders 144/145

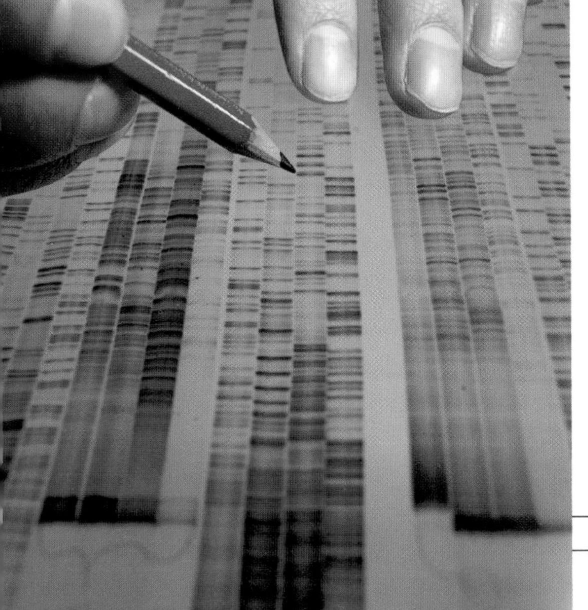

THE HUMAN GENOME – MAPPING OUT HEREDITY

"Sequencing gels" (**left**) are the basic technology used to map the genetic base sequences of human genes. Each gel has four rows of "ladder rungs" with each row standing for one of the four DNA nucleotide bases – adenine (A), thymine (T), cytosine (C), and guanine (G). The order of the bases along the DNA molecule provides the instructions for a particular protein.

When a human gene is appropriately chemically manipulated and then spread along the gel, the sequence of "ladder rung" bands on the gel can tell a researcher the order of bases on the gene. A huge international research project, the Human Genome Project, is now in operation. Its aim is to list the entire base sequence of all the DNA molecules of the human genome – several billion bases. When completed, the project will have discovered the genes for the making of every protein in the human body.

Bibliography

Agur, Anne M.R. (ed.) *Grant's Atlas of Anatomy* Wilkins and Wilkins, Baltimore, Maryland, 1991

Alexander, R. McNeill *The Human Machine* Natural History Museum Publications, London, 1992

American Medical Association Home Medical Library Reader's Digest, Pleasantville, New York, 1992

Atlas of the Body and Mind Mitchell Beazley, London, 1976

Bevan, James *A Pictorial Handbook of Anatomy and Physiology* Mitchell Beazley/Reed International Books, London, 1994

Birren, James E. and Warner K. Schraie *Handbook of the Psychology of Aging* Van Nostrand, New York, 1977

Blakemore, Colin *The Mind Machine* BBC Books, London, 1988

Bodanis, David *Being Human* Century Publishing, London, 1984

Carlson, Neil R. *Physiology of Behaviour* Allyn and Bacon, Newton, Massachusetts, 1986

Carola, Robert, John P. Harley and Charles R. Noback *Human Anatomy and Physiology,* McGraw-Hill, Inc., New York, 1992

Charness, N. (ed.) *Aging and Human Performance* John Wiley and Sons, Chichester, 1985

Cunningham, John D. *Human Biology* Harper & Row, New York, 1989

Dixon, Bernard *Health and the Human Body* Perseus Press, London, 1986

Dorit, Robert L., Warren F. Walker Jr. and Robert D. Barnes *Zoology* Saunders College Publishing, Philadelphia, Pennsylvania, 1991

Dox, Ida, Biagio John Melloni and Gilbert M. Eisner *Melloni's Illustrated Medical Dictionary* Wilkins and Wilkins, Baltimore, Maryland, 1979

Downer, John *Supersense* BBC Books, London, 1988

Farndon, John *The All Color Book of the Body* Arco Publishing, Inc., New York, 1985

FitzGerald, M.J.T. *Neuroanatomy Basic and Applied* Baillière Tindall, London, 1985

Gallahue, David L. and John C. Ozmun *Understanding Motor Development* Brown & Benchmark, Madison, Wisconsin, 1995

Gold, Jay J. and John B. Josinovich (eds.) *Gynaecologic Endocrinology* Harper, Hagerstown, Maryland, 1980

Graham, Robert B. *Physiological Psychology* Wadsworth Publishing Company, Belmont, California, 1990

The Guinness Book of Records Guinness Publishing Limited, London, 1994

Guyton, Arthur C. *Textbook of Medical Physiology* W.B. Saunders Company, Philadelphia, Pennsylvania, 1981

Hobson, J.Allan *Sleep* Scientific American, Inc., New York, 1989

The Human Body Arch Cape Press, New York, 1989

Jennett, Sheila *Human Physiology* Churchill Livingstone, Edinburgh, 1989

Kirkpatrick, C.T. *Illustrated Handbook of Medical Physiology* John Wiley and Sons, Chichester, 1992

Kuby, Janis *Immunology* Freeman, New York, 1994

McArdle, William D., Frank I. Katch and Victor L. Katch *Exercise Physiology* Lea & Febiger, Malvern, Pennsylvania, 1991

Man's Body an Owner's Manual Paddington Press Ltd., New York, 1976

Parker, Steve *The Body Atlas* Dorling Kindersley, London, 1993

——*How the Body Works* Dorling Kindersley, London, 1994

Roberts, Jean *Mastering Human Biology* Macmillan, London, 1991

Roitt, Ivan M., Jonathan Brostoff and David K. Male *Immunology* Churchill Livingstone, Edinburgh, 1985

Ronan, Colin A. *Science Explained* Henry Holt, New York, 1993; Doubleday, London, 1993

Rose-Neil, Wendy *The Complete Handbook of Pregnancy* Warner Books, London, 1993

Russell, Peter *The Brain Book* Routledge, London, 1989

Stockley, Corinne, Chris Oxlade and Jane Wertheim *The Usborne Illustrated Dictionary of Science* Usborne, London, 1988

Strickberger, Monroe W. *Evolution* Jones and Bartlett Publishers, Boston, Massachusetts, 1990

Tortora, Gerard J. and Nicholas P. Anagnostakos *Principles of Anatomy and Physiology* Canfield Press, San Fransisco, California, 1978

Whitfield, Philip *From So Simple a Beginning: The Book of Evolution* Macmillan, New York, 1993

Whitfield, Philip and Ruth Whitfield *Why Does My Heart Beat?* British Museum (Natural History), London, 1988

Williams, Peter L., Roger Warwick, Mary Dyson and Lawrence H. Bannister (eds.) *Gray's Anatomy* Churchill Livingstone, Edinburgh, 37th ed., 1989

Suggested journals and periodicals

Nature Macmillan Magazines, London

New Scientist IPC Magazines Ltd., London

Scientific American Scientific American, Inc., New York

UNDER PRESSURE

The force with which blood is pushed against the walls of blood vessels is the blood pressure. While pressure is usually measured in pounds per square inch or Newtons per square meter (Pascals), blood pressure is measured in millimeters of mercury, or mmHg. This refers to the height to which a column of mercury can be raised by the pressure. The typical pressure in the main arteries caused by the contraction of the heart's left ventricle is enough to raise a column of mercury by 120 mm.

ACID TEST

The acidity or alkalinity of a solution is measured using the pH scale. A pH of 1 is very acid and a pH of 14 is very alkaline; a pH of 7 is neutral. The pH of pure water is 7. The pH of the blood is between 7.35 and 7.45, making it slightly alkaline. The term pH stands for "power of hydrogen" and the scale reflects the number of free hydrogen ions and the number of free hydroxyl ions in a solution. Each whole number change on the scale represents a tenfold change in the number of ions.

MEASURE FOR MEASURE

In the metric system a millimeter (mm) is a thousandth of a meter ($\frac{1}{25}$ inch); a micrometer (μm) is a millionth of a meter ($\frac{1}{25,000}$ inch); and a nanometer (nm) is a billionth of a meter ($\frac{1}{25,000,000}$ inch).

In this book, a billion is 1,000 million, or 10^9; a trillion is a million million, or 10^{12}.

Index

age numbers in *italic* efer to the captions and ox text.

Acknowledgments

l = left; *r* = right; *c* = center; *t* = top; *b* = bottom

Picture credits

2*t* A.B. Dowsett/Science Photo Library; 2*b* John Barlow; 5*t* Alan Brooke/Telegraph Colour Library; 5*b* John Barlow; 10*t* Mark Conlin/Telegraph Colour Library; 10*b* Peter Aprahamtan/Sharples Stress Engineers Ltd/Science Photo Library; 12/13 Mark Conlin/Telegraph Colour Library; 13 The Ronald Grant Archive; 16 John Barlow; 18/19 John Barlow; 19*l* Wellcome Institute Library, London; 19*r* Peter Aprahamtan/Sharples Stress Engineers Ltd/Science Photo Library; 20/21 John Barlow; 23 Corey Ross/Professional Sport; 24 Wellcome Institute Library, London; 24/25 Dr. Philip J. Benson; 26 Philip Quirk/Wildlight Photo Agency; 28 Adam Hart-Davis/Science Photo Library; 28/29 Alan Brooke/Telegraph Colour Library; 30 John Barlow; 30/31 Michael Fogden/Oxford Scientific Films; 31 Avon Rubber P.L.C.; 32 Michael Fogden/Oxford Scientific Films; 34 Dr. R. Clark & M. Goff/Science Photo Library; 34/35 Barry Lewis/Network; 36 John Barlow; 36/37 Rover Group; 39 Anthony Suau/Network; 40/41 Aspect Picture Library; 42 John Barlow; 44 G. Vandystadt/Allsport; 48/49 The Ronald Grant Archive; 52*t* NASA; 52/53*b* John Barlow; 54/55 Ron Haviv/Saba-Rea/Katz; 57 Institute of Opthalmology; 58/59 Michael Fogden/Oxford Scientific Films; 59*t* © Breslich & Foss from Visual Magic (John Murray); 59*b* John Barlow; 60/61 Barros & Barros/The Image Bank; 64/65 John Barlow; 66*r* John Barlow; 68 The Bridgeman Art Library; 70 Robert Harding Picture Library; 71 NASA; 73 Gamma Sport/Frank Spooner Pictures; 74/75 Nobby Clark; 75*t* National Library of Medicine, U.S.A.; 75*b* AT&T Archives; 77 James Holmes/Janssen Pharmaceutical Ltd/Science Photo Library; 78 John Barlow; 78/79 Zefa Picture Library; 80/81 F. Rombout/Bubbles; 82/83 Robert Harding Picture Library; 83 Helen A. Lisher; 84/85 Jean Michel Turpin/Frank Spooner Pictures; 85 John Barlow; 86 Peter Underwood; 87 John Barlow; 88*l* Dr. Ray Clark/Science Photo Library; 88*r* John Barlow; 90/91 Paul Lowe/Network; 92/93 Alan Becker/The Image Bank; 93 Gray Mortimore/Allsport; 94 Dr. Ray Clark/Science Photo Library; 96 John Barlow; 106/107 D.C. Lowe/Tony Stone Images; 108–112 John Barlow; 116*t* E.T. Archive; 116*c&b* John Barlow; 119*l* John Barlow; 119*r* François Gohier/Ardea; 120 John Barlow; 124 Bill Dobbins/Allsport; 125 Wellcome Institute Library, London; 126/127 A.B. Dowsett/Science Photo Library; 128/129 Tom van Sant/Geosphere Project, Santa Monica/Science Photo Library; 130 Arnulf Husmo/Tony Stone Images; 134 John Barlow; 136 Chicago Historical Society; 136/137 Scott Frances/Esto/Arcaird; 138/139 Randy Wells/Tony Stone Images; 140 John Massey Stewart; 144 J.C. Revy/Science Photo Library; 145 John Barlow; 147 E.T. Archive; 148/149 Dr. Arnold Brody/Science Photo Library; 150/151*t* John Barlow; 150/151*b* Jerry Mason/Science Photo Library; 152*t&c* John Barlow; 152*b*–155 J. Feingersh/Zefa Picture Library; 156 Martel/Rapho/Network; 156/157 CNRI/Science Photo Library; 159 Richardson/Custom Medical Stock Photo/Science Photo Library; 162/163 John Barlow; 163 G.A. Maclean/Oxford Scientific Films; 166/167 John Shaw/NHPA; 171 John Barlow; 174/175 F. Guenet/Imapress/Camera Press; 176/177 Zefa Picture Library; 182/183 Zigy Kaluzny/Gamma/Frank Spooner Pictures; 184 Larry Lefever/Zefa Picture Library; 185 Sinclair Stammers/Science Photo Library

If the publishers have unwittingly infringed copyright in any illustration reproduced, they would pay an appropriate fee on being satisfied to the owner's title.

Illustration credits

David Ashby 144/145, 156/157, 158/159; John Bavosi 170/171; Richard Bonson 112/113; Sarah Bowers 178/179; Richard Coombes 140/141; Michael Courtney 66/67, 98/99, 122/123, 174/175; Bill Donahoe 42/43, 72/73, 112/113, 164/165, 176/177; Andrew Farmer 12/13, 14/15, 52/53, 94/95, 114/115, 136/137, 138/139, 144/145, 182/183 and all connection icons; Chris Forsey 22/23, 26/27, 36/37, 44/45, 56/57, 64/65, 70/71, 78/79, 80/81, 92/93, 108/109, 118/119, 128/129, 166/167, 172/173; David Gifford 124/125; Ed Gillah 18/19; Mick Gillah 38/39, 44/45, 114/115; Greensmith Associates 138/139; Gary Hinks 126/127, 146/147, 148/149; Richard Hook 50/51; Frank Kennard 14/15, 26/27, 28/29, 56/57, 60/61, 62/63, 98/99, 120/121; Mainline Design 16/17, 28/29, 34/35, 40/41, 46/47, 52/53, 54/55, 58/59, 60/61, 64/65, 66/67, 74/75, 76/77, 80/81, 82/83, 84/85, 90/91, 96/97, 102/103, 110/111, 124/125, 130/131, 132/133, 134/135, 158/159, 160/161, 170/171, 178/179, 180/181; Tom McArthur 184/185; Annabel Milne 22/23, Ed Musy 18/19, Lillith Pollock 178/179; Paul Richardson 58/59, 174/175; Peter Sarson 20/21, 24/25, 54/55, 94/95, 122/123, 140/141; Mike Saunders 30/31, 38/39, 40/41, 42/43, 48/49, 50/51, 62/63, 68/69, 84/85, 132/133, 134/135, 142/143; Sue Sharples 162/163 Les Smith 42/43, 100/101, 104/105, 122/123, 172/173; UCLH Cochlear Implant Programme 62/63

Marshall Editions would like to thank the followi for their assistance in the compilation of this book

Authors: Steve Parker 12–31, 34–87
Philip Whitfield 90–115, 118–151, 154–185

Copy editors: Isabella Raeburn, Maggi McCormick
Managing editor: Lindsay McTeague
Editorial director: Sophie Collins
DTP editors: Mary Pickles, Pennie Jelliff
DTP assistance: Heather Magrill
Editorial assistant: Jolika Feszt
Index: Caroline S. Sheard
Assistant art editor: Lynn Bowers
Art assistant: Eileen Batterberry
Picture manager: Zilda Tandy
Production: Sarah Hinks

British Telecom for the loan of red telephone; David Mellor for the loan of frying pan, stainless steel knife, sifter, plate, glass, and cutlery; Kristin Baybars for the loan of toy giraffe; The Algerian Coffee Shop for the loan of cafetière; Pravins Jewellers Ltd for the loan of watch; Racing Green for the loan of man's top; Olympus Sport for the loan of man's sports clothes and trainers; Jubilee Sports for the loan of rugby ball; Revelations for the loan of leather goods; C.J. Graphics for the loa of magic markers; James Bodenham & Co. for the loan of pot-pourri and toiletries.

George Bridgeman and Jo Tomlin at Guy's Hospital anatomy department. David Weeks "Magician" for card tricks. Dr. Deirdre O'Gallagh for prescription drugs.